STUDY GUIDE

JENNIFER SHANOSKI
Merritt College

CHEMISTRY

A Molecular Approach

Second Edition

NIVALDO J. TRO

Courseware Portfolio Management Director: Jeanne Zalesky
Executive Courseware Portfolio Manager: Terry Haugen
Courseware Director, Content Development: Jennifer Hart
Development Editor: Erin Mulligan
Courseware Analyst: Coleen Morrison
Portfolio Management Assistant: Lindsey Pruett
Portfolio Management Assistant: Shercian Kinosian
VP, Product Strategy & Development: Lauren Fogel
Content Producers: Lisa Pierce, Mae Lum
Managing Producer: Kristen Flathman
Director, Production & Digital Studio: Laura Tommasi
Editorial Content Producer: Jackie Jakob
Director, Production & Digital Studio: Katie Foley
Senior Mastering Media Producer: Jayne Sportelli
Rights and Permissions Manager: Ben Ferrini
Rights and Permissions Management: Cenveo Publisher Services
Photo Researcher: Eric Schrader
Production Management and Composition: codeMantra
Design Managers: Marilyn Perry, Maria Guglielmo Walsh
Cover and Interior Designer: Jeff Puda
Contributing Illustrators: Lachina
Printer/Binder: LSC Communications
Manufacturing Buyer: Maura Zaldivar-Garcia
Product Marketer: Elizabeth Bell
Executive Field Marketing Manager: Chris Barker

Cover Art: Quade Paul

ISBN 10: 0-134-46068-5; ISBN 13: 978-0-134-46068-0

www.pearsonhighered.com

Table of Contents

Preface

To the general chemistry student:

Chemistry is one of the subjects that college students dread. I see it on my students' faces at the beginning of a new semester: fear and anxiety about the class. In my opinion, this attitude is primarily derived from a belief that chemistry is unimportant, abstract, and inherently difficult.

The importance of chemistry is in its ubiquity—molecules, atoms, and ions are everywhere and make up everything; the study of chemistry is the study of molecules, atoms, and ions, so it is the study of everything. And what does the study of chemistry entail? Mathematics, while important, is not chemistry, and this is what most students do not realize upon first glance. Mathematical relationships are certainly important in a general chemistry course, but they are a means to an end. The end is an understanding of the physical principles that govern the world (and universe) that we live in.

This study guide is aimed at simplifying your exploration of general chemistry concepts so that you can find and explore the underlying physical principles, which provide meaning and context to your studies. Each chapter is divided into six sections. First, the learning objectives of each chapter are provided. These are the skills that you should have mastered upon completion of the chapter. Second, a brief summary of the chapter is given so that you have context for the topics and how they fit together. Third, an extensive outline of the chapter is given; the outline should be used as a review of your textbook reading and not in its place. In each outline, example problems are worked out with explanations for the logic and physical meaning of the steps involved and the solutions for each problem. Fourth, fill-in-the-blank problems are provided. You should use these to test your familiarity with the main ideas and definitions of the chapter. The fifth section of each chapter contains problems for you to work out. These problems should be used as a final test of your knowledge—if you are able to complete these problems on your own, you have mastered the mechanics of the concepts presented in the chapter. Finally, and most importantly, the sixth section of each chapter is the concept questions. These questions require that you thoroughly think through the concepts presented and look at the underlying principles behind those concepts. Often, these questions require that you link together multiple topics and/or find the connections between seemingly disparate topics. If you can answer these questions thoroughly, you truly understand the material. Answers are provided, but it is more important that you understand your answer and its reasonableness than it is to get the right number. Check your work by going through your approach; a solid understanding is always better than a correct number.

The likelihood of your success, in this and other courses, is based on your plan for learning the material. I give the series of steps listed below to my own students each semester; following these steps will almost certainly lead to a passing grade and, more importantly, will generally lead you to a true understanding that you can take away with you when the course is done. You will see that this study guide is listed as an endpoint for your strategy—it is not meant to replace either your textbook or your lectures.

1. Read the chapter before going to the lecture.
   You don't have to read for content; even skimming over the pages and looking at pictures and graphs will help. Your goal here is to get an idea of the concepts and language that you will be learning next; it is often the language of chemistry that students struggle with most, and the more you expose yourself to that language, the less likely it is to be a barrier.

2. Attend lecture, take notes, participate in class, and ask questions.
   You are in class to learn, and your instructor is there to help you—so use the resources that are available to you. If concepts remain unclear after lecture, go to your instructor's office and ask questions.

3. Read the material again after class.
   This time use a pen and some paper…create an outline of the chapter and go through the example problems. Writing concepts down and doing problems as you go will help to ensure that you retain the information. You can also make notes about concepts that are confusing or unclear; ask your instructor or your fellow students about concepts with which you are struggling.

4. Work on problems at the end of the chapter.
   The surest way to understand concepts is to practice using them in many ways. One of the most difficult parts of chemistry is the incorporation of math into physical problems—story problems. You need to practice taking apart these problems and figuring out what is important to solving them. It is a good idea to look at the physical meaning behind the problem and the answer—think about it in real world terms.

5. Form a study group.
   There is no better way to learn than to teach, and there is no better way to see if you really understand something than to try to explain it. Getting a group of your peers together to discuss chemistry is one of the smartest ways that you can study.

6. Use the study guide as a review and a final test of your understanding.
   Once you have looked over the material and feel comfortable with the concepts and problems, use this study guide to help clarify difficult concepts or to test your skills. The fill-in-the-blank questions, problems, and concept questions are all meant to be a final test—you can use these as a mock exam to see if you have truly learned the material.

I believe strongly that everyone can really learn chemistry, and once it is learned, everyone can not only excel in a general chemistry course, but also start to see the chemistry around them as interesting and exciting. Learning new material is difficult, however, and requires time and patience. Don't expect that you can study for four hours before an exam and be successful; pace yourself so that you can learn the material instead of simply trying to regurgitate facts. You will find that the slow and steady approach works best for this type of material.

I think that chemistry is a fascinating and rewarding field of study—I see it everywhere and I am always amazed at its underlying simplicity. I don't expect that you will see chemistry as simple and fascinating right from the beginning because as a student it is also challenging and frustrating. But over the course of your studies, if you look carefully and learn thoroughly, the beauty and simplicity will shine through.

I would like to thank Merritt College for supporting me while writing this book and giving me the opportunity to share my love of chemistry with my students; I am especially grateful to my fellow chemistry instructor, Ray Chamberlian, who is always encouraging and enthusiastic. I would like to thank my students for their ever-growing curiosity and desire to learn. Finally, I would like to thank my husband, Engelbert, for encouraging me in all of my projects, my son, Vicente, for patiently sharing some time with me, and my daughter, Sophie, for napping while I work.

Jennifer Shanoski, Ph.D.
Merritt College
Oakland, California

Chapter E: Essentials: Units, Measurement, and Problem Solving

<u>Key Learning Outcomes:</u>

- Convert between Temperature Scales

- Report Scientific Measurements to the Correct Digit of Uncertainty

- Determine the Number of Significant Digits in a Number

- Follow Significant Figure Rules in Calculations

- Calculate the Density of a Substance

- Use Conversion Factors to Convert Quantities from One Unit to Another

- Solve Problems Using Equations

<u>Chapter Summary:</u>

In this chapter, you will be introduced to many of the tools you will use throughout the remainder of this text. You will start with an introduction to units generally and the metric system specifically. Next, you will learn how to assess the reliability of a measurement, how to indicate the reliability using significant figures, and how to keep track of that information during calculations. You will then learn about density and energy—two measurable quantities. Unit conversion problems will then be introduced, with an emphasis on strategies for solving these, as well as equation, problems.

<u>Chapter Outline:</u>

1. The Metric Mix-up: A $125 Million Unit Error

    a. The unit of a number is a critical piece of information.

2. The Units of Measurement

    a. Units are standard quantities that are used to specify the meaning of a measurement.

    b. In science, the metric system is used, and the standard units are the International System of Units (SI). Four of the seven SI base units and their abbreviations (in parentheses) are given below:

        i. Length is a measure of distance. The standard unit for length is the meter (m).

        ii. Mass is a measure of the quantity of matter in an object; mass is different from weight, which measures the gravitational pull on an object. The standard unit for mass is the kilogram (kg).

        iii. Time is a measure of periods between events. The standard unit for time is the second (s).

        iv. Temperature is a measure of the average kinetic energy of the atoms or molecules that compose matter. The standard unit for temperature is kelvin (K).

            1. The Kelvin scale is often referred to as the absolute scale because the lowest possible temperature is 0 K or absolute zero. This is the temperature at which all molecular motion stops.

            2. In order to convert a temperature to the Kelvin scale, you must begin with units of °C.

a. Conversion between °C and °F:

$$°C = \frac{\left(°F - 32\right)}{1.8}$$

b. Conversion between °C and K:

$$K = °C + 273.15$$

EXAMPLE:

The average temperature of the human body is 98.6 °F. Convert this into °C and K.

We can convert to the Celsius scale using the equation given:

$$°C = \frac{\left(98.6 \; °F - 32\right)}{1.8} \quad \rightarrow \quad 37.0 \; °C$$

And we can now use this result to calculate the temperature in the Kelvin scale:

$$K = 37.0 + 273.15 \quad \rightarrow \quad 310.2 \; K$$

c. Prefix Multipliers

    i. Using the metric system, we can change the value of a unit by multiplying by powers of 10 and incorporating a prefix. For example:

        1. kilo (k) is 10^3 or 1000.

        2. centi (c) is 10^{-2} or 0.01.

        3. milli (m) is 10^{-3} or 0.001.

        4. micro (μ) is 10^{-6} or 0.000001.

        5. nano (n) is 10^{-9} or 0.000000001.

    ▶ You can translate the prefix multipliers directly into numbers so that 15 km is really just $15 \, (\times 10^3)$ m, which we write simply as 15×10^3 m or, in proper scientific notation, as 1.5×10^4 m.

EXAMPLE:

Convert the following measurements into numbers between 1 and 10 using prefix multipliers:

a. 1908750 m

Rewriting in scientific notation: 1.90875×10^6 m. 10^6 corresponds to mega, so we have 1.90875 Mm.

b. 0.00000000000989 s

Rewriting in scientific notation: 9.89×10^{-12} s. 10^{-12} corresponds to pico, so we have 9.89 ps.

c. 6.8526×10^{-6} g

10^{-6} corresponds to micro, so we have 6.8526 μg.

d. 0.06548 m

Rewriting in scientific notation: 6.548×10^{-2} m. 10^{-2} corresponds to centi, so we have 6.548 cm.

e. 4.548×10^3 g

10^3 corresponds to kilo, so we have 4.548 kg.

---

d. Derived Units

   i. Derived units are units that come from combinations of the base units; we will see many examples in the coming chapters.

      1. Volume is a measure of space. Volume is an extensive property, which means that it depends on the sample size or the amount of substance. The SI base unit of volume is m^3. In general, units of liter (L) are used.

$$1 \text{ L} = 1000 \text{ mL} = 1000 \text{ cm}^3 = 0.001 \text{ m}^3$$

      2. Density is the ratio of mass to volume ($d = m/V$). Density is an intensive property, which means that it is independent of the sample size or amount of substance; density is therefore characteristic of particular materials. The SI base unit of density is kg/m^3. The common units for density are g/mL for solids and liquids and g/L for gases.

EXAMPLE:

A solid sample of a metal is found to have a mass of 13.45 g and a volume of 4.98 mL. Calculate the density of the metal and determine its possible identity using the table of known densities given below.

Metal	Density (g/mL)
Aluminum	2.70
Iron	7.87
Silver	10.5
Lead	11.4

The density is simply the mass divided by the volume: 13.45 g/4.98 mL = 2.70 g/mL. We can see from the table that this is the same density as aluminum, so we can conclude that the metal slug may be aluminum.

---

3. The Reliability of a Measurement

   a. The number of digits in a reported measurement indicates the certainty associated with that measurement.

      i. The last digit reported is always assumed to be the uncertain digit.

   b. Uncertainty comes from measurement devices, so the number of digits reported is determined by the measurement device. The rule in measurements is to report all digits given explicitly and then guess one more; the uncertain digit (the last digit) will be the one that you guessed.

   c. Precision is a measure of reproducibility; that is, how close various measurements are to one another. Accuracy is a measure of how close measurements are to the actual or true value.

      i. Random errors are those errors that have an equal probability of being higher and lower than the true value. These errors can be reduced by averaging multiple measured values.

ii. Systematic errors tend to cause results that are either too high or too low and do not average out with multiple measurements. Systematic errors often cause results that are inaccurate, but they do not, in themselves, affect precision.

▶ Random errors will give a less precise result, while systematic errors give a less accurate result.

4. Significant Figures in Calculations

a. Uncertainty is conveyed in measurements and calculations by keeping track of the number of significant figures. The rules for determining the number of significant figures (sig. figs.) are:

i. All nonzero digits are significant.

ii. All interior zeros are significant.

iii. All leading zeros are not significant.

iv. All trailing zeros after the decimal point are significant.

v. Trailing zeros in numbers without a decimal point are ambiguous and should be avoided by using scientific notation.

b. Exact numbers are numbers that come from counting discrete objects, are defined quantities, or are integral numbers in an equation. Exact numbers have no uncertainty and should be treated as though they have unlimited significant digits.

EXAMPLE:

Determine the number of significant figures in the following numbers:

a. 0.548704

This number has six significant figures. The first zero is a leading zero and is not, therefore, significant. The zero between the 7 and the 4 is significant because it comes between two nonzero integers.

b. 4580000

This number is ambiguous. It could have three, four, five, six, or seven significant figures. If we rewrite this number in scientific notation as 4.58×10^6, then it would have three significant figures; in that case, the four zeroes would simply be placeholders. Written another way as 4.5800×10^6, it would have five significant figures.

c. 654089.0

This number has seven significant figures. The zero between the 4 and the 8 is significant because it comes between two nonzero integers. The zero at the end of the number is significant because it is a trailing zero in a number with a decimal point.

d. 0.0000548

This number has three significant figures. All of the zeros in this number are leading zeros and serve only as placeholders.

---

c. When carrying out calculations, the following rules should be followed in regards to significant figures:

i. When multiplying or dividing, the result will have the same number of significant figures as the factor with the fewest significant figures.

ii. When adding or subtracting, the result will have the same number of decimal places as the quantity with the fewest decimal places. This means that the uncertainty in the answer will have the same uncertainty as the most uncertain term in the calculation.

iii. At the end of a calculation, you should round your answer. If the digit after the last significant digit is less than or equal to four, round down; if the digit after the last significant digit is greater than or equal to five, round up.

▶ Rounding off before the end of the calculation will introduce errors.

EXAMPLE:

Carry out the following operations:

a. 65.587×654.0

We first calculate the answer, and then we can determine the number of significant figures:

$$65.587 \times 654.0 = 42893.898$$

We see that 65.587 has five sig. figs. and 654.0 has four sig. figs. Our answer should then have four sig. figs., so it is 4.289×10^4. Do not write 42890 because the trailing zero is ambiguous.

b. $648/0.02$

The answer is $648/0.02 = 32400$.

648 has two sig. figs. and 0.02 has one sig. fig. so our answer should have one sig. fig: 3×10^4.

c. $(4.58 \times 10^{-6}) \times (5.89 \times 10^8)$

We will take a slightly different approach here. First, let us collect the powers of ten together so that we can easily simplify that part of the equation (using exponent rules) and avoid calculator errors.

$$(4.58 \times 5.89) \times (10^{-6} \times 10^8) = (4.58 \times 5.89) \times (10^{-6+8}) = (4.58 \times 5.89) \times 10^2$$

Now we can continue the calculation:

$$26.9762 \times 10^2 = 2.69762 \times 10^3$$

Since both 4.58 and 5.89 have three sig. figs., our answer will have three sig. figs.: 2.70×10^3.

d. $5.687 - 0.000087$

We carry out the calculation: $5.687 - 0.000087 = 5.686913$. Since this is a subtraction problem, we will use the decimal place to determine the number of sig. figs. in our answer. The uncertainty in 5.687 is in the thousandths digit, whereas the uncertainty in 0.000087 is in the millionths digit. Our answer will therefore only be reported to the thousandths digit: 5.687.

e. $5897 - 12$

The answer is $5897 - 12 = 5885$. Since the uncertainty in both 5897 and 12 is in the ones digit, our answer will be reported to the ones digit as 5885.

f. $(9.687 \times 10^5) + (8.568 \times 10^4)$

Here we again want to isolate the powers of ten, but we will make both numbers multiplied by the same power of ten in order to add or subtract and easily determine the significant figures:

$$(96.87 \times 10^4) + (8.568 \times 10^4) = (96.87 + 8.568) \times 10^4$$

We can now carry out the addition and determine the significant digits without worrying about the power of ten:

$$(96.87 + 8.568) \times 10^4 = (105.438) \times 10^4$$

Since the uncertainty of 96.87 is in the hundredths place and the uncertainty of 8.568 is in the thousandths digit, our answer will be reported to the hundredths digit: 105.44×10^4. We can convert this to proper scientific notation to obtain our final answer: 1.0544×10^6.

g. $6.549 + 8.4 \times 12.2$

In this problem, we have more than one operation to carry out, so we will follow the order of operations and keep track of the uncertain digit as we go along.

First, we will multiply: $8.4 \times 12.2 = 102.48$. Since this is a multiplication problem, the number of sig. figs. will be the same as the term with the fewest sig. figs. So there should be two: $1\underline{0}2.48$. We will use an underline to indicate the last sig. fig. since we wait to round off until the end of the problem. We now have:

$6.549 + 1\underline{0}2.48 = 109.029$

Since this is an addition problem, we will use the decimal place to determine the sig. figs. in our answer. The uncertainty in 6.549 is in the thousandths place, while the uncertainty in $1\underline{0}2.48$ is in the tens place. Our answer will therefore be reported to the tens place: 110.

---

5. Density

   a. The density (d) of a substance is its mass (m) over its volume (V):

$$d = \frac{m}{V}$$

   b. An intensive property is one that is independent of the amount of a substance.

      i. Density is an intensive property.

   c. An extensive property is one that depends on the amount of substance in a sample.

      i. Mass is an extensive property.

   d. The units of density are those of mass over volume.

      i. The SI unit for density is kg/m^3, but it is rarely used.

      ii. Solids and liquids commonly have density units of g/mL or g/cm^3.

      iii. Gases commonly have density units of g/L.

EXAMPLE:

A solid sample of a metal is found to have a mass of 13.45 g and a volume of 4.98 mL. Calculate the density of the metal, and determine its possible identity using the following table of known densities.

Metal	Density (g/mL)
Aluminum	2.70
Iron	7.87
Silver	10.5
Lead	11.4

The density is simply the mass divided by the volume: 13.45g/4.98 mL = 2.70 g/mL. We can see from the table that this is the same density as aluminum, so we can conclude that the metal slug may be aluminum.

---

6. Energy and Its Units

   a. Energy is the capacity to do work, and work is a force applied over a distance.

   b. Most physical and chemical changes are accompanied by a change in energy.

c. The total energy is always equal to the sum of kinetic energy (energy associated with motion) and the potential energy (energy associated with position or composition).

d. Thermal energy is associated with temperature; which is a measure of the average kinetic energy of the constituents of a substance. The link between temperature and kinetic energy is important to remember for future chapters.

e. The law of conservation of energy states that energy cannot be created or destroyed but can change form and flow from one object to another.

f. Systems with high potential energy tend to change in a way that lowers their potential energy.

    i. In relative terms, low potential energy is more stable than high potential energy.

    ▶ In chemistry, "more stable" is synonymous with "lower potential energy."

g. The units of energy can be deduced from the definition of kinetic energy (KE):

$$KE = \frac{1}{2}mv^2$$

where m is the mass (in kg) and v is the velocity (in m/s).

    i. The joule (J) is the SI unit of energy.

$$1\ J = 1\ \frac{kg \cdot m^2}{s^2}$$

    ii. Calories (cal), nutritional calories (Cal), and kilowatt-hours (kWh) are also commonly used.

        1. 1 cal = 4.184 J

            a. The calorie is defined as the amount of energy required to raise the temperature of 1 g of a substance by 1 °C (at a pressure of 1 atm).

        2. 1 Cal = 1000 cal (as used in the food industry)

        3. $1\ kWh = 3.60 \times 10^6\ J$ (as used for electricity)

h. In this book, we will view the transfer of energy from the point of view of the system that we are studying.

    i. The surroundings will be defined as anything that interacts with the system.

    ii. A loss of energy from the system will be given a negative sign.

    iii. A gain of energy by the system will be given a positive sign.

i. An exothermic process is one in which the system loses heat, and so the change in energy is a negative number.

j. An endothermic process is one in which the system gains heat, and so the change in energy is a positive number.

7. Converting between Units

a. Dimensional analysis uses units to guide problem solving.

b. A conversion factor is a fractional quantity that is equivalent to one and relates equivalent quantities with two different units. For example, 1 mile = 1.61 km. This equivalency can be expressed as:

$$\frac{1\ mi.}{1.61\ km} = 1 \quad OR \quad \frac{1.61\ km}{1\ mi.} = 1$$

c. Multiplication by a conversion factor changes the units but not the actual quantity.

d. Unit conversion factors are always of the form:

$$\text{Given unit} \times \frac{\text{desired unit}}{\text{given unit}} = \text{desired unit}$$

where the ratio of the desired to given units is the conversion factor.

EXAMPLE:

How many miles will I have run in a half marathon (21.1 km)?

Here we see that the given unit of km needs to be converted to the desired unit of mi using the conversion factor given above:

$$21.1 \text{ km} \times \frac{1 \text{ mi}}{1.61 \text{ km}} = 13.1056 \text{ mi}$$

Our final answer should have three sig. figs., so we report our answer as 13.1 mi.

8. Problem-Solving Strategies

a. The general strategy for solving quantitative problems in chemistry is:

i. Identify the type of problem (conversion or equation).

ii. Make a list of the information given.

iii. Determine what you are looking for.

iv. Develop a plan for going from the given information to the desired answer.

v. Check your answer to be sure that it makes sense in the context of the problem and that you have reported the correct units and number of significant figures.

EXAMPLE:

Speed limits are typically reported in units of miles/hour. In science, speeds are often given in units of m/s. What is your speed in m/s if you are traveling 55 mph?

This is a conversion problem.

First, we identify the starting point of the problem: the initial quantity of 55 mi/hr.

Then, we identify the desired ending point: the equivalent quantity with units of m/s.

Next, we map out the path that will take us from the starting point to the ending point using conversion factors that we know (or can look up):

$$\frac{\text{mi.}}{\text{hr}} \xrightarrow{\text{mi. to km}} \frac{\text{km}}{\text{hr}} \xrightarrow{\text{km to m}} \frac{\text{m}}{\text{hr}} \xrightarrow{\text{hr to min}} \frac{\text{m}}{\text{min}} \xrightarrow{\text{min to s}} \frac{\text{m}}{\text{s}}$$

And now we can include the numbers and conversion factors:

$$\frac{55 \text{ mi}}{\text{hr}} \times \frac{1.61 \text{ km}}{1 \text{ mi}} \times \frac{1000 \text{ m}}{1 \text{ km}} \times \frac{1 \text{ hr}}{60 \text{ min}} \times \frac{1 \text{ min}}{60 \text{ s}} = 24.597 \frac{\text{m}}{\text{s}}$$

Our final answer should have two sig. figs., so we report our answer as 25 m/s.

EXAMPLE:

Saltwater, with a density of 1027 kg/m^3, is denser than pure deionized water. Convert the density of saltwater to the more conventional units of g/mL.

This is a conversion problem.

First, we identify the starting point of the problem: the initial quantity of 1027 kg/m^3.

Then, we identify the desired ending point: the equivalent density in units of g/mL.

Next, we map out our problem-solving pathway:

$$\frac{kg}{m^3} \xrightarrow{kg\ to\ g} \frac{g}{m^3} \xrightarrow{m\ to\ cm} \frac{g}{cm \cdot m^2} \xrightarrow{m\ to\ cm} \frac{g}{cm^2 \cdot m} \xrightarrow{m\ to\ cm} \frac{g}{cm^3} \xrightarrow{cm^3\ to\ mL} \frac{g}{mL}$$

▶ Note that we had to convert m^3 to cm^3 by converting m to cm three times since m^3 = m × m × m.

And now we include the numbers and conversion factors:

$$\frac{1027\ kg}{m^3} \times \frac{1000\ g}{kg} \times \frac{1\ m}{100\ cm} \times \frac{1\ m}{100\ cm} \times \frac{1\ m}{100\ cm} \times \frac{cm^3}{mL} = 1.027 \frac{g}{mL}$$

The calculated answer already has the correct number of sig. figs., so we are finished!

---

b. Order-of-magnitude estimates can be used as a fast method for checking your answer. An order-of-magnitude estimate looks at powers of ten in calculations.

EXAMPLE:

Do an order-of-magnitude calculation for the conversion of the density of saltwater from 1027 kg/m^3 to units of g/mL as described in the example above.

We will set up the problem in the same way but round off all numbers and look only at the powers of ten:

$$\frac{10^3\ kg}{m^3} \times \frac{10^3\ g}{kg} \times \frac{1\ m}{10^2\ cm} \times \frac{1\ m}{10^2\ cm} \times \frac{1\ m}{10^2\ cm} \times \frac{cm^3}{mL} = \frac{10^6\ g}{10^6\ mL} = 1 \frac{g}{mL}$$

We see from this that we should have an answer that is a number times 10^0 (or 1), which is in agreement with what we calculated above.

---

9. Solving Problems Involving Equations

a. Some problems require the use of an equation in solving, but the strategy above still applies.

EXAMPLE:

What is the density (in g/mL) of a brick of iron that measures 1.0 cm long, 2.0 cm wide, and 1.0 cm deep and weighs 15.72 g?

First, we will make a list of information given:

-   dimensions: 1.0 cm × 2.0 cm × 1.0 cm

-   weight: 15.72 g

Next, we determine what we are looking for: density in g/mL.

We have an equation for density ($d = m/V$), which we will use in solving this problem. So we need to find the mass and the volume in order to compute density.

First, we will consider volume. We are not given the volume, but we can get the volume from the dimensions.

$$\text{Volume} = \text{length} \times \text{width} \times \text{height} = 1.0 \text{ cm} \times 2.0 \text{ cm} \times 1.0 \text{ cm} = 2.0 \text{ cm}^3$$

Since the mass is given in the problem, we can simply plug the numbers into our density equation:

$$d = \frac{m}{V} = \frac{15.72 \text{ g}}{2.0 \text{ cm}^3}$$

The units in our calculated density are not the same as those requested, so we need to do a conversion:

$$\frac{15.72 \text{ g}}{2.0 \text{ cm}^3} \times \frac{1 \text{ cm}^3}{1 \text{ mL}} = \frac{15.72 \text{ g}}{2.0 \text{ mL}}$$

Finally, we will simplify, determine significant figures, and report our answer. The density of the iron sample is 7.9 g/mL.

Fill in the Blank:

1. Energy associated with position or composition is _____ energy.

2. The _____ states that energy cannot be created or destroyed; it may only change form or flow from one body to another.

3. A(n) _____ property is one that does not depend on the size of the sample or system.

4. The _____ of a substance is a measure of its average kinetic energy and is measured in units of _____.

5. The _____ of a measurement is a measure of how close it is to the actual value; the _____ is a measure of how close a set of measurements are to one another.

Problems:

1. Indicate the number of significant figures in the following numbers:
   a. 2305
   b. 1.20×10^4
   c. 0.00345
   d. 1.00230
   e. 1250

2. Carry out the following calculations being sure to keep track of significant figures:
   a. $3.854 + 52353$
   b. $2345 - 2324.023$
   c. 23966×32.0
   d. $92486 \div 893$
   e. $(2597 + 249) - 8673 \times 87 + 16.01$

3. The density of air is 1.184 kg/m^3 at 25 °C. Calculate the mass (in g) of 1 gallon of air.

4.   Perform the following unit conversions:

   a.   89000 mg to pounds

   b.   8.789 km to inches

   c.   0.489 weeks to seconds

   d.   67.580 mL to dm^3

5.   You measure the volume of a sample of aluminum using the method of displacement. Your measurements are 5.45 mL, 5.58 mL, 5.49 mL, and 5.51 mL. Calculate the average value, and comment on your accuracy if the actual value is reported to be 5.5 mL.

6.   You decide that you are going to put up a fence around your house. You go to the hardware store and find out that the fence costs $50 per yard. You then measure your yard and find that it is 40 feet wide and 150 feet long. How much will it cost to buy the fencing that you will need?

7.   A light-year is the distance that light will travel (in a vacuum) in one year. Given that the speed of light is 3.0×10^8 m/s, how many miles will light travel in one year?

8.   A small fish tank will hold 29 gallons of water. If the tank leaks water at a rate of 1.0 mL/min., how long (in days) will it take for all of the water to drain from a full tank?

9.   Consider a chunk of concrete that measures 1.0 ft wide, 1.0 ft long, and 4 in. tall. This type of concrete has a density of 2403 kg/m^3.

   a.   What is the mass (in lb) of the chunk of concrete?

   b.   Will the chunk of concrete float in 100 lb of water? Explain.

   c.   Will the chunk of concrete float in 100 lb of water if it is first broken up into small (<1 lb) pieces? Explain.

Concept Questions:

1.   A student collects from her classmates the following measurements of the length of a stick: 4.57 cm, 4.67 cm, 4.64 cm, 4.58 cm, and 4.6 cm.

   a.   Did all of the students use the same measuring device? Explain how you arrived at your answer.

   b.   Comment on the accuracy and precision of the class results.

2.   Dimensional analysis is used all the time in real life, but we usually don't recognize it as such. State one example from your life where you have used dimensional analysis.

3.   When wood burns, the chemical potential energies of wood and oxygen (the reactants of the combustion) are, collectively, higher than the potential energy of carbon dioxide and water (the products of the combustion). This is why heat is released when we burn wood in our fireplace. Explain, on a microscopic or molecular level, what happens to this excess energy.

Chapter 1: Atoms

Key Learning Outcomes:

- Classify Matter by State and Composition

- Distinguish between Laws and Theories

- Apply the Law of Definite Proportions

- Apply the Law of Multiple Proportions

- Work with Atomic Numbers, Mass Numbers, and Isotope Symbols

- Calculate Atomic Mass

- Convert between Mass and Amount (Number of Moles)

- Use the Mole Concept

Chapter Summary:

In this chapter, you will be introduced to atoms, the fundamental building blocks of all matter. You will learn how to classify matter based on its physical state and by its composition. After an introduction to the scientific method, you will learn how scientists progressed from early ideas of matter to modern atomic theory as summarized by John Dalton. Subatomic particles, protons, neutrons, and electrons will be compared and contrasted. Once you have learned how to identify the number of each subatomic particle in a given atom or ion, you will be able to determine its average atomic mass. Next, you will learn about the mole, which is a unit of quantity that will be used throughout the remainder of this book. Finally, the chapter concludes with a brief discussion of the origin of atoms.

Chapter Outline:

1. A Particle View of the World: Structure Determines Properties

    a. Matter is anything that occupies space and has mass.

    b. Matter is particulate, meaning it is composed of particles.

    c. The structure of matter's particles determines its properties.

        i. Atoms are the basic particles of matter.

        ii. Molecules are atoms bound together into a single unit.

    d. Chemistry is the science that seeks to understand the properties of matter by studying the structure of the particles that compose it.

2. Classifying Matter: A Particulate View

    a. A substance is a specific instance of matter.

    b. Substances can be classified by phase or by composition.

        i. The phase of matter is classified by its physical state. There are three states of matter: solid, liquid, and gas.

            1. A solid consists of a large collection of particles (atoms, molecules, or ions) held close together and maintains a rigid shape. Solids can be orderly (crystalline) or lack long-range order (amorphous).

            2. A liquid consists of a large collection of particles held close together but able to move relative to each other. A liquid is therefore a relatively incompressible fluid that can assume the shape of its container.

3. A gas consists of a large collection of particles that are very far apart. A gas is thus a compressible fluid. Gases assume the shape and volume of the containers that hold them.

ii. Depending on the variability of its composition, matter is classified as a pure substance or a mixture.

1. A pure substance has no variability in its composition. No matter what the size of the sample is or how it is prepared, it will always be composed of the same "building blocks."

a. A pure substance can be classified as a compound or an element. An element is a substance that contains only one type of atom. A compound is a substance that contains atoms of two or more chemical elements in fixed proportions; it can be separated into elements by chemical means.

2. A mixture can have a variable composition. A mixture can be prepared in an infinite number of ways; each time it is made, it will be different on the microscopic scale.

a. A mixture can be classified as either homogeneous or heterogeneous. A homogeneous mixture is uniform throughout; a sample of the mixture taken from one area will be identical to a sample of the mixture taken from another area. A heterogeneous mixture has variations throughout the sample.

In general, a homogeneous mixture forms when the constituent parts can move freely in random directions; this constant, random motion results in mixing on the molecular/atomic level. For example, gaseous water and gaseous carbon dioxide will mix to form a homogeneous mixture, while solid water and solid carbon dioxide will form a heterogeneous mixture.

b. A mixture can be separated based on differences in the physical and chemical properties of its components.

When deciding if a sample is a pure substance or a mixture, think about the recipe for its preparation. A pure substance has a definite recipe, whereas a mixture can have an infinite number of recipes. For example, water will always be made of two hydrogen atoms and one oxygen atom, so water is a pure substance; saltwater can have a pinch of salt or a bucket of salt in a given amount of water, so saltwater is a mixture.

EXAMPLE:

Identify the following as pure substances or mixtures. If it is a substance, determine whether it is an element or a compound. If it is a mixture, indicate if it is a homogeneous or heterogeneous mixture.

a. Salt and sugar that have been ground down to a fine powder

This is a mixture; it contains two distinct chemical compounds: sugar and salt. This mixture is heterogeneous because no matter how small the crystals have become, the lack of dynamic motion in the solid state will ultimately prevent uniform dispersion of the two solids.

b. Glucose

This is a pure substance. Since it contains more than one type of atom (C, H, O), it is a compound.

c. Stainless steel

This is a mixture of various elements that have been combined into an alloy. Since the elements are not chemically bound to one another, they are not constituents of a compound. This is a homogeneous mixture that has the same microscopic composition throughout.

d. Exhaled breath

A breath contains a number of different molecules and atoms, which means that it is a mixture. Since the mixture is a fluid that is in constant motion, the mixture is homogeneous or the same throughout.

---

3. The Scientific Approach to Knowledge

    a. Chemistry is an empirical science, meaning that it is based on experimentation and observation.

    b. The scientific method begins with an observation that leads to a tentative explanation (hypothesis). The hypothesis is repeatedly tested by experimentation and then either accepted, revised, or discarded.

    c. Scientific laws are generalizations about nature that summarize past observations and predict future ones (the *what*).

    d. Theories are models based on well-established hypotheses and validated by experimental evidence. By explaining the generalizations stated in scientific laws, theories give insight into how nature works (the *how* and *why*).

        i. Atomic theory is the root of the particulate view of matter and will be discussed later in this chapter.

    e. It is important to realize that science is constantly expanding and changing because a law or theory cannot be "proven" absolutely. The "strongest" theories are those that have been consistently supported by experimental data over time, but any theory that fails to account for new observations must be modified or discarded.

    f. Science is quantifiable—it can be expressed numerically.

        i. A unit is a standard, agreed-upon quantity used to specify a given measurement.

            1. Two systems of units are common: the metric system and the English system.

            2. Scientists use the International System of Units (SI), which is based on the metric system.

4. Early Ideas about the Building Blocks of Matter

    a. Matter was considered to be composed of fire, earth, air, and water until the sixteenth century, which brought about modern science.

5. Modern Atomic Theory and the Laws That Led to It

    a. The law of conservation of mass states that matter is not created or destroyed in a chemical reaction. A chemical reaction results in the rearrangement of atoms, but the total mass is not changed.

    b. The law of definite proportions states that all samples of a given compound have the same proportion of their constituent elements no matter how the compound is prepared or where it comes from.

EXAMPLE:

A sample of methane, CH_4, is found to have 46.12 g of carbon and 3.87 g of hydrogen. Another sample is analyzed and found to have 51.25 g of carbon and 4.30 g of hydrogen. Analyze these results in the context of the law of definite proportions.

First, we will analyze the first sample. We simply take the ratio of the mass of carbon to the mass of hydrogen and simplify:

$$\frac{46.12 \text{ g C}}{3.87 \text{ g H}} = 11.9$$

So the C/H mass ratio is 11.9:1.

The second sample will be inspected in a similar manner:

$$\frac{51.25 \text{ g C}}{4.30 \text{ g H}} = 11.9$$

Again, the C/H mass ratio is 11.9:1.

We see from this example that methane has an 11.9:1 mass ratio of carbon to hydrogen despite the sample size, which is consistent with the law of definite proportions.

---

    c.   The law of multiple proportions states that when two elements combine to form two different compounds, the masses of one element that combines with 1 g of the other element can be expressed as a ratio of small whole numbers.

EXAMPLE:

Many different iron oxides are commonly encountered in nature. Wustite contains 2.62 g of iron for every 1.00 g of oxygen, while magnetite contains 3.49 g of iron for every 1.00 g of oxygen. Show that this ratio is consistent with the law of multiple proportions.

We will construct a ratio of the masses of iron and find the smallest whole-number ratio:

$$\frac{3.49 \text{ g Fe in magnetite}}{2.62 \text{ g Fe in wustite}} = 1.33$$

This is not a simple whole number, so we must multiply the top and bottom by a whole number that will give us a whole-number ratio:

$$1.33 \cdot \frac{3}{3} = \frac{4}{3}$$

We see that 4:3 is a simple whole-number ratio consistent with the law of multiple proportions.

---

    d.   John Dalton's atomic theory centers on the following four ideas:

        i.   Each element is composed of tiny, indestructible particles called atoms.

        ii.   All atoms of a given element have properties (including mass) that differentiate them from atoms of other elements.

        iii.   Atoms combine in whole-number ratios to form compounds.

        iv.   Atoms of one element cannot change into atoms of another element via chemical means. New substances are formed in a chemical reaction by atoms rearranging the way they are bound to other atoms.

           Not all of these postulates remain valid today; modern atomic theory takes into account nuclear reactions, isotopes, and nonstoichiometric compounds.

6.   The Discovery of the Electron

    a.   When a high electrical voltage is applied between the electrodes of a cathode ray tube, cathode rays travel from the negatively charged cathode to the positively charged anode.

    b.   J. J. Thomson discovered that cathode rays are composed of electrons, which are very small, negatively charged particles contained within atoms. He measured the electron's charge-to-mass ratio as -1.76×10^8 C/g, where C represents Coulomb.

    c.   The charge of an electron was determined by Robert Millikan using charged droplets of oil suspended in an electric field. The charge of a single electron is -1.60×10^{-19} C. The mass of an electron is 9.10×10^{-28} g.

7.   The Structure of the Atom

    a.   Ernest Rutherford used positively charged alpha particles aimed at gold foil to show the existence of the nucleus.

        i.   Alpha particles are one type of ionizing radiation or radioactivity (to be discussed in Chapter 19).

    b.   The nuclear theory of an atom states that:

        i.   Most of an atom's mass and all of its positive charge are contained in a small, dense core called the nucleus.

        ii.   Most of the volume of an atom is empty space; throughout this space, negatively charged electrons are dispersed.

        iii.   The number of electrons is equal to the number of protons so that charge neutrality is maintained.

    c.   James Chadwick showed that the nucleus also contains neutrons, which are small, massive, neutral particles.

8.   Subatomic Particles: Protons, Neutrons, and Electrons in Atoms

    a.   Both protons and neutrons are found in the nucleus of an atom and have approximately the same mass: 1 atomic mass unit (amu).

        i.   An atomic mass unit is defined as 1/12 the mass of a carbon-12 atom.

    b.   Electrons are found outside of the nucleus and have a mass of 0.00055 amu.

▶   In practice, we say that protons and neutrons each have a mass of 1 amu and electrons have zero mass—this allows for easy estimates of the atomic mass.

    c.   The charge of an electron in *relative* units is −1, and the charge of a proton in *relative* units is +1.

    d.   Elements are defined by the number of protons in the nucleus. This is called the atomic number (Z) and is the number of the element on the periodic table.

    e.   Each element has a unique chemical symbol that is a one- or two-letter abbreviation of its name. Note that the abbreviations are sometimes derived from non-English names; for example, the symbol for gold is Au, which is from the Latin word for gold, *aurum*.

        i.   In a chemical symbol, the first letter is always capitalized, and the second letter (if there is one) is always lowercase.

    f.   Not all atoms of a given element have the same number of neutrons.

        i.   Atoms with the same number of protons but a different number of neutrons are called isotopes.

        ii.   The natural abundance of an element's particular isotope is the percent of a naturally occurring sample of the element that is a particular isotope.

EXAMPLE:

A sample of 10^{10} atoms of chlorine is collected. Chlorine has two stable isotopes: chlorine-35 and chlorine-37. The natural abundance of chlorine-35 is 75.77%, and the natural abundance of chlorine-37 is 24.23%. How many atoms of the sample are chlorine-35, and how many are chlorine-37?

The natural abundance is the percentage of atoms in a sample that is a particular isotope, so we can simply calculate the percentage of the 10^{10} chlorine atoms that is each isotope by multiplying the total number of atoms by the percentage (expressed as a decimal):

Cl-35: $10^{10} \times 0.7577 = 7.577 \times 10^9$ chlorine-35 atoms in the sample

Cl-37: $10^{10} \times 0.2423 = 2.423 \times 10^9$ chlorine-37 atoms in the sample

---

g. The mass number (A) is the sum of protons and neutrons in an atom.

h. The full symbol for an element uses the chemical symbol (X), the mass number (A), and the atomic number (Z):

$$_Z^A X$$

i. Since inclusion of the atomic number and the element symbol is redundant, the full symbol is often abbreviated in one of two ways:

$$^A X \text{ or } X - A$$

▶ Note: The textbook does not use the $^A X$ notation, but you may see it used elsewhere, including in this study guide.

EXAMPLE:

Give the full atomic symbol for each of the following:

a. chlorine-35

Chlorine's symbol is Cl, it has 17 protons, and the mass number given is 35; the full symbol is $_{17}^{35}Cl$.

b. aluminum-27

Aluminum's symbol is Al, it has 13 protons, and the mass number given is 27; the full symbol is $_{13}^{27}Al$.

c. calcium-44

Calcium's symbol is Ca, it has 20 protons, and the mass number given is 44; the full symbol is $_{20}^{44}Ca$.

d. lithium-6

Lithium's symbol is Li, it has 3 protons, and the mass number given is 6; the full symbol is $_3^6Li$.

---

j. Ions are atoms that do not have an equal number of electrons and protons.

   i. Cations are positively charged ions; they have fewer electrons than protons.

   ii. Anions are negatively charged ions; they have more electrons than protons.

▶ Ions are formed when electrons (and NOT protons) are added/removed. Note: Changing the number of protons would change the identity of the atom.

▶ The following table summarizes the important aspects of each of the three subatomic particles.

Name	Abbreviation	Determines the…	Relative Mass (amu)	Relative Charge
Proton	p	identity	1	+1
Neutron	n	isotope	1	0
Electron	e^-	charge	0	−1

EXAMPLE:

State the number of electrons, protons, and neutrons in each of the following examples:

a. $^{23}Na^+$

Sodium is atomic number 11 on the periodic table, so the given ion has 11 protons. The number of neutrons is given by the mass number (23 in this case) minus the number of protons, so it has 12 neutrons. The number of electrons in a neutral atom is equal to the number of protons, but in this case we have a 1+ ion, so there is one less electron than proton; thus, there are 10 electrons.

b. $^{40}Ca^{2+}$

Calcium is atomic number 20 on the periodic table, so the given ion has 20 protons. The number of neutrons is given by the mass number (40 in this case) minus the number of protons, so it has 20 neutrons. The number of electrons in a neutral atom is equal to the number of protons, but in this case we have a 2+ ion, so there are two less electrons than protons; thus, there are 18 electrons.

c. $^{54}Fe^{3+}$

Iron is atomic number 26 on the periodic table, so the given ion has 26 protons. The number of neutrons is given by the mass number (54 in this case) minus the number of protons, so it has 28 neutrons. The number of electrons in a neutral atom is equal to the number of protons, but in this case we have a 3+ ion, so there are three less electrons than protons; thus, there are 23 electrons.

d. $^{18}O^{2-}$

Oxygen is atomic number 8 on the periodic table, so the given ion has 8 protons. The number of neutrons is given by the mass number (18 in this case) minus the number of protons, so it has 10 neutrons. The number of electrons in a neutral atom is equal to the number of protons, but in this case we have a 2– ion, so there are two more electrons than protons; thus, there are 10 electrons.

e. $^{37}Cl^-$

Chlorine is atomic number 17 on the periodic table, so the given ion has 17 protons. The number of neutrons is given by the mass number (37 in this case) minus the number of protons, so it has 20 neutrons. The number of electrons in a neutral atom is equal to the number of protons, but in this case we have a 1– ion, so there is one more electron than protons; thus, there are 18 electrons.

9. Atomic Mass: The Average Mass of an Element's Atoms

   a. The atomic mass is listed on the periodic table and is the relative average mass of that atom's isotopes weighted by their natural abundance.

$$\text{Atomic mass} = \sum_n (\text{fraction of isotope } n) \times (\text{mass of isotope } n)$$

EXAMPLE:

Neon has three stable isotopes: neon-20, neon-21, and neon-22. Neon-20 has a mass of 19.9924 amu and a natural abundance of 90.60%. Neon-21 has a mass of 20.9938 amu and a natural abundance of 0.26%. Neon-22 has a mass of 21.9914 amu and a natural abundance of 9.20%. Calculate the average atomic mass of neon.

In order to calculate the average mass, we will add together the masses of each of the isotopes multiplied by their natural abundance converted to a decimal.

$$(19.9924 \text{ amu} \times 0.9060) + (20.9938 \text{ amu} \times 0.0026) + (21.9914 \text{ amu} \times 0.0920) = 20.19 \text{ amu}$$

If we look on the periodic table, we see that the atomic mass is listed as 20.18 amu, which is very close to our calculated answer. Note that the difference is probably a result of more or less precise values of the mass used in our calculation versus the ones used for the periodic table.

  b. The masses of atoms and the percent abundances of their isotopes are determined using mass spectrometry.

 10. Atoms and the Mole: How Many Particles?

  a. The particles that compose matter are too small to count, so we need to determine amounts by weighing samples.

  b. Avogadro's number, 6.022×10^{23}, gives the number of particles (electrons, atoms, ions, molecules, etc.) in 1 mole of material.

  c. One mole is defined as the number of atoms in exactly 12 g of pure carbon-12.

  d. The mole is an SI base unit and is abbreviated as mol.

  e. Avogadro's number is a conversion factor between moles (chemical amount) and the number of particles (a count).

EXAMPLE:

Calculate the number of particles in each of the following samples:

a. 3.468 moles of gold atoms

$$3.468 \text{ mol Au atoms} \times \frac{6.022 \times 10^{23} \text{ Au atoms}}{1 \text{ mol Au atoms}} = 2.088 \times 10^{24} \text{ Au atoms}$$

b. 2.45 moles of Ca^{2+} ions

$$2.45 \text{ mol Ca}^{2+} \text{ ions} \times \frac{6.022 \times 10^{23} \text{ Ca}^{2+} \text{ ions}}{1 \text{ mol Ca}^{2+} \text{ ions}} = 1.48 \times 10^{24} \text{ Ca}^{2+} \text{ ions}$$

c. 1.87×10^{-13} moles of CCl_4 molecules

$$1.87 \times 10^{-13} \text{ mol CCl}_4 \text{ molecules} \times \frac{6.022 \times 10^{23} \text{ CCl}_4 \text{ molecules}}{1 \text{ mol CCl}_4 \text{ molecules}} = 1.13 \times 10^{11} \text{ CCl}_4 \text{ molecules}$$

---

  f. The molar mass of an element is the mass of 1 mol of its atoms and is numerically equal to the element's atomic mass in amu.

  ▶ You can check your work when calculating the molar mass of an atom by determining the mass number—the two should be roughly the same.

EXAMPLE:

How many gold atoms are there in 5.89 g of solid Au?

In solving this problem, we follow exactly the same procedure as in the previous example, but we include one additional step. First, we convert mass to moles (using molar mass) and then moles to number of atoms (using Avogadro's number):

$$5.89 \text{ g Au atoms} \times \frac{1 \text{ mol Au Atoms}}{196.97 \text{ g Au Atoms}} \times \frac{6.022 \times 10^{23} \text{ Au atoms}}{1 \text{ mol Au Atoms}} = 1.80 \times 10^{22} \text{ Au atoms}$$

---

11. The Origins of Atoms and Elements

   a. The birth of the universe is described by the Big Bang Theory.

      i. The Big Bang Theory asserts that the beginning of the universe came about as a hot dense collection of matter that rapidly expanded.

Fill in the Blank:

1. The amount of material containing 6.022×10^{23} particles is a(n) _____.

2. The value of an element's molar mass in grams per mole is numerically equal to the element's _____.

3. Avogadro's number is the number of atoms contained in exactly 12 g of _____.

4. The _____ states that all samples of a given compound have the same proportions of their constituent elements.

5. Most of the volume of an atom is empty space, throughout which are dispersed _____.

6. Cathode ray tubes were used in the discovery of _____.

7. The law of conservation of mass states that in a chemical reaction, matter is never _____.

8. Most of an atom's mass and all of its positive charge are contained in the atom's _____.

9. A good hypothesis makes predictions that can be confirmed or refuted by further _____.

10. The number of _____ in an atom defines its identity.

11. The percentage of a particular isotope of an atom found in a natural sample of that element is called the _____.

12. _____ are brief statements that summarize past observations and predict future ones.

13. The three states of matter are _____, _____, and _____.

14. A(n) _____ mixture is one in which the composition is uniform throughout the sample.

15. When two elements combine to form different compounds, the masses of one element that combine with 1 g of the other element can be expressed as a(n) _____.

16. One of the postulates of John Dalton's atomic theory is that each element is composed of tiny, indestructible particles called _____.

17. A(n) _____ is a substance that cannot be chemically broken down into simpler substances.

18. A(n) _____ property is one that is displayed only when the composition is changed through a chemical change.

19. The percent abundance of elements is measured using _____.

20. _____ are positively charged ions.

21. A(n) _____ is two or more atoms joined together via chemical bonds in a specific geometric orientation.

22. The SI unit for mass is the _____.

Problems:

1. Classify each of the following as a mixture or a pure substance:

    a. Gasoline

    b. Milk

    c. Gravel

    d. Nail polish remover

    e. A silver ring

    f. Vegetable soup

2. Classify each of the following as a homogeneous or heterogeneous mixture:

    a. Freshly squeezed lemonade

    b. Ocean water

    c. Lemonade drink made from a powder mix and water

    d. Oil and vinegar salad dressing

    e. Gas in a scuba diver's tank

    f. Hydrogen peroxide solution

3. A sample of glucose is found to have 34.92 g of carbon, 5.87 g of hydrogen, and 46.56 g of oxygen. Another sample is found to have 0.4471 g of carbon, 0.07510 g of hydrogen, and 0.5962 g of oxygen. Show that these results are consistent with the law of definite proportions.

4. Complete the following table:

Atomic Symbol	Typical Ion Formed	Number of Electrons	Number of Protons	Number of Neutrons
^{40}Ca	Ca^{2+}			
	Be^{2+}			5
^{79}Se		36		
		18	17	20

5. Identify the anion or cation that will most likely form from each of the following elements:

    a. Na

    b. O

    c. Mg

    d. Al

    e. S

    f. Cl

$.07(6.941) + .93(6.941)$
$= 6.941$

$\dfrac{7}{100} \times (6.941) + (.93)(6.941)$

$\dfrac{93}{100} : .93$

6. Lithium has two isotopes, ^6Li and ^7Li, with natural abundances of 7% and 93%, respectively. What is the average atomic mass of lithium?

7. Magnesium has three naturally occurring isotopes: ^{24}Mg (23.985042 amu and 78.99% abundance), ^{25}Mg (24.965837 amu and 10.00% abundance), and ^{26}Mg (25.962593 amu and 11.01% abundance). What is the average mass of a magnesium atom?

$(23.985042 \text{ amu} \times .7899) + (24.965837 \times .1) + (.1101 \times 25.962593)$

$\approx 24.3 \text{ amu}$

8. Using the masses of the proton, neutron, and electron given in your book, calculate the mass of 1.0 mole of sodium ions and the mass of one mole of sodium atoms. Based on your answer, do you think that it is reasonable to report the mass of one mole of sodium ions as equal to the mass of one mole of sodium atoms (as is generally done)? Why or why not?

9. How many silver atoms are there in a 10.5 g sample of solid silver?

10. What is the mass of 5.89×10^{30} atoms of Al?

Concept Questions:

1. One of the arguments against teaching evolution in public schools is that it is just a theory. Why is this a weak argument against the inclusion of evolution in a science curriculum?

2. Dry air is a mixture of nitrogen (76%), oxygen (23%), argon (1%), and trace amounts of other substances.

   a. Air can be separated into its components. Is this a physical or chemical change? Explain.

   b. Is the mixture homogeneous or heterogeneous? Explain.

3. The mass of a substance is the amount of matter in that substance. We define 12 amu to be the mass of an atom of carbon-12. Other elements do not have integer masses; for example, a hydrogen-1 atom has a mass of 1.007825 amu. Explain why you think that this might be.

4. Silicon has three naturally occurring isotopes: ^{28}Si, ^{29}Si, and ^{30}Si. Their natural abundances are 92.3%, 4.7%, and 3.0%, respectively. Without doing any calculations, estimate the average molar mass of silicon to two significant figures. Explain the reasoning behind your answer.

5. The law of conservation of mass states that matter is not created or destroyed in a chemical reaction. In one particular reaction, 100.0 g of water decompose to form 11.1 g of hydrogen gas according to the following equation:

$$2H_2O \rightarrow 2H_2 + O_2$$

According to the law of conservation of mass, how much oxygen (in g) must be produced in this reaction? Explain the reasoning behind your answer.

6. The masses listed on the periodic table are weighted averages, as discussed in your book. When we do calculations in chemistry, these are the masses that we use. Why is it reasonable to use the average mass when carrying out calculations despite the fact that we don't know the exact composition of the isotopes in a macroscopic sample? Will it always be acceptable to use the average masses?

7. Enriched uranium has a higher concentration of uranium-235 than is found in nature. One method for obtaining enriched uranium involves ionizing a raw sample of uranium (containing U-235 and the more abundant U-238) and then separating the components using their mass-to-charge ratio. Explain how this is done using what you have learned about mass spectrometry in this chapter. Which isotope will be collected first in the experiment? Explain how you arrived at your answer.

8. The atomic mass (in amu) is numerically equivalent to the molar mass (in g). Does this imply that 1 g is equal to 1 amu? Explain your answer.

9. The mass of a substance is the amount of matter in that substance. We define 12 amu to be the mass of an atom of carbon-12. Other elements do not have integer masses; for example, a hydrogen-1 atom has a mass of 1.007825 amu. Explain why you think that this might be.

10. The masses listed on the periodic table are weighted averages, as discussed in your book. When we do calculations in chemistry, these are the masses that we use. Why is it reasonable to use the average mass when carrying out calculations despite the fact that we don't know the exact composition of the isotopes in a macroscopic sample? Will it always be acceptable to use the average masses?

Chapter 2: The Quantum-Mechanical Model of the Atom

Key Learning Outcomes:

- Relate the Wavelength and Frequency of Light

- Calculate the Energy of a Photon

- Relate Wavelength, Energy, and Frequency to the Electromagnetic Spectrum

- Use the de Broglie Relation to Calculate Wavelength

- Relate Quantum Numbers to One Another and to Their Corresponding Orbitals

- Relate the Wavelength of Light to Transitions in the Hydrogen Atom

Chapter Summary:

In this chapter, you will explore the subject of quantum mechanics, which provides a model for atomic structure and behavior. You will begin by looking at electromagnetic radiation (light) and the electromagnetic spectrum. Properties of light such as frequency, wavelength, and amplitude will be explained, and then you will learn the dependence of these properties on one another. The wave nature of light will be discussed in terms of interference, and then the particle properties of light will be explored using the photoelectric effect. The connection between light and matter will be introduced using atomic spectroscopy, and the first theory to explain emission and absorption spectra, the Bohr model, will be discussed. In order to explore a more correct description of electron behavior, you will then look at the wave-particle duality of electrons. Your understanding of electron wave properties will be used to introduce quantum-mechanical orbitals, which are specified using four quantum numbers and provide a probability distribution map for electrons in atoms. You will then learn how to calculate the energy associated with an orbital using spectroscopic data. Finally, we will explore each of the different orbital shapes in order to understand the physical characteristics of the atoms that result from them.

Chapter Outline:

1. Schrödinger's Cat

    a. Very small particles can exist in multiple states simultaneously.

    b. Behaviors of very small things are affected by our observations of them. The observation causes multiple states to collapse into a single observable.

    c. The quantum-mechanical model explains how electrons exist in atoms and how these electrons determine the chemical and physical properties of elements.

2. The Nature of Light

    a. Some properties of light are best described using its wave characteristics.

        i. Light is electromagnetic radiation that consists of electric and magnetic fields oscillating perpendicular to one another while propagating through space.

        ii. The speed of light in a vacuum is a constant, $c = 3.00 \times 10^8$ m/s.

        iii. Electromagnetic waves are characterized by their amplitude, wavelength, and frequency.

            1. The amplitude is the intensity of light.

            2. The wavelength (λ) is the distance between two crests of a light wave.

            ▶ Wavelength and amplitude are both related to energy but are independent of each other.

3. The frequency (v) is the number of cycles that pass a point in a given period of time. The units of frequency are s^{-1} or Hz, where $1 \text{ Hz} = 1 \text{ s}^{-1}$.

4. The wavelength and frequency are inversely related through the speed of light:

$$v = \frac{c}{\lambda}$$

EXAMPLE:

Calculate the frequency of red light with a wavelength of 700. nm.

We will first convert the wavelength of light from nm to m:

$$700. \text{ nm} \times \frac{1 \text{ m}}{10^9 \text{ nm}} = 7.00 \times 10^{-7} \text{ m}$$

We can now use this to calculate the frequency:

$$v = \frac{c}{\lambda} = \frac{3.00 \times 10^8 \text{ m/s}}{7.00 \times 10^{-7} \text{ m}} = 4.29 \times 10^{14} \text{ s}^{-1}$$

5. The electromagnetic spectrum includes all wavelengths of light.

6. The regions of the electromagnetic spectrum have different names.

    a. In order of increasing wavelength or decreasing frequency: gamma rays, X-rays, ultraviolet light, visible light, infrared radiation, microwaves, and radio waves.

  iv. Interference occurs when waves interact with one another.

1. Constructive interference occurs when waves travel in phase with one another. When the crests of waves overlap, their amplitudes are additive.

2. Destructive interference occurs when waves travel out of phase with respect to one another. The waves cancel each other out.

  v. Diffraction is the phenomenon of a wave bending around an obstacle or slit.

  vi. The diffraction of light through two closely spaced slits results in an interference pattern that represents regions of constructive and destructive interference. This is an inherent property of waves.

  b. Some properties of light are best described using its particle characteristics.

    i. The photoelectric effect occurs when electrons are emitted from metal surfaces upon their exposure to light.

1. Electrons are emitted from a metal when light of sufficient energy is shined upon its surface.

    a. The minimum energy of light required to emit electrons from the surface of a metal is called the threshold energy.

    b. When the intensity of the light is increased, more electrons are emitted. When the intensity is decreased, fewer electrons are emitted. In both cases, the electron energies for a given wavelength are the same.

2. If light with an energy lower than the threshold frequency is shined on a metal, electrons will not be emitted no matter the light intensity.

ii.   Light comes in packets (called photons or quanta) that have a pa quantized:

$$E = h\nu = \frac{hc}{\lambda}$$

where h is Planck's constant and is equal to 6.626×10^{-34} J·s.

EXAMPLE:

What is the energy of red light with a wavelength of 700. nm?

We calculated the frequency of this light in the last problem, so we can easily calculate the energy:

$$E = (6.626 \times 10^{-34} \, J \cdot s) \times (4.29 \times 10^{14} \, s^{-1}) = 2.84 \times 10^{-19} \, J$$

Note that visible light will always have energy that is on the order of 10^{-19} J, as in this problem.

---

iii.   The threshold frequency is related to the binding energy of the electron to the metal (also called the metal's work function), ϕ:

$$h\nu = \phi$$

▶   The greater the binding energy (i.e., the more tightly held is an electron), the greater is the frequency (thus, energy) of light needed to eject it.

iv.   When energy greater than the binding energy is incident on the surface of a metal, electrons are emitted with a kinetic energy (KE) that is equal to the difference between the energy of the incident light and the binding energy of the particular metal:

$$KE_{ejected\ electron} = h\nu - \phi$$

▶   The idea here is that electrons need a certain energy in order to "escape" the atom. Any additional energy makes them move faster (kinetic energy).

c.   Wave-particle duality of light refers to the fact that light is best described as a wave or as a particle depending on the circumstances.

EXAMPLE:

What is the threshold frequency of a metal that ejects electrons with a kinetic energy of 1.58×10^{-19} J when 400. nm light is shined upon its surface?

We are given the $KE_{ejected\ electron}$ and the wavelength of the incident light. We can convert the latter from nm to m:

$$400 \text{ nm} \times \frac{1 \text{ m}}{10^9 \text{ nm}} = 4.00 \times 10^{-7} \text{ m}$$

We can use these data to calculate the binding energy of the metal:

$$KE = \frac{hc}{\lambda} - \varphi$$

$$1.58 \times 10^{-19} J = \frac{(6.626 \times 10^{-34} J \cdot s)(3.00 \times 10^8 \, m/s)}{4.00 \times 10^{-7} m} - \varphi$$

$$\varphi = 3.39 \times 10^{-19} \, J$$

ally, we can calculate the threshold frequency from the binding energy:

$$(3.39 \times 10^{-19}\,\text{J}) = (6.626 \times 10^{-34}\,\text{J·s})\ \nu$$

$$\nu = 5.12 \times 10^{14}\,\text{s}^{-1}$$

---

3. Atomic Spectroscopy and the Bohr Model

    a. After energy is added to an atom, it is often released in the form of light.

    b. Atomic spectroscopy is the study of the electromagnetic radiation that is absorbed and/or emitted by atoms.

    c. Line spectra are emission (or absorption) spectra of atoms with the emitted (or absorbed) wavelengths of light separated from each other.

        i. Atomic line spectra are unique for different elements and are always the same for a given element. They serve as an "atomic fingerprint."

    d. Continuous spectra have no sudden interruptions in the intensity of light as a function of the wavelength.

    e. The Rydberg equation predicts the line spectrum for hydrogen:

$$\frac{1}{\lambda} = R\left(\frac{1}{m^2} - \frac{1}{n^2}\right)$$

        where n and m are integers and R is the Rydberg constant $= 1.097 \times 10^7\,\text{m}^{-1}$.

    f. The Bohr model was the first atomic model introduced to explain line spectra.

        i. In the Bohr model of the atom, electrons were described as orbiting the nucleus at particular distances; these distances were quantized (at particular energy values) so that electrons could only jump from one orbit to another but could not be in between.

            1. Energy is absorbed when an electron goes from a lower-energy orbit to a higher-energy orbit.

            2. Energy is emitted when an electron goes from a higher-energy orbit to a lower-energy orbit.

    g. The Bohr model was replaced by a model that considers electrons as waves.

      ▶ It is important to realize that electrons do NOT travel in orbits—this was the first promising model set forth but was replaced with a more thorough and correct description of electron behavior: the quantum-mechanical description.

    h. Each element on the periodic table has a unique spectrum associated with it.

        i. A flame test utilizes the emission spectrum to identify ions in a sample.

    i. An absorption spectrum consists of dark lines on a bright background. When white light is passed through a sample, certain wavelengths are absorbed.

    j. In emission, an electron makes a transition from high to low energy. In absorption, an electron makes a transition from low to high energy.

4. The Wave Nature of Matter: The de Broglie Wavelength, the Uncertainty Principle, and Indeterminacy

    a. Electron beams that travel through two closely spaced slits exhibit a diffraction pattern.

        i. The diffraction pattern occurs even when electrons are sent through the slits one at a time.

        ii. The diffraction pattern occurs from single electrons interfering with themselves.

        iii. The wave nature of an electron is an inherent property of an individual electron.

b. The de Broglie relation mathematically relates the wave properties (wavelength: λ) to the particle properties (momentum: mv) of matter.

$$\lambda = \frac{h}{mv}$$

▶ While this relation applies to all objects—large and small—it holds that as the object's mass increases, its wavelength decreases.

EXAMPLE:

Calculate the speed of an electron (in m/s) whose wavelength is 8.54 µm.

We will first rearrange the de Broglie relation to solve for velocity (a vector quantity that is equal in magnitude to the speed):

$$v = \frac{h}{m\lambda}$$

We can convert the electron's wavelength into SI units of m:

$$8.54 \ \mu m \times \frac{1 \ m}{10^6 \ \mu m} = 8.54 \times 10^{-6} \ m$$

The mass of an electron is given on the inside back cover of your book, allowing us to solve for the velocity:

$$v = \frac{(6.626 \times 10^{-34} \ J \cdot s)}{(9.109 \times 10^{-31} \ kg)(8.54 \times 10^{-6} m)} = 85.2 \ m/s$$

c. Complementary properties are those that exclude one another; the more certain you know one of the properties, the less certain you are of the value of the other.

i. If we try to measure the position of an electron in a two-slit experiment, no interference pattern will be observed because the observation causes the states to collapse into one state.

ii. It is not possible to measure the wave properties (interference) and particle properties (position) simultaneously because they are complementary.

iii. The Heisenberg principle is a statement of the uncertainty associated with complementary properties:

$$\Delta x \times m\Delta v \geq \frac{h}{4\pi}$$

Where Δx is the uncertainty in the position, m is the object's mass, Δv is the uncertainty in the velocity, and h is Planck's constant.

d. The wave nature of electrons makes them *nondeterministic*, which means that their trajectories are unknowable.

i. In quantum mechanics, trajectories are replaced with probability distribution maps that show where electrons are likely to be found.

5. Quantum Mechanics and the Atom

a. Energy and position are complementary properties.

b. Electron positions are described using atomic orbitals, which are probability distribution maps.

c. The energy of an electron's wave function is given by the Schrödinger equation:

$$\hat{H}\Psi = E\Psi$$

where $\hat{H}$ is the Hamiltonian operator, which represents the total energy dependencies, E is the energy, and ψ is the wave function that describes the wave nature of the electron. The wave function has angular and radial components.

d. A plot of the wavefunction squared (ψ^2) represents an atomic orbital, which is a position probability distribution map of the electron(s).

e. Electrons in orbitals are specified by four quantum numbers.

   i. The principal quantum number, n, is also called the principal level or shell.

      1. The principal level determines the overall size and energy of the orbital for hydrogen:

$$E = -2.18 \times 10^{-18} \text{ J}\left(\frac{1}{n^2}\right) \quad n = 1, 2, 3, 4, \dots$$

EXAMPLE:

What is the energy of the $n = 2$ level for the hydrogen atom?

$$E = -2.18 \times 10^{-18} \text{ J}\left(\frac{1}{2^2}\right) = -5.45 \times 10^{-19} \text{ J}$$

---

      2. The energy is negative because an electron close to the nucleus is at a lower energy than when it is far away (according to Coulomb's law). Zero energy is defined as infinite separation between electrons and protons.

      3. As the principal quantum number increases, the spacing between subsequent levels decreases.

   ii. The angular momentum quantum number, l, is also called the azimuthal quantum number, the sublevel, or the subshell.

      1. The value of l determines the shape of the orbital.

      2. l can take on values of 0, 1, 2, 3...$(n - 1)$. For $n = 1$, l must equal zero; for $n = 2$, l can equal either zero or one; etc.

      3. Letters are often used to replace the number of l.

         a. $l = 0$ is an s orbital.

         b. $l = 1$ is a p orbital.

         c. $l = 2$ is a d orbital.

         d. $l = 3$ is an f orbital.

   iii. The magnetic quantum number, m_l, gives the orientation of the orbital in space.

      1. The magnetic quantum number can have values of: $-l, -(l-1), ..., (l-1), l$.

   iv. The spin quantum number (m_s) specifies the orientation of the electron spin.

      1. The spin quantum number can have values of $+\frac{1}{2}$ (spin up) or $-\frac{1}{2}$ (spin down).

EXAMPLE:

What are the first three quantum numbers for an electron in the 4d orbital?

The principal quantum number, n, is 4, and the angular momentum quantum number, l, is 2 (d orbital). The magnetic quantum number, m_l, can be -2, -1, 0, $+1$, or $+2$ (that is, there are five spatial orientations of d orbitals); we cannot determine which without more information.

f. The wavelengths of light that appear in atomic emission spectra correspond to electronic transitions between atomic orbitals. The energy of the light is equal to the energy of the emitted/absorbed photon:

$$\Delta E_{\text{atomic transition}} = -E_{\text{photon}}$$

i. The addition of energy to an atom/ion results in excitation: Electrons move from a lower-energy orbital to a higher-energy orbital.

ii. When an electron moves from a higher-energy orbital to a lower-energy orbital, energy is released (in the form of light) as the electron relaxes.

EXAMPLE:

What wavelength of light (in nm) is emitted when an electron in a hydrogen atom moves from the $n = 3$ shell to the $n = 2$ shell?

First, we will calculate the energy associated with each level:

$$E_2 = -2.18 \times 10^{-18} \text{ J} \left(\frac{1}{2^2} \right) = -5.45 \times 10^{-19} \text{ J}$$

$$E_3 = -2.18 \times 10^{-18} \text{ J} \left(\frac{1}{3^2} \right) = -2.42 \times 10^{-19} \text{ J}$$

The difference in energy, $E_{\text{final}} - E_{\text{initial}}$, is:

$$E_2 - E_3 = (-5.45 \times 10^{-19} \text{ J}) - (-2.42 \times 10^{-19} \text{ J}) = (-5.45 + 2.42) \times 10^{-19} \text{ J} = -3.03 \times 10^{-19} \text{ J}$$

The energy change associated with this transition is equal in magnitude to the energy of the emitted photon. The wavelength of emitted light is then:

$$E = \frac{hc}{\lambda}$$

$$\lambda = \frac{(6.626 \times 10^{-34} \text{ J·s})(3.00 \times 10^{8} \text{ m})}{3.03 \times 10^{-19} \text{ J}} = 6.56 \times 10^{-7} \text{ m} = 656 \text{ nm}$$

6. The Shapes of Atomic Orbitals

   a. The shape of an atomic orbital is determined primarily by the angular momentum quantum number.

      i. The shape of an atomic orbital is actually a probability distribution map.

         1. Typically, the orbital shapes are drawn to represent the region in which the probability of finding an electron is 90%.

      ii. The square of the wave function (ψ^2) is the probability (per unit volume) of finding an electron in a given region of space.

    b.   We plot the probability of finding an electron at a given distance using the radial distribution function as a function of the distance from the nucleus.

        i.   The radial distribution function is the total probability of finding an electron in a thin spherical shell at a distance, r, from the nucleus.

        ii.   The probability of finding an electron at the nucleus is zero because the size of a shell on the nucleus, effectively a mathematical point compared to the volume of the atom, is zero.

    c.   A node is a point where the wave function (ψ)—and therefore the probability of finding an electron (ψ^2)—is equal to zero.

        i.   There are (n − 1) nodes—either spherical or angular—in each shell.

        ii.   There are l angular nodes (planes or cones) in each subshell.

    d. s orbitals have an angular momentum quantum number of zero ($l = 0$).

        i. s orbitals are spherically symmetric.

        ii. s orbitals have no angular nodes and have ($n − 1$) radial nodes.

    e. p orbitals have an angular momentum quantum number of one ($l = 1$).

        i.   There are three p orbitals in each shell (where $n > 2$), one each with $m_l = -1, 0$, or $+1$.

        ii. p orbitals have two lobes and one angular node.

    f. d orbitals have an angular momentum quantum number of two ($l = 2$).

        i.   There are five d orbitals in each shell (where $n > 3$) with $m_l = -2, -1, 0, +1$, or $+2$.

        ii.   Four of the d orbitals have four lobes, and the fifth d orbital has two lobes with a ring where an electron is most likely to be found.

        iii. d orbitals have two angular nodes.

    g. f orbitals have an angular momentum quantum number of three ($l = 3$).

        i.   There are seven f orbitals in each shell (where $n > 4$) with $m_l = -3, -2, -1, 0, +1, +2$, or $+3$.

    h.   The phase of a wave is the sign of the wave's amplitude—positive or negative.

Fill in the Blank:

1. The phenomenon of light bending around an obstacle or slit is called _____.

2. The _____ determines the orientation of an orbital.

3. _____ is a type of energy embodied in oscillating electric and magnetic fields.

4. The _____ of light is the number of cycles that pass through a stationary point in a given period of time.

5. A(n) _____ is a point where the wave function goes through zero.

6. An increase in the intensity of light that causes electrons to be emitted causes _____ electrons to be emitted.

7. The _____ is the minimum frequency of light that must be incident on a metal surface in order to emit electrons.

8. The highest energy region of the electromagnetic spectrum is _____.

9. _____ occurs when two waves travel in phase with one another.

10. Energy is emitted from an atom when an electron _____.

11. The _____ of light is the distance in space between any two analogous points of a wave.

12. The _____ explains how electrons exist in atoms and how these electrons determine the chemical and physical properties of elements.

13. _____ relates the uncertainty in an electron's position to the uncertainty in its velocity.

14. The series of wavelengths that are emitted from an excited species is called a(n) _____.

15. The more certainly you know a given property, the less certain you know its _____ property.

16. There are _____ p orbitals with magnetic quantum numbers of _____ for a given value of n.

17. The speed of light is a constant in a(n) _____.

18. The _____ determines the overall size and approximate energy of an orbital.

19. The path of an electron is _____; it can only be described statistically.

20. The interference pattern that occurs when electrons are passed through two closely spaced slits is a result of the electron's _____.

21. The wave function squared can be plotted in three dimensions and represents the _____.

22. An orbital is specified by three interrelated _____.

23. The _____ primarily determines the shape of the orbital.

24. The _____ is the observation that many metals emit electrons when light is shined upon them.

25. As the energies of orbitals increase, the distance between subsequent levels _____.

26. The energy of an emitted photon is equal to _____ of the atomic orbitals.

Problems:

1. Sketch the radial distribution plots for the 2s and 3s orbitals.

    a. On your plots, label the axes appropriately.

    b. Identify the radial nodes on your graphs.

    c. Use your plots to determine if an electron in the 3s orbital can be closer to the nucleus than an electron in the 2s orbital. Explain.

2. The threshold energy of chromium is 7.00×10^{-19} J.

    a. What wavelength of light will cause an electron to be emitted with zero kinetic energy?

    b. What is the threshold frequency for chromium metal?

    c. What do you expect to observe when 500 nm light is incident on the metal surface?

    d. What do you expect to observe when the intensity of the light at the threshold frequency is increased tenfold?

    e. What will the kinetic energy of the electron be when light that has a frequency twice that of the threshold frequency is incident on chromium?

    f. When an electron is ejected from a metal, it is ejected from the highest occupied orbital. In chromium, the $n = 4$ level is the highest populated level. Will the transition from $n = 1$ to $n = 4$ require more or less energy than the threshold energy?

3. How many different electrons can be in the $n = 4$ level? What will be the four quantum numbers of these electrons?

4. All matter has wave and particle characteristics, but we don't often consider wave properties of large things.

   a. Calculate the de Broglie wavelength of a 1-lb ball moving at a speed of 90.0 mph.

   b. Based on your answer in part a, explain why the wave properties are unimportant in describing the behavior of the ball.

   c. Recently, scientists have measured the de Broglie wavelengths of large carbon molecules called fullerenes. Current detection limits do not allow for the detection of wavelengths for larger molecules. Explain how these experiments represent the current transition between quantum and classical behavior.

   d. What do you expect to happen when the two-slit experiment is carried out using baseballs? Explain the reasoning behind your answer.

   e. Zero-point energy is the minimum energy that a quantum-mechanical system can have (the energy can't actually be zero). What will be the zero-point energy of a baseball?

5. A blue advertising sign emits light with a wavelength of 465 nm.

   a. What is the frequency of the light?

   b. What is the energy of 1.00 mol of photons of this light?

   c. When the power of the light is reduced, what about the light has changed?

6. Photons of light can be split using special materials. When a single photon is split into two, the total energy must be conserved. Derive an expression that will allow you to determine the energy of the two photons emitted when ultraviolet light of 200 nm is split into two equivalent photons.

Concept Questions:

1. The uncertainty principle states that the position and velocity (or momentum) of a particle cannot be determined exactly at the same time.

   a. Why, despite uncertainty, is it possible for astronomers to track the position of satellites with a very high precision?

   b. The uncertainty principle necessarily discredits the possibility that electrons orbit the nucleus. Explain why it is impossible for an electron to be in an orbit if uncertainty is true.

2. In this chapter, radial distribution plots were drawn for you; in these plots, all angles were considered at a given radius. We could also draw angular distribution plots for the various orbitals by plotting the probability of finding an electron at a given angle for all radii. In doing this, we would need to consider two different angles to define all of three-dimensional space: the angle defined by motion away from the z-axis in the x-z plane and the angle defined by motion away from the x-axis in the x-y plane. Draw such orbitals for the 1s, 2s, and 2p orbitals. Hint: Consider the number of angular nodes present for each orbital in order to situate your graphs.

3. Radio waves are permitted, by law, to be "everywhere" around us.

   a. How can you tell that radio waves are everywhere?

   b. Why are radio waves the chosen frequency range for so many common applications?

4. The two-slit experiment results in an interference pattern that depends on the distance between the slits and the wavelength of the incident light/electrons.

   a. How will the interference pattern change as the slits are moved apart from one another?

   b. How will the interference pattern change as the wavelength of light is increased?

   c. In this chapter, the experiment that attempted to identify the positions of electrons in the two-slit experiment was discussed. We can also gain a better understanding of the position by decreasing the size of the slits. Use the uncertainty principle to explain how making the slits narrower will affect the interference pattern.

5. A metal sample was irradiated with blue light, and 1.25 mol of electrons were ejected with a kinetic energy of 5.23×10^{-19} J.

    a. How will shining ultraviolet light of the same intensity affect the number of electrons ejected? Explain.

    b. How will shining ultraviolet light of the same intensity affect the kinetic energy of the electrons ejected? Explain.

    c. How will shining blue light of greater intensity affect the number of electrons ejected? Explain.

    d. How will shining blue light of greater intensity affect the kinetic energy of the electrons ejected? Explain.

6. In this chapter, you saw that the spacing between energy levels gets smaller as the energy of the levels increases. The levels become infinitely close together when the value of n approaches infinity. Explain how this relates to the idea that kinetic energy is not quantized.

7. Radioactive material is often detected using a Geiger counter, which detects the electromagnetic radiation that is emitted upon radioactive decay. It is often possible to hear single clicks, each click representing a single decay process. Explain how this experiment reinforces the idea that light has particle properties.

8. Electron microscopes allow one to image very small things. You can search for electron microscope images on the Internet to see some amazing pictures. Based on the idea that one can only use waves that are comparable to the size of the particle to be imaged, do you expect electrons to have longer or shorter wavelengths than visible light (used in an ordinary microscope)?

9. Magnesium has a work function of 355 kJ/mol, while sodium has a work function of 220 kJ/mol. Both metals emit electrons upon the absorption of violet light. One of the metals absorbs green light, while the other does not. Which metal absorbs green light? Explain.

Key Learning Outcomes:

- Write Electron Configurations

- Write Orbital Diagrams

- Differentiate between Valence Electrons and Core Electrons

- Write Electron Configurations from the Periodic Table

- Predict the Charge of Ions

- Use Periodic Trends to Predict Atomic Size

- Write Electron Configurations for Ions

- Apply Periodic Trends to Predict Ion Size

- Apply Periodic Trends to Predict Relative Ionization Energies

- Predict Metallic Character Based on Periodic Trends

Chapter Summary:

In this chapter, you will be introduced to periodic properties of the elements. Your understanding of the periodic table will begin by learning to write electron configurations and examining how these configurations compare for elements in the same and different columns (groups or families) on the periodic table. Once you have a full grasp of electron configurations, you will learn how to predict ion formation and then explore periodic trends in atomic size, ionization energy, and metallic character.

Chapter Outline:

1. Aluminum: Low-Density Atoms Result in Low-Density Metal

    a. Aluminum has a low density (2.70 g/cm^3) because the density of an aluminum atom is low.

    b. The density of elements tends to increase down a column on the periodic table because the mass increases more than the volume.

    c. A periodic property is one that can be predicted based on the element's position on the periodic table.

    d. Quantum-mechanical theory explains the electronic structure of an atom, which then determines an atom's properties.

2. The Periodic Law and the Periodic Table

    a. The periodic law states that when elements are arranged in order of increasing mass, certain properties recur periodically.

    b. The modern periodic table is arranged according to atomic number and not atomic mass.

    c. The periodic table can be divided into main-group elements, transition elements (or transition metals), and inner transition elements.

        i. Main-group elements have properties that are largely predictable based on position and are in columns labeled with a number and the letter A.

        ii. Transition elements have less predictable properties based on their position on the periodic table and are designated with a number and the letter B.

        iii. An alternate numbering system from the A/B system uses numbers 1–18 across the columns on the periodic table.

3. Electron Configurations: How Many Electrons Occupy Orbitals

    a. Electron configurations show the atomic orbitals occupied for an element.

        i.   The ground-state electron configuration is the lowest energy configuration.

        ii. In an electron configuration, the energies and positions of electrons are described by stating the value of n, the symbol of l, and the number of electrons in the orbital as a superscript.

    b. Orbital diagrams symbolize electrons as arrows in boxes that represent orbitals.

        i.   The direction of the arrow represents the electron's spin.

            1. Electrons all have the same amount of spin, but it is quantized with spin up or down.

            2. The electron spin is a fourth quantum number, m_s, and can equal $+\frac{1}{2}$ (spin up, ↑) or $-\frac{1}{2}$ (spin down, ↓).

        ii. The Pauli exclusion principle states that no two electrons in an atom can have the same four quantum numbers.

            1. Each orbital can hold a maximum of two electrons: each with opposite spin.

    c. In a hydrogen atom (and other one-electron species), the energy of an orbital depends only on the value of the principal quantum number, n.

        i.   The energies of orbitals in a given shell (n) are the same; they are degenerate.

    d. In multi-electron atoms, the energy depends on the principal quantum number, n, and the orbital angular momentum quantum number, l.

        i.   The lower the value of l, the lower the energy of a subshell:

        $$E(s; l = 0) \;<\; E(p; l = 1) \;<\; E(d; l = 2) \;<\; E(f; l = 3)$$

        ii. This orbital splitting is the result of differences in the potential energy of electrons in different sublevels.

            1. The potential energy, E_p, associated with charged particles is described using Coulomb's law:

            $$E = \frac{1}{4\pi\varepsilon_o}\frac{q_1 q_2}{r}$$

            where ε_o is a physical constant, q_1 is the charge of one particle, q_2 is the charge of another particle, and r is the distance between the two particles.

                a. The potential energy is negative for oppositely charged species, positive for like charged species, and depends on the distance between the charges.

        iii. The presence of multiple electrons in an atom results in the shielding of the nucleus by some of the electrons. Electrons closer to the nucleus block the outer electrons from experiencing the full charge of the nucleus.

            1. The effect of shielding is quantified by the effective nuclear charge, Z_{eff}, which is the charge "felt" by an electron.

            2. The energy of s orbitals is lowest (relative to other orbitals in the same shell) because electrons in s orbitals have a higher probability of being closest to the nucleus. Relative to p, d, or f orbitals, s orbitals have the greatest penetration, meaning that electrons in this sublevel have a greater probability of being closer to the nucleus, so they experience a larger Z_{eff} and shield outer electrons more effectively.

        iv. The splitting of levels is, in essence, a result of differences in the spatial arrangement of electrons within a given sublevel.

   e. The aufbau (or building up) principle states that electrons occupy orbitals from lowest to highest energy.

   f. Hund's rule states that electrons fill orbitals of equal energy (i.e., orbitals of subshell) singly with parallel spins before pairing up.

EXAMPLE:

Write electron configurations and orbital diagrams for an atom of the following elements.

a. S

Sulfur ($Z = 16$) is in the third period and group 6A; an S atom has 16 electrons. We will fill up all of the $n = 1$ (2 electrons) and $n = 2$ (8 electrons) orbitals, then fill the $3s$ orbital, and finally have four electrons in the $3p$ orbital:

$$1s^2 2s^2 2p^6 3s^2 3p^4$$

Notice that the two unpaired electrons in the $3p$ orbital are in the same direction—they have parallel spin according to Hund's rule.

b. Sr

Strontium ($Z = 38$) is in the fifth period and group 2A; an Sr atom has 38 electrons. We will fill up all of the $n = 1$, $n = 2$, $n = 3$ and $n = 4$ s and p orbitals as well as the $n = 3d$ orbitals:

$$1s^2 2s^2 2p^6 3s^2 3p^6 4s^2 3d^{10} 4p^6 5s^2$$

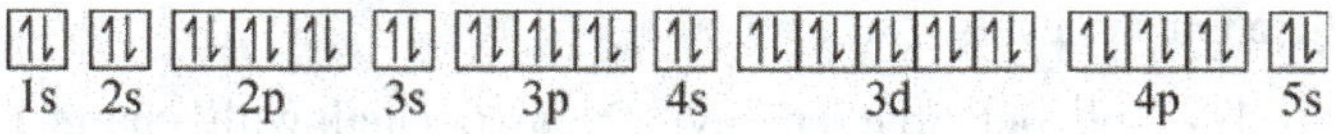

The orbitals are listed according to increasing energy, and all of the orbitals are filled.

c. Ti

Titanium ($Z = 22$) is a transition metal; a Ti atom has 22 electrons. We fill the $n = 1$, $n = 2$, and $n = 3$ s and p orbitals, then fill the $4s$ orbital and partially fill the $3d$ orbital:

$$1s^2 2s^2 2p^6 3s^2 3p^6 4s^2 3d^2$$

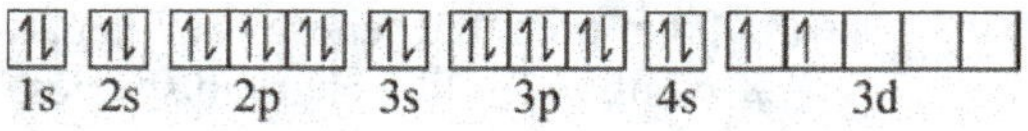

We see from the orbital diagram that titanium has two unpaired electrons in the $3d$ orbital and that these electrons have parallel spin.

d. P

Phosphorus ($Z = 15$; a P atom has 15 electrons) is in group 5A. It has filled $1s$, $2s$, $2p$, and $3s$ orbitals; it has a partially filled $3p$ orbital:

$$1s^2 2s^2 2p^6 3s^2 3p^3$$

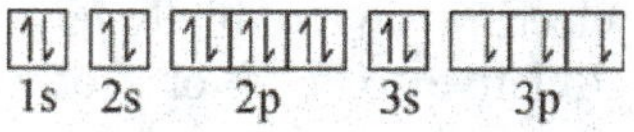

Again we see that the partially filled $3p$ orbital has unpaired electrons with parallel spin. In this orbital diagram, the unpaired electrons have been drawn with their spins directed down to emphasize the fact that the spin direction does not matter; only the fact that the spins are in the same direction matters. By custom, we draw unpaired electrons pointing up, however.

4. Electron Configurations, Valence Electrons, and the Periodic Table

    a. The electrons in the outermost shell of an atom are called the valence electrons; elements in a group have the same number of valence electrons.

        i. Valence electrons for main-group elements are those in the outermost principal quantum level (or outermost shell).

        ii. Valence electrons for transition metals are those in the outermost shell plus the outermost d orbital electrons.

    b. All other (inner shell) electrons in an atom are called core electrons.

EXAMPLE:

Determine the number of valence electrons for each of the elements in the previous example.

a. We can see from the orbital diagram that sulfur (group 6A) has six valence electrons in the outermost $n = 3$ shell.

b. From the orbital diagram, we see that strontium (a group 2A metal) has two electrons in the outermost shell ($n = 5$) and therefore has two valence electrons.

c. Because titanium (group 4B) is a transition metal, we include the $3d$ electrons in the valence electron count. Titanium therefore has four valence electrons.

d. Phosphorus (group 5A) has five valence electrons since there are five electrons in the $n = 3$ shell.

    c. The periodic table can be separated into blocks labeled s, p, d, and f that represent the valence subshell that is being filled as one moves across a period; the number of the columns in each block is equal to the maximum number of electrons in that sublevel; that is, $s = 2$, $p = 6$, $d = 10$ and $f = 14$.

    d. For main-group elements, the number of the group is equal to the number of valence electrons.

    e. The row number is equal to the principal quantum level for the main-group elements.

    f. We can abbreviate electron diagrams using the noble gas abbreviation for the inner (core) electron configuration. We write the symbol for the noble gas in square brackets, followed by the outer electron configuration.

    g. For transition metals, the d orbital will be in the $n = $ (row $-$ 1) level, and there are anomalies introduced in orbital filling because ns and $(n - 1)d$ orbitals are very close in energy and change relative to one another as they are filled.

EXAMPLE:

Use the periodic table to write electron configurations for the following elements using noble gas abbreviations.

a. F

Fluorine ($Z = 9$) is in the second row of the periodic table and is located in the p block. Since fluorine is in group 7A, it will have seven valence electrons. Helium is the noble gas that comes before fluorine and will be the noble gas core. The electron configuration is $[\text{He}]2s^2 2p^5$.

b. As

Arsenic ($Z = 33$) is in the fourth row of the periodic table and is located in the p block. Since arsenic is in group 5A, it has five electrons in the $n = 4$ shell. Argon is the noble gas core for arsenic. Since the $3d$ orbital is not part of the argon core, we will need to write the full $3d$ shell in the electron configuration. The electron configuration is $[Ar]4s^23d^{10}4p^3$.

c. Ca

Calcium ($Z = 20$) is in the s block of the fourth row of the periodic table, with argon as the noble gas that immediately precedes it on the periodic table. Since calcium is in group 2A, it has two valence electrons and an electron configuration of $[Ar]4s^2$.

d. Zr

Zirconium ($Z = 40$) is a transition metal (d block element) in the fifth row of the periodic table. Krypton is the noble gas that immediately precedes it on the periodic table, and the d orbital electrons will be in the (row − 1)d or $4d$ orbital. The electron configuration of zirconium is $[Kr]5s^24d^2$.

e. Mo

Molybdenum ($Z = 42$) is in the fifth row of the periodic table and will have an electron configuration that is very similar to that of zirconium. We could expect the electron configuration of molybdenum to be $[Kr]5s^24d^4$, but because of the close energy spacing of the $5s$ and $4d$ orbitals, the electron configuration is actually $[Kr]5s^14d^5$, which results in two half-filled shells.

---

5. Electron Configurations and Elemental Properties

    a. Elements in a family (or group) have similar properties because they have the same number of valence electrons.

        i. Column 8A is the noble gas family. The noble gases are relatively inert.

    b. When a quantum level is full, the overall potential energy of the electrons in that level is low, so they are not prone to undergoing change.

    c. Elements without a noble gas configuration undergo reactions in order to attain a noble gas configuration.

    d. Elements on the periodic table can be classified as metals, nonmetals, or metalloids.

        i. A metal is a substance that is a good conductor of heat and electricity, usually a solid at room temperature, malleable, ductile, and tends to lose electrons in chemical reactions.

            1. Metals tend to lose electrons to obtain a noble gas configuration.

        ii. A nonmetal has more varied properties; it may be a gas, liquid, or solid but is usually a poor conductor of heat and electricity and tends to gain electrons in chemical reactions.

            1. Nonmetals tend to gain electrons to obtain a noble gas configuration.

        iii. A metalloid has properties that are in between metals and nonmetals. Semiconductors are metalloids and can be conductors or insulators, depending on the conditions in which they are used.

EXAMPLE:

Identify the following elements as metals, nonmetals, or metalloids.

a. Cr

Chromium is element 24 and is found to the left of the "staircase" on the periodic table; it is therefore a metal.

b. O

Oxygen is element 8 and is found to the right of the "staircase" on the periodic table; it is therefore a nonmetal.

c. He

Helium is element 2 and is found to the right of the "staircase" on the periodic table; it is therefore a nonmetal.

d. Si

Silicon is element 14 and is found on the "staircase" on the periodic table; it is therefore a metalloid.

e. Ga

Gallium is element 31 and is found to the left of the "staircase" on the periodic table; it is therefore a metal.

---

   e. Each column within the main group is called a family or group of elements.

      i. The alkali metals are group 1 (1A) on the periodic table and react with water to form a basic solution.

         1. Alkali metals all have ns^1 electron configurations and form 1+ ions.

      ii. The alkaline earth metals are group 2 (2A) on the periodic table and react with water to form a basic solution in a less vigorous way than the alkali metals.

         1. Alkaline earth metals all have ns^2 electron configurations and form 2+ ions.

      iii. The halogens are group 7 (7A) on the periodic table and are reactive nonmetals.

         1. Halogens all have ns^2np^5 electron configurations and form 1− ions.

   f. Main-group elements tend to form predictable ions based on their electron configurations and the configuration of their nearest noble gas.

      i. Metals tend to form ions with a charge that is equal to their group number. These ions are all cations (positively charged).

      ii. Nonmetals tend to form ions with a charge that is equal to their group number minus 8. These ions are all anions (negatively charged).

   g. The tendency for a specific ion to form does NOT mean that the formation is energetically favorable. (We will discuss this further when we discuss bond formation in Chapter 5.)

6. Periodic Trends in Atomic Size and Effective Nuclear Charge

   a. The nonbonding atomic radius, or van der Waals radius, is the distance between the centers (nuclei) of adjacent nonbonding atoms.

   b. The bonding atomic radius or covalent radius for nonmetals is one-half the distance between two nuclei that are bonded together.

   c. The bonding atomic radius for metals is one-half the distance between two atoms next to each other in a crystal lattice.

   d. The atomic radius is the average radius value based on many measurements of many different elements and compounds.

   e. The atomic radius increases down a group and decreases across a period in the periodic table.

      i. The radius increases down a column because the valence electrons are at higher energy and are further away from the nucleus on average.

      ii. The radius decreases across a row because the effective nuclear charge, Z_{eff}, increases across a row

$$Z_{eff} = Z - S$$

where Z is the nuclear charge (number of protons) and S is the shielding.

1. The further away the electron is from the nucleus, the more electrons there are present to shield the charge of the protons in the nucleus.

   a. Core electrons effectively shield the electrons in the outermost shell.

   b. Outer shell electrons only partially shield each other.

f. The radii of transition metals increase down a group (for the first two periods only) but remain approximately constant across a row.

   i. Electrons in d orbitals all experience approximately the same nuclear charge.

   ii. Elements following the lanthanides (in period 6) have atomic radii that are smaller than expected because electrons in $4f$ orbitals are not effective at shielding; this is termed the "lanthanide contraction."

EXAMPLE:

List the following atoms in order of increasing atomic radius: Ca, F, Ba, Se, and Cl.

The element that is furthest down and to the left on the periodic table will have the largest radius, so Ba is the largest of these. Calcium is the element that is the next lowest and to the left, so it is the second largest. The element that is furthest to the right and the top of the periodic table will have the smallest radius, so F is the smallest of these. Chlorine is also very far to the right and is just under fluorine, so it is the next smallest. Finally, we see that selenium is below and to the left of chlorine, making it larger while it is in the same row and to the right of calcium, making it smaller. The order is then F < Cl < Se < Ca < Ba.

7. Ions: Electron Configurations, Magnetic Properties, Radii, and Ionization Energy

   a. Electron configurations of ions are the same as their neutral counterparts with the appropriate number of electrons added (anions) or subtracted (cations).

      i. Electrons are added to the lowest energy sublevel and removed from the highest energy sublevel.

         1. For main-group elements, the electrons are removed in the reverse order that they are added.

         2. For transition metals, the electrons are not removed in the reverse order that they are added.

            a. s and d orbitals are very close in energy.

            b. As d orbitals are filled, they are stabilized with respect to the s orbitals.

               i. ns electrons are removed before $(n-1)d$ electrons.

   b. When an atom has unpaired electrons, it is called paramagnetic; it is attracted to a magnetic field.

   c. An atom with all of its electrons paired is called diamagnetic; it is not attracted to a magnetic field.

EXAMPLE:

Write electron configurations for the following ions:

a. Br⁻

   Bromine has an electron configuration of $[Ar]4s^2 3d^{10} 4p^5$ as a neutral atom. The 1– ion has one additional electron, so the electron configuration becomes $[Ar]4s^2 3d^{10} 4p^6$.

b. Cd^{2+}

Cadmium has an electron configuration of $[Kr]5s^2 4d^{10}$ as a neutral atom. The 2+ ion has lost two electrons, which are removed from the $5s$ orbital, so the electron configuration becomes $[Kr]4d^{10}$.

c. O^{2-}

A neutral oxygen atom has an electron configuration of $[He]2s^2 2p^4$. Oxygen becomes a 2– ion upon the addition of two electrons to give an electron configuration of $[He]2s^2 2p^6$.

---

d. The size of an ion relative to its neutral counterpart depends on the charge.

  i. A cation is much smaller than the neutral atom from which it is formed; this is due to the increase in Z_{eff} experienced by the outer shell electrons.

  ii. An anion is much larger than the neutral atom from which it is formed; this is because the value of Z_{eff} has not changed, while the number of electrons (negative charge) has increased.

EXAMPLE:

Rank the following in order of increasing ionic radius: Ca^{2+}, Ba^{2+}, Cs^+, and F^-.

We will first focus on the cations listed. Ca^{2+} will be the smallest since it is highest up on the periodic table and has a 2+ charge. Barium and cesium are in the same row of the periodic table, and a neutral barium atom will have a smaller radius than cesium's neutral atomic radius. This difference will be exaggerated by the charge difference of their ions: since barium is a 2+ ion, its radius will decrease more, with respect to the neutral atom, than cesium with a 1+ charge. The order for increasing radius is therefore $Ca^{2+} < Ba^{2+} < Cs^+$. We now need to consider where F^- fits into this. Although anions increase in size relative to their neutral counterparts, F^- will still be the smallest of the species listed because it has its valence electrons in the $n = 2$ shell, whereas the smallest of the cations, Ca^{2+}, has a full $n = 3$ shell. The ions are thus ordered according to increasing radius: $F^- < Ca^{2+} < Ba^{2+} < Cs^+$.

---

e. The ionization energy is the energy required to remove an electron from an atom in the gas phase:

$$X(g) + energy \rightarrow X^+(g) + 1e^-$$

  i. Ionization energies are always positive. That is, energy is always required in order to remove an electron from a neutral atom.

  ii. Ionization energy decreases down the periodic table.

    1. As the distance between an electron and the nucleus increases, the energy of the electron increases and the amount of energy required to remove it decreases.

  iii. Ionization energy generally increases across a row on the periodic table; it is a maximum for group 8A elements.

    1. As the effective nuclear charge increases, more energy is required to remove an electron.

    2. The ionization energy decreases slightly when the ion formed has a full or half-full shell because of the energy associated with populating a higher level or pairing up electrons.

  iv. For any element, the first ionization energy is less than the second ionization energy, which is less than the third ionization energy and so on.

    1. Successive ionization energies increase due to the increase in Z_{eff}; increasing energy is required to remove an electron from a positively charged ion.

2. Large jumps in the ionization energy value occur when an electron is removed from an atom or ion with a full or half-full shell.

EXAMPLE:

List the following neutral species in order of increasing (first) ionization energy: K, Mg, P, S, and O.

In general, we expect the ionization energy to increase across a period and up a group, so we tentatively order these species as: $K < Mg < P < S < O$. We need to recognize, however, that phosphorus has a half-filled shell and will therefore have a slightly higher ionization energy than sulfur, which is next to it on the periodic table. The order is therefore $K < Mg < S < P < O$.

---

8. Electron Affinities and Metallic Character

    a. The electron affinity is the energy change that results from an atom gaining an electron in the gas phase:

    $$A(g) + 1e^- \rightarrow A^-(g) + (\text{energy})$$

    Note that energy is usually, but not always, released.

    i. Down a group, the trends in electron affinity are not as regular.

    ii. Across a row the trend generally increases as expected.

        1. The electron affinity is greatest (i.e., most exothermic) for group 7A elements.

    b. Metallic character decreases across a row and increases down a column.

EXAMPLE:

Rank the following in order of increasing electron affinity: N, S, F, Al, Mg.

The more metallic an element is, the lower its electron affinity generally is. The closer the element is to having a full shell (the more valence electrons it has), the higher its electron affinity is. So we have $Mg < N < Al < S < F$. Note that the EA of N is smaller than that of Al because N has a half-filled shell.

EXAMPLE:

Identify each of the following as a noble gas, an alkali metal, an alkaline earth metal, or a halogen.

a. Ba

    Barium is found in group 2 (2A), so it is an alkaline earth metal.

b. Cl

    Chlorine is found in group 17 (7A), so it is a halogen.

c. Ar

    Argon is found in group 18 (8A), so it is a noble gas.

d. Li

    Lithium is found in group 1 (1A), so it is an alkali metal.

e. Rb

    Rubidium is found in group 1 (1A), so it is an alkali metal.

---

c. Main-group metals tend to lose electrons to form cations with the same number of electrons as the nearest noble gas.

▶ The number of electrons lost (i.e., the charge of main-group metal cations) is generally equal to the group number.

d. Main-group nonmetals tend to gain electrons to form anions with the same number of electrons as the nearest noble gas.

▶ The number of electrons gained (i.e., the charge of main-group metal cations) is generally equal to 8 minus the group number.

EXAMPLE:

Identify the anion or cation that will form from each of the following elements:

a. Ca

The nearest noble gas to calcium is argon, which has 18 electrons. When calcium, with 20 protons, has 18 electrons, it will form a 2+ ion and will therefore be Ca^{2+}.

b. S

The nearest noble gas to sulfur is argon, which has 18 electrons. When sulfur, with 16 protons, has 18 electrons, it will form a 2– ion and will therefore be S^{2-}.

c. I

The nearest noble gas to iodine is xenon, which has 54 electrons. When iodine, with 53 protons, has 54 electrons, it will form a 1– ion and will therefore be I^-.

d. Sr

The nearest noble gas to strontium is krypton, which has 36 electrons. When strontium, with 38 protons, has 36 electrons, it will form a 2+ ion and will therefore be Sr^{2+}.

Fill in the Blank:

1. The lowest energy state of an atom is its _____.

2. _____ are charged species that are much smaller than their corresponding atoms.

3. An element with unpaired electrons is _____.

4. Metallic character _____ across the periodic table and _____ down the periodic table.

5. The principal quantum number of the *d* orbital being filled for a transition metal is equal to _____.

6. The second ionization energy is _____ than the first ionization energy.

7. Electron affinity is the energy change associated with _____.

8. The orientation of an electron's spin is quantized: it is either _____ or _____.

9. The Pauli exclusion principle states that no two electrons of an atom can have the same four _____.

10. The effective nuclear charge differs from the actual nuclear charge due to _____.

11. Hund's rule states that when filling degenerate orbitals, electrons fill them singly first with parallel _____.

12. Alkali metals have an electron configuration of _____.

13. A(n) _____ property is one that is predictable based on an element's position on the periodic table.

14. Noble gases are very stable and unreactive; under extreme conditions _____ and _____ can be made to react with fluorine.

15. The atomic radius is the _____ based on measurements of a large number of elements and compounds.

16. The _____ is the energy required to remove an electron from an atom or ion in the gaseous state.

17. _____ are the electrons in the outermost shell and are important in chemical bonding.

18. The van der Waals radius represents the radius of an atom when it is _____.

19. The atomic radius _____ across a row and _____ down a column.

20. Degenerate orbitals have the same _____.

21. Nonmetals tend to gain electrons to form _____.

22. _____ are good conductors of heat and electricity, often shiny, ductile, malleable, and tend to lose electrons when undergoing a chemical change.

Problems:

1. Carbon is one of the most important elements to us because it is the most abundant element in living organisms.

    a. What is the full electron configuration for carbon?

    b. What are the four quantum numbers for the highest energy electron in carbon?

    c. Do you expect carbon atoms to be magnetic? Explain.

    d. Do you expect the radius of carbon to be larger or smaller than that of nitrogen? Explain.

    e. Do you expect the ionization energy of carbon to be greater or less than that of nitrogen? Explain.

    f. Do you expect the carbide ion (C^{4-}) to have a radius that is larger or smaller than the radius of a neutral carbon atom? Explain.

    g. Do you expect the electron affinity of carbon to be a positive or negative number? Explain.

2. Consider the following atoms: B, O, F, He, Na, K, Rb.

    a. Write the electron configurations and orbital diagrams for each of these species.

    b. How many valence electrons will each of the species listed have?

    c. Which of these species, if any, is paramagnetic?

    d. Rank these elements in order of increasing effective nuclear charge.

    e. Rank these elements in order of increasing atomic radius.

    f. Rank these elements in order of increasing ionization energy.

    g. Rank these elements in terms of their reactivity with $F_2(g)$.

3. Consider the following atoms and ion: P, Ca, Ga, Ca^{2+}, and Cl.

    a. Write the electron configurations and orbital diagrams (using noble gas abbreviations) for each of these species.

    b. How many valence electrons will each of the species listed have?

    c. Which of these species, if any, is paramagnetic?

    d. Estimate the effective nuclear charge for each of these species.

    e. Rank these species in order of increasing atomic radius.

    f. Rank these species in order of decreasing ionization energy.

g. Rank these species in order of increasing metallic character.

h. In one to two sentences, explain why the ionization energy follows the particular periodic trend that it does. Relate the ionization energy trend to the size of the atom/ion.

i. Which of these species do you expect to have the largest ionization energy? Explain why in terms of the explanation that you provided in part h.

4. Consider the following set of isoelectronic species: S^{2-}, Cl^-, Ar, K^+, Ca^{2+}.

a. Write the electron configurations for all species.

b. Rank these species according to increasing atomic radius.

c. Rank these species according to increasing electron affinity.

d. Rank these species in order of increasing ionization energy.

e. How do you expect the chemical reactivity of these species to compare? Explain your answer.

5. The ground state configuration is the lowest energy configuration of an element.

a. Write the ground state configuration for Zn.

b. Will Zn be paramagnetic or diamagnetic? Explain.

c. Based on the ground state configuration for Zn, what will the charge of a zinc ion be when it reacts with chlorine to form zinc chloride?

d. What are the four quantum numbers for the highest energy electron in the ground state configuration of Zn?

e. Write an excited-state configuration of Zn.

f. Will the ionization energy of the ground state be higher or lower than the ionization energy of the excited state that you wrote in part e? Explain.

6. The first ionization energy of lithium is 520 kJ/mol. What is the electron affinity for a lithium ion?

7. Using the size of their atomic radii, explain why you think that some noble gases (Kr, Xe, and Rn) will react to form compounds while others will not.

8. List two metals, two metalloids, and two nonmetals.

9. Which group on the periodic table consists of the halogen family? What properties do the halogens all share?

10. Identify the anion or cation that will most likely form from each of the following elements:

a. Na

b. O

c. Mg

d. Al

e. S

f. Cl

Concept Questions:

1. Sketch the radial distribution plots for the 2s and 3s orbitals of hydrogen.

a. Using the radial distribution plots that you have drawn, explain the periodic trend for atomic radius as you move down the periodic table.

b. Using the radial distribution plots that you have drawn, explain the periodic trend for ionization energy as you move down the periodic table.

   c.   Is it possible for an electron in the 3s orbital to be closer to the nucleus than an electron in the 2s orbital? Explain.

2.   In Chapter 2, we saw that many transition metals can have variable charges in ionic compounds. Based on what you have learned in this chapter, explain why you think this is so.

3.   The ionization energy for any species is always positive; that is, energy is always required to remove an electron from a neutral species. Sodium metal, however, reacts violently with chlorine gas to form sodium chloride.

   a.   Explain how this is possible and why the reaction is exothermic.

   b.   Do you expect the reaction of magnesium with chlorine to be more or less exothermic? Explain.

4.   Figure 8.14 shows the first ionization energy (IE) versus the atomic number for all elements through xenon.

   a.   The differences in IE for subsequent elements are much larger in the first two rows of the periodic table than for subsequent rows. Explain why you think this might be.

   b.   Notice that there is a jump in the region for the period 4 transition elements. Explain why this jump occurs for the IE of Mn.

   c.   The decrease in IE that occurs between group 5A and 6A disappears for period 5. Explain why you think this might be.

5.   In Chapter 7, we saw the line spectrum of hydrogen.

   a.   Using what you have learned about orbital energies in this chapter, qualitatively describe how the line spectra of multi-electron atoms will be different and how they will be the same.

   b.   Experimentally, we observe that the spacing of energy levels gets smaller as the atomic number increases. Explain this observation using the ideal of effective nuclear charge and atomic radius.

6.   According to Hund's rule, electrons are placed in orbitals singly with parallel spin before they are paired up.

   a.   Recall that a spinning electron creates a magnetic field. Explain why it is more energetically favorable for the spins to be parallel than for them to be randomly oriented.

   b.   The energy associated with pairing electrons in an orbital is often called the pairing energy. Explain two reasons for the existence of pairing energy.

   c.   In this chapter, you learned that some transition metals fill orbitals in a seemingly anomalous way. Based on the fact that s and d orbitals are close in energy, use the pairing energy to explain how this seemingly anomalous behavior actually follows the aufbau principle.

Chapter 4: Molecules and Compounds

<u>Key Learning Outcomes:</u>

- Write Molecular and Empirical Formulas

- Use Lewis Symbols to Predict the Chemical Formula of an Ionic Compound

- Write Formulas for Ionic Compounds

- Name Ionic Compounds

- Name Ionic Compounds Containing Polyatomic Ions

- Name Molecular Compounds

- Calculate Formula Mass

- Use Formula Mass to Count Molecules by Weighing

- Calculate Mass Percent Composition

- Use Mass Percent Composition as a Conversion Factor

- Use Chemical Formulas as a Conversion Factor

- Obtain an Empirical Formula from Experimental Data

- Calculate a Molecular Formula from an Empirical Formula and Molar Mass

- Determine an Empirical Formula from Combustion Analysis

<u>Chapter Summary:</u>

In this chapter, you will be introduced to molecules and compounds. You will begin by learning about various types of chemical bonds: ionic, covalent, and metallic. Then you will learn how these substances are represented using formulas and molecular models. Lewis theory will be used to explain both ionic and covalent bonding. A systematic naming scheme, chemical nomenclature, will be described to allow you to name binary ionic and covalent compounds as well as to determine the formulas from the names. Next, the mole concept will be applied to compounds allowing for the use of formulas and percent composition to be used as conversion factors. You will then see how experimental data can be used to determine a chemical formula. Finally, you will be introduced to organic compounds.

<u>Chapter Outline:</u>

1. Hydrogen, Oxygen, and Water

    a. When two or more elements combine to form a compound, a new substance is formed.

    b. The physical and chemical properties of compounds are generally different from the properties of the elements that compose them.

2. Types of Chemical Bonds

    a. A chemical bond holds atoms together and is the result of the electrostatic interactions between positive and negative charges.

        i. Chemical bonds form because they lower the potential energy of the particles that compose the atoms.

    b. An ionic bond generally occurs between metals and nonmetals and involves the transfer of electrons.

        i. When electrons are transferred between substances, a negatively charged ion (usually a nonmetal) and a positively charged ion (usually a metal) are formed.

        ii.   The positively and negatively charged ions are electrostatically attracted to each other, forming a neutral, three-dimensional array, or lattice, in the solid phase.

    c.   A covalent bond typically occurs between two nonmetals and involves the sharing of electrons.

        i.   Covalent bonds usually occur between nonmetal elements and are a result of the lowering of (potential) energy that comes from the sharing of electrons between two positively charged nuclei.

        ii.   Covalently bonded compounds are composed of individual molecules and are called molecular compounds.

EXAMPLE:

Identify the type of bonding present in each of the following compounds.

a.   NaCl

   Sodium is a metal, and chlorine is a nonmetal. This is an ionic compound with ionic bonds.

b. CH_4

   Carbon and hydrogen are both nonmetals. This is a covalent compound with covalent bonds.

c.   MgS

   Magnesium is a metal, and sulfur is a nonmetal. This is an ionic compound with ionic bonds.

d. N_2O_4

   Nitrogen and oxygen are both nonmetals. This is a covalent compound with covalent bonds.

e.   NO

   Nitrogen and oxygen are both nonmetals. This is a covalent compound with covalent bonds.

   3.   Representing Compounds: Chemical Formulas and Molecular Models

    a.   Chemical formulas give the elements and the numbers of atoms of each element in a compound.

        i.   A chemical formula gives the symbols for each element present and a subscript that refers to the number of atoms of that element.

        ii.   The element that is most metallic (furthest left and lower on the periodic table) is generally listed first, and the element that is least metallic (furthest right and higher on the periodic table) is generally listed last.

        iii.   The empirical formula is the simplest whole-number ratio of elements in a compound; it provides a *relative* number of atoms of each element type.

        iv.   The molecular formula gives the actual number of atoms of each element in one molecule of a compound. The molecular formula is related to the empirical formula—the subscripts in the empirical formula are multiplied by a whole number in order to give the molecular formula.

EXAMPLE:

State the number of atoms of each type of element in the compounds below.

a. Na_2SO_4

   There are 2 sodium atoms, 1 sulfur atom, and 4 oxygen atoms in a formula unit of sodium sulfate.

b. HCO_2H

There are 2 hydrogen atoms, 1 carbon atom, and 2 oxygen atoms in a molecule of formic acid.

c. $Mg_3(PO_4)_2$

There are 3 magnesium atoms, 2 phosphorous atoms, and 8 oxygen atoms in a formula unit of magnesium phosphate.

d. $C_6H_{12}O_6$

There are 6 carbon atoms, 12 hydrogen atoms, and 6 oxygen atoms in this fructose molecule.

EXAMPLE:

Given the following molecular formulas, provide the empirical formula for each molecular compound.

a. Benzene, C_6H_6

Both subscripts are 6, so if we divide them both by 6, we get the empirical formula CH

b. Dinitrogen tetroxide, N_2O_4

Both subscripts are multiples of 2, so we can divide them by 2 to get the empirical formula NO_2

c. Hydrogen peroxide, H_2O_2

Both subscripts are multiples of 2, so we can divide them by 2 to get the empirical formula HO

d. Glucose, $C_6H_{12}O_6$

All of the subscripts are multiples of 6, so we can divide them by 6 to get the empirical formula CH_2O

---

     v.   The structural formula gives information about atom connectivity; in an expanded structural formula, lines are used to represent covalent bonds, whereas in a condensed formula, bonds are eliminated but atom attachment is evident.

    vi.   Molecular models give a visual representation of compounds.

        1.   Ball-and-stick models represent atoms as spheres connected by sticks representing bonds.

        2.   Space-filling models have the space between atoms filled in to represent a more accurate picture of how we think compounds look.

4.   The Lewis Model: Representing Valence Electrons with Dots

    a.   In the Lewis model, valence electrons are represented as dots around an atom's symbol to give a Lewis structure.

        i.   Dots are placed on each of the four sides of an element's atomic symbol. There is a maximum of two dots per side, and dots remain unpaired if possible.

        ii.   An octet is eight electrons and represents a full outer shell.

            1.   Noble gases have full octets, except for helium which forms a duet (two-paired valence electrons).

        iii.   The octet rule refers to the fact that stable electron configurations usually have eight electrons. This rule applies to main-group elements in period (row) 2 and, with many exceptions, to elements in the third period and beyond.

      b.   Bonding occurs between atoms—via electron transfer or sharing—in order to form stable valence electron configurations.

           i.   Bonds occur so that atoms can acquire full electron shells.

  5.   Ionic Bonding: Lewis Symbols and Lattice Energies

      a.   The basic unit of an ionic compound is the formula unit, the smallest, electrically neutral collection of atoms.

           i.   The formula unit exists as part of a larger lattice—it does not exist by itself.

      b.   In ionic compounds, electrons are transferred from a metal to a nonmetal such that the resulting ions acquire a stable electron configuration.

EXAMPLE:

Draw Lewis structures in order to predict the ionic compound that forms from the following metals and nonmetals.

a.   K and F

Potassium has one valence electron, and fluorine has seven valence electrons:

$K\cdot \quad :\ddot{F}\cdot$

In order for both to have a full valence shell, potassium must lose one electron and fluorine will gain one electron to form K^+ and F^-:

$K^+ \left[:\ddot{F}: \right]^-$

b.   Mg and Cl

Magnesium has two valence electrons, and chlorine has seven valence electrons:

$\cdot Mg\cdot \quad :\ddot{C}l\cdot$

In order for both species to have a full valence shell, two chlorine atoms need to be associated with each magnesium atom:

$\left[:\ddot{C}l: \right]^- Mg^{2+} \left[:\ddot{C}l: \right]^-$

c.   Li and O

Lithium has one valence electron, and oxygen has two valence electrons:

$Li\cdot \quad \cdot\ddot{O}\cdot$

In order for both species to have a full valence shell, two lithium atoms need to be associated with each oxygen atom:

$Li^+ \left[:\ddot{O}: \right]^{2-} Li^+$

---

      c.   The lattice energy is the energy associated with forming a crystalline lattice from ions in the gas phase.

           i.   The lattice energy decreases with increasing cation and anion size because the distance between the ions increases (r in Coulomb's law).

          ii.   The lattice energy increases with increasing ion charge because of an increase in attraction (q_1 and q_2 in Coulomb's law).

      d.   The Lewis model shows an ionic solid as a lattice of ions held together by nondirectional coulombic forces.

          i.   In melting, the coulombic forces must be overcome; this requires a lot of energy, which explains the high melting points of ionic solids.

          ii.   In lattices, electrons are localized on a given ion so ionic solids do not conduct electricity.

          iii.   Solutions of ions conduct electricity because they consist of charged ions moving freely in solution.

6.   Ionic Compounds: Formulas and Names

      a.   Ionic compounds consist of positive and negative ions interacting in a crystal lattice.

      b.   In writing formulas for ionic compounds, the following rules must be obeyed:

          i.   Ionic compounds always contain positive and negative ions.

          ii.   Since a compound is a neutral substance, the sum of positive charge (from cations) and negative charge (from anions) must be zero.

          ▶   This is the key to writing formulas for ionic compounds. You will need to determine the charge of the ions (using the periodic table or, in the case of polyatomic ions, from memory) and figure out how to make the overall charge zero by combining the ions.

          iii.   The formula unit is the basic unit of an ionic compound. It gives the smallest neutral whole-number ratio of ions.

EXAMPLE:

Give the formula for the ionic compound that forms between the two elements given:

a.   Mg and O

The charge of a magnesium ion is 2+, and the charge of an oxygen ion is 2−. In order to have a charge-neutral compound, the formula for the ionic compound must be MgO.

b.   Sr and Cl

The charge of a strontium ion is 2+, and the charge of a chloride ion is 1−. In order to have a charge-neutral compound, the formula for the ionic compound must be $SrCl_2$.

c.   Al and S

The charge of an aluminum ion is 3+, and the charge of a sulfur ion is 2−. In order to have a charge-neutral compound, the formula for the ionic compound must be Al_2S_3.

d.   Ba and N

The charge of a barium ion is 2+, and the charge of a nitrogen ion is 3−. In order to have a charge-neutral compound, the formula for the ionic compound must be Ba_3N_2.

---

      c.   Naming Ionic Compounds

          i.   Common names exist for some compounds and require memorization.

          ii.   A systematic method allows any compound to be unambiguously named.

          iii.   Ionic compounds are always named with two words: "cation anion."

              1.   Metals cations that have only one type of cation (groups 1, 2 and Al in group 3A) are named using the name of the metal. The charge of the metal is not indicated.

              2.   Nonmetal anions are named using the base name of the nonmetal with the suffix -ide.

EXAMPLE:

Provide names for the following ionic compounds.

a. AgCl

   Silver chloride

b. Al_2O_3

   Aluminum oxide

c. Na_2S

   Sodium sulfide

d. Ca_3P_2

   Calcium phosphide

e. Rb_2O

   Rubidium oxide

3. Metals that have multiple possible cation charges (lower main-group metals in groups 3A–5A and many transition metals) are named using the name of the metal followed by the charge in parentheses and then the base name of the nonmetal with the suffix -ide.

▶   You must determine the charge of the metal using the charge of the anion and working backwards.

EXAMPLE:

Provide names for the following ionic compounds.

a. $FeCl_3$

   There are 3 Cl^- ions; therefore, for charge balance, iron is Fe^{3+}: Iron (III) chloride

b. CuO

   There is one O^{2-} ion; therefore, for charge balance, copper is Cu^{2+}: Copper (II) oxide

c. MnS_2

   There are two S^{2-} ions; therefore, for charge balance, manganese is Mn^{4+}: Manganese (IV) sulfide

d. CrF_2

   There are two F^- ions; therefore, for charge balance, chromium is Cr^{2+}: Chromium (II) fluoride

4. Ionic compounds may also contain polyatomic (molecular) ions, whose names and formulae must be committed to memory. Compounds containing polyatomic ions are named using the name of the metal (following by the charge if necessary) followed by the name of the polyatomic ion.

   a. A common polyatomic cation is ammonium (NH_4^+).

   b. Many of the polyatomic anions are oxyanions, which contain oxygen and one other atom.

   c. Once the -ate form of a polyatomic ion is memorized, all other versions of that ion can be derived from it:

      i.   Add an oxygen atom to the -ate form to get the per-ate form.

      ii.   Remove an oxygen atom from the -ate form to get the -ite form.

      iii.   Remove two oxygen atoms from the -ate form to get the hypo-ite form.

  d. All forms of a given polyatomic ion have the same charge. For example, perchlorate, chlorate, chlorite, and hypochlorite ions all have a charge of 1.

  e. A simple way to remember some of the polyatomic ions is to memorize CO_3^{2-}, NO_3^-, PO_4^{3-}, SO_4^{2-}, and ClO_3^-. All other polyatomic ions will be of the form of the ion that is nearest in the same group on the periodic table. For example, the oxyanion of arsenic will be like that of phosphorus: AsO_4^{3-} is arsenate.

EXAMPLE:

Provide names for the following ionic compounds.

a. $Pb(NO_3)_2$

    Lead (II) nitrate

b. $Cu(NO_2)_2$

    Copper (II) nitrite

c. K_2SO_3

    Potassium sulfite

d. $Mg(ClO)_2$

    Magnesium hypochlorite

e. $(NH_4)_2SO_4$

    Ammonium sulfate

---

5. Hydrates are ionic compounds that have a certain number of water molecules associated with each formula unit.

    a. Hydrates are named using the ionic compound name followed by (number prefix)-hydrate.

EXAMPLE:

Provide names for the following hydrates.

a. $MgSO_4 \cdot 9H_2O$

    Magnesium sulfate nonahydrate

b. $BaCl_2 \cdot 2H_2O$

    Barium chloride dihydrate

c. $KHCO_3 \cdot 2H_2O$

    Potassium bicarbonate dihydrate

d. $CoF_2 \cdot H_2O$

    Cobalt (II) fluoride monohydrate

---

7. Covalent Bonding: Simple Lewis Structures

    a. Bonding pairs are the electrons shared between two atoms.

        i. A bonding pair of electrons is represented as a dash (—).

        ii. A single covalent bond is comprised of one electron pair.

        iii. A double bond has two electron pairs shared between two atoms.

        iv. A triple bond has three electron pairs shared between two atoms.

    b. Lone pairs (or nonbonding pairs) are electron pairs associated with a single atom.

    c. Covalent bonds are directional (as opposed to ionic bonds), which means that individual molecules can form.

        i. Interactions between molecules in the bulk phase are much weaker than the covalent bonds between the atoms within the molecules themselves, which means that molecular substances have much lower melting and boiling points than do ionic compounds.

    d. Covalent bonding is also present in extended covalent network solids and macromolecular structures.

8. Molecular Compounds: Formulas and Names

    a. Nonmetals can combine with each other in variable ways; the numbers of each type of element are not obvious from the group number as in ionic compounds.

    b. Common names are given for many molecular compounds, but this method requires extensive memorization.

    c. The systematic method for writing and naming molecular compounds is based on the number of each type of atom.

        i. The first element listed in the molecular formula is generally the more metallic (the one that is farthest left and lowest down on the periodic table).

        ii. The number of atoms of each element is denoted using a prefix (mono-, di-, tri-, tetra-, penta-, hexa-, hepta-, octa-, nona-, deca-, etc.)

        iii. The name is "prefix-(first element name) prefix-(second element base name)-ide."

            1. If there is only one atom of the more metallic element, the prefix mono- is dropped.

EXAMPLE:

Provide names for the following molecular compounds.

a. NO_2

    Nitrogen dioxide

b. CCl_4

    Carbon tetrachloride

c. CO_2

    Carbon dioxide

d. P_4O_{10}

    Tetraphosphorus decoxide

---

9. Formula Mass and the Mole Concept for Compounds

    a. The formula mass (also called formula weight) of a compound is the sum of the atomic masses (in amu) of the atoms in the empirical formula of the compound.

      i.   Ionic compounds do not exist as discrete molecules, so the term "molecular mass" should not be applied to them, though it often is. For ionic compounds, we will assume that a mol refers to a mole of formula units.

    b.   The molecular mass (also called molecular weight) of a molecule is its average mass (in amu); it is the sum of the atomic masses (in amu) of the atoms in the molecular formula.

    c.   Since the mass in amu of a single atom is numerically equal to the mass in grams of 1 mole of those atoms, the same is true for compounds.

    d.   The molar mass of a compound is equal to the mass in grams of 1 mole of that compound.

EXAMPLE:

Calculate the molar mass of the following compounds.

a. K_3PO_4

In this case we have three potassium atoms, one phosphorous atom, and four oxygen atoms. The mass of each element is given on the periodic table. Adding up the masses gives:

$$3 \text{ mol K} \times \frac{39.10 \text{ g K}}{1 \text{ mol K}} + 1 \text{ mol P} \times \frac{30.97 \text{ g K}}{1 \text{ mol P}} + 4 \text{ mol O} \times \frac{16.00 \text{ g O}}{1 \text{ mol O}} = 212.27 \text{ g/mol } K_3PO_4$$

b. $Ca(OH)_2$

In this case we have one calcium atom, two oxygen atoms, and two hydrogen atoms. Adding up the masses given on the periodic table we have:

$$1 \text{ mol Ca} \times \frac{40.08 \text{ g Ca}}{1 \text{ mol Ca}} + 2 \text{ mol O} \times \frac{16.00 \text{ g O}}{1 \text{ mol O}} + 2 \text{ mol H} \times \frac{1.008 \text{ g H}}{1 \text{ mol H}} = 74.10 \text{ g/mol } Ca(OH)_2$$

c. $(NH_4)_2CO_3$

In this case we have two nitrogen atoms, eight hydrogen atoms, one carbon atom, and three oxygen atoms. Adding up the masses given on the periodic table, we have:

$$2 \text{ mol N} \times \frac{14.01 \text{ g N}}{1 \text{ mol N}} + 8 \text{ mol H} \times \frac{1.008 \text{ g H}}{1 \text{ mol H}} + 1 \text{ mol C} \times \frac{12.01 \text{ g C}}{1 \text{ mol C}}$$

$$+ 3 \text{ mol O} \times \frac{16.00 \text{ g O}}{1 \text{ mol O}} = 96.09 \text{ g/mol } (NH_4)_2CO_3$$

    e.   Molar mass can be thought of as a conversion factor between moles of a compound and grams of that compound.

EXAMPLE:

Determine the number of carbon atoms in 150.0 g of $CaCO_3$.

First, we must calculate the molar mass of $CaCO_3$. We do this by finding the sum of the masses of the atoms that comprise it:

$$40.08 \text{ g/mol} + 12.01 \text{ g/mol} + 3 \times 16.00 \text{ g/mol} = 100.09 \text{ g/mol}$$

Now we can map out the path to find the answer:

$$g \text{ } CaCO_3 \xrightarrow{\text{g } CaCO_3 \text{ to mol } CaCO_3} \text{mol } CaCO_3 \xrightarrow{\text{mol } CaCO_3 \text{ to mol C atoms}} \text{mol C atoms} \xrightarrow{\text{mol C atoms to C atoms}} \text{C atoms}$$

Now we will use conversion factors to solve the problem:

$$150.0 \text{ g CaCO}_3 \times \frac{1 \text{ mol CaCO}_3}{100.09 \text{ g CaCO}_3} \times \frac{1 \text{ mol C atoms}}{1 \text{ mol CaCO}_3} \times \frac{6.022 \times 10^{23} \text{ C atoms}}{1 \text{ mol C atoms}} = 9.025 \times 10^{23} \text{ C atoms}$$

10. Composition of Compounds

    a. The mass percent of an element in a compound is the fraction of a compound's total mass that comes from one element present in that compound (multiplied by 100):

$$\text{mass percent of X} = \frac{\text{mass of X in 1 mol of compound Y}}{\text{mass of 1 mol of compound Y}} \times 100\%$$

EXAMPLE:

Calculate the mass percent of oxygen in sulfuric acid, H_2SO_4.

First, we will calculate the molar mass of sulfuric acid. We do this by finding the sum of the masses of the atoms that comprise this covalent compound:

$$2 \text{ mol H} \times \left(\frac{1.008 \text{ g H}}{1 \text{ mol H}} \right) + 1 \text{ mol S} \times \left(\frac{32.07 \text{ g S}}{1 \text{ mol S}} \right) + 4 \text{ mol O} \times \left(\frac{16.00 \text{ g O}}{1 \text{ mol O}} \right) = 98.09 \text{ g/mol}$$

Then we can calculate the mass of oxygen that is present in sulfuric acid:

$$\frac{4 \text{ mol O}}{1 \text{ mol H}_2SO_4} \times \frac{16.00 \text{ g O}}{1 \text{ mol O}} = 64.00 \text{ g O/mol H}_2SO_4$$

And now we can use the formula given to calculate the mass percent:

$$\frac{64.00 \text{ g O/mol H}_2SO_4}{98.09 \text{ g H}_2SO_4/\text{mol H}_2SO_4} \times 100\% = 65.25\% \text{ O}$$

So sulfuric acid is 65.25% oxygen by mass.

    b. Mass percent can be viewed as a conversion factor between an element's mass and the compound's total mass.

EXAMPLE:

Using the percent mass of oxygen calculated in the previous example, determine the mass of oxygen present in a 5.43 kg sample of sulfuric acid.

This is a one-step conversion problem using percent by mass as the conversion factor:

$$5.43 \text{ kg H}_2SO_4 \times \frac{65.25 \text{ g O}}{100.00 \text{ g H}_2SO_4} = 3.54 \text{ kg O}$$

We can also solve this problem by multiplying the total mass of the compound by the percentage of that mass that is oxygen to get the same result:

$$5.43 \text{ kg} \times 65.25/100 = 3.54 \text{ kg oxygen}$$

    c. In a chemical formula, we are given a ratio of atoms to molecules (or formula units) or moles of atoms to moles of molecules (or formula units). Subscripts do not represent relative masses. For example, 1 mole of water molecules (H_2O) has twice as many moles of hydrogen as oxygen, but the mass of hydrogen in water is only one-eighth the mass of oxygen.

11. Determining a Chemical Formula from Experimental Data

    a. A compound can be decomposed into its elements, and the masses of the elements can be measured. From this process, the relative number of moles of each element can be calculated, and the empirical formula can be determined.

EXAMPLE:

Fructose is one of the sugars found in fruit. Elemental analysis of fructose gave the following mass percent composition: 40.0% carbon, 6.72% hydrogen, and 53.28% oxygen. What is the empirical formula of fructose?

First, we will assume that we have 100.0 g of fructose and determine the number of moles of each element in that 100.0 g sample.

Since we have 100.0 g of fructose and 40.0% of the mass is carbon, we have 40.0 g of carbon:

$$40.0 \text{ g C} \times \frac{1 \text{ mol C}}{12.01 \text{ g C}} = 3.33 \text{ mol C}$$

Since we have 100.0 g of fructose and 6.72% of the mass is hydrogen, we have 6.72 g of hydrogen:

$$6.72 \text{ g H} \times \frac{1 \text{ mol H}}{1.008 \text{ g H}} = 6.67 \text{ mol H}$$

Since we have 100.0 g of fructose and 53.28% of the mass is oxygen, we have 53.28 g of oxygen:

$$53.28 \text{ g O} \times \frac{1 \text{ mol O}}{16.00 \text{ g O}} = 3.33 \text{ mol O}$$

We now need to find the simplest whole-number ratio of each of these elements. By dividing all three by the smallest number of moles (3.33 mol in this case), we find that the ratio of C to H to O is 1:2:1. The empirical formula is therefore CH_2O.

---

    b. All covalent compounds with the same empirical formula will have the same mass percentages, so the molecular formula cannot be determined without the molar mass of the compound.

EXAMPLE:

The molar mass of fructose is 180.16 g/mol. Using the empirical formula found in the example above, determine the molecular formula of fructose.

We start out by calculating the molar mass of the empirical formula from above:

$$1 \text{ mol C} \times \frac{12.01 \text{ g C}}{1 \text{ mol C}} + 2 \text{ mol H} \times \frac{1.008 \text{ g}}{1 \text{ mol H}} + 1 \text{ mol O} \times \frac{16.00 \text{ g O}}{1 \text{ mol O}} = 30.03 \text{ g/mol } CH_2O$$

Now by dividing the molar mass by the empirical formula molar mass, we find the integer multiplier:

$$\frac{180.16 \text{ g/mol}}{30.03 \text{ g/mol}} = 5.999 \sim 6$$

Thus, we multiply all of the subscripts of the empirical formula by 6 in order to find the molecular formula, which is $C_6H_{12}O_6$.

---

    c. Combustion analysis can also be used in order to determine the empirical formula of a compound.

EXAMPLE:

A certain hydrocarbon (compound containing only carbon and hydrogen) is burned in excess oxygen. Analysis of the combustion products showed that 24.38 g of water and 30.29 g of carbon dioxide were produced in the reaction. What is the empirical formula of the hydrocarbon?

We need to calculate the number of moles of carbon and the number of moles of hydrogen in the compound, that is, C_xH_y. Once we have this, then the problem becomes simple.

If we assume that all of the hydrogen in H_2O originated from the hydrocarbon, we can calculate the number of moles of hydrogen reacted:

$$24.38 \text{ g } H_2O \times \frac{1 \text{ mol } H_2O}{18.02 \text{ g } H_2O} \times \frac{2 \text{ mol } H}{1 \text{ mol } H_2O} = 2.706 \text{ mol } H$$

If we assume that all of the carbon in CO_2 originated from the hydrocarbon, we can calculate the number of moles of carbon reacted:

$$30.29 \text{ g } CO_2 \times \frac{1 \text{ mol } CO_2}{44.01 \text{ g } CO_2} \times \frac{1 \text{ mol } C}{1 \text{ mol } CO_2} = 0.6883 \text{ mol } C$$

Now we will divide both numbers by the smaller of the two (0.6883 in this case) to get a whole-number ratio. We see from this that the ratio of C:H is 1:4, so the empirical formula is CH_4.

12. Organic Compounds

    a. Organic compounds are a broad class of compounds that contain mostly carbon with a few other elements.

        i. Carbon forms four bonds to other atoms.

       ii. Carbon can form single, double, or triple covalent bonds.

Fill in the Blank:

1. The average mass of a molecule is the _____.

2. The _____ is the smallest, electrically neutral collection of ions.

3. A(n) _____ bond typically occurs between a metal and a nonmetal.

4. The _____ gives the number of atoms of each element in a molecule of a compound.

5. Oppositely charged species are attracted to one another via _____ forces.

6. _____ are compounds that contain only two elements.

7. A(n) _____ group is a characteristic atom or group of atoms attached to a hydrocarbon.

8. The _____ of an element is that element's percentage of a compound's total mass.

9. A(n) _____ bond occurs between two nonmetals.

10. Many molecular elements exist as _____ molecules.

11. Organic compounds that contain only carbon and hydrogen are called _____.

12. Atoms that have a full outer shell have a(n) _____ of electrons.

13. When electrons are transferred from one species to another, the ions form _____ bonds.

14. A(n) _____ is the sharing or transfer of electrons to attain stable electron configurations.

15. _____ represent molecules in which valence electrons are represented as dots.

16. A pair of electrons that is shared between two atoms is called a(n) _____.

17. The energy associated with forming a crystalline lattice of alternating positively and negatively charged ions is called the _____.

18. The average length of a bond between two particular atoms in a large number of compounds is the _____.

19. A(n) _____ bond is one in which electrons are shared unequally.

Problems:

1. List five elements that are found in nature as diatomic molecules.

2. Provide formulas for the following compounds:

    a. Sodium bromide

    b. Magnesium sulfide

    c. Disulfur tetroxide

    d. Carbonic acid

    e. Hydroiodic acid

    f. Perchloric acid

    g. Selenium dinitride

    h. Iron(III) oxide

    i. Silver chloride

    j. Silicon tetrachloride

    k. Calcium bicarbonate

    l. Cobalt(III) phosphate monohydrate

3. Provide names for the following compounds:

    a. P_2O_5

    b. Rb_2O

    c. TiO_2

    d. BeI_2

    e. N_2O_4

    f. $Fe_2(SO_4)_3$

    g. XeF_4

    h. H_3PO_3

    i. $LiOH$

    j. CuF_2

    k. $CuSO_4 \cdot 5H_2O$

    l. $HClO_2$

4. Calculate the formula mass for the compounds given in question 3.

5. What is the mass percent of carbon in pentanol, $CH_3CH_2CH_2CH_2CH_2OH$?

6. The daily nutritional allowance for salt is about 2.5 g. How many sodium atoms are there in a day's worth of table salt, $NaCl$?

7. How many carbon atoms are there in a 5.0-g sample of fructose, $C_6H_{12}O_6$?

8. A hydrate of copper(II) chloride is heated until all of the water is driven off. The sample had a mass of 6.10 g before heating and 3.41 g after heating. What is the formula of the hydrate?

9. A mixture of $CaCO_3$ and $(NH_4)_2CO_3$ is 61.9% CO_3 by mass. What is the percent of $CaCO_3$ in the mixture?

10. A sample of a certain molecular compound is analyzed and found to have 36.45 g of C, 6.12 g of H, and 32.4 g of O.

    a. What is the mass percent of each element in the compound?

    b. How much oxygen (in g) is present in a 12.74 kg sample of the compound?

    c. What is the empirical formula of the compound?

    d. Given that the molar mass of the compound is 148 g/mol, determine the molecular formula.

11. Sodium bicarbonate is a common baking ingredient. Calculate the mass percent of each element in sodium bicarbonate. Using the mass percent, determine the number of grams of oxygen present in a 5.0-g sample of sodium bicarbonate.

12. A certain hydrocarbon is placed in a sealed container with pure oxygen and burned. After the reaction is complete, all of the hydrocarbon and oxygen have been consumed. Quantitative analysis of the products showed that 52.10 g of water and 84.85 g of carbon dioxide were produced. What is the empirical formula of the hydrocarbon, and how many grams of oxygen were present in the container before the reaction was carried out?

13. A recent report has suggested that children are becoming more susceptible to tooth decay because they are drinking bottled water, which lacks fluoride. Studies have shown that the optimal concentration of fluoride in drinking water is 1.1 mg/L, given that the average amount of water consumed in one day is six 8-ounce glasses. How many grams of sodium fluoride will need to be consumed in order to obtain the same benefit as drinking fluoridated water?

14. Atrazine is a pesticide that is very controversial because of its environmental impact; it is especially dangerous for amphibians. The IUPAC name for atrazine is 1-chloro-3-ethylamino-5-isopropylamino-2,4,6-trizine, and its molecular formula is $C_8H_{14}ClN_5$.

    a. What is the molar mass of atrazine?

    b. How many molecules are in 10.0 g of atrazine?

    c. How many hydrogen atoms are in 10.0 g of atrazine?

    d. What is the mass percent of carbon in atrazine?

15. Each of the following substances contains both ionic and covalent bonds; indicate which bond is which and draw the Lewis structures for each.

    a. $BaCO_3$

    b. KNO_3

    c. Na_2SO_4

    d. $LiIO$

    e. $Mg(ClO_3)_2$

Concept Questions:

1. What is wrong with the following statement? "The chemical formula of ammonia indicates that the compound contains 3 grams of hydrogen for each gram of nitrogen."

2. As we learned in this chapter, ionic compounds exist in a crystal lattice, a large network of anions and cations, and thus are solids at room temperature. Conversely, molecular compounds exist as discrete entities—molecules—and may be gases, liquids, or solids at room temperature. Explain why you think this is so based on the principles of bonding that exist in ionic and covalent compounds.

3. Molecular compounds have empirical and molecular formulas, while ionic compounds only have empirical formulas associated with them. Explain why you think that this difference exists.

4. The number of compounds that have been synthesized in the laboratory is on the order of millions. Most of these compounds are molecular instead of ionic. Suggest one reason why the number of molecular compounds is larger than the number of ionic compounds.

5. The structures of morphine (a) and codeine (b) are shown below:

    (a) 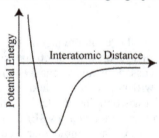   (b)

    In these structures, the intersection of two lines indicates that a carbon atom is present.

    These two compounds have very different properties. Identify any functional groups that you recognize in these structures. What aspect of their structure is responsible for the differences in their reactivity?

6. In the beginning of this chapter, you were told that there is competition between attractions and repulsions when forming a bond. We can combine these statements into a single graph of potential energy:

    a. Identify the region that represents the repulsions of electrons.

    b. Identify the region that represents the attraction of electrons to protons.

    c. Identify the region that represents the average bond energy.

    d. Notice that the potential energy approaches zero as the distance between atoms increases. Explain this behavior.

Chapter 5: Chemical Bonding I

Key Learning Outcomes:

- Classify Bonds: Pure Covalent, Polar Covalent, or Ionic

- Write Lewis Structures for Covalent Compounds

- Write Lewis Structures for Polyatomic Ions

- Write Resonance Lewis Structures

- Assign Formal Charges to Assess Competing Resonance Structures

- Draw Resonance Structures and Assigning Formal Charge for Organic Compounds

- Write Lewis Structures for Compounds Having Expanded Octets

- Use VSEPR Theory to Predict the Basic Shapes of Molecules

- Predict Molecular Geometries Using VSEPR Theory and the Effects of Lone Pairs

- Predict the Shapes of Larger Molecules

- Use Molecular Shape to Determine Polarity of a Molecule

Chapter Summary:

In this chapter, covalent bonding will be explored using Lewis dot structures. Electronegativity will be explained, and the polarity of covalent bonds that result from electronegativity differences will be described. A further expansion of covalent bonding will be discussed using resonance structures, formal charge considerations, and exceptions to the octet rule. Bond energies will be explained, and the connection between reaction enthalpies and average bond energies will be explored. The Lewis bonding model will then be expanded with a method to rationalize molecular shapes called valence shell electron pair repulsion (VSEPR) theory, which uses the repulsions of valence shell electron pairs to predict common electronic and molecular geometries. Using these geometries and your knowledge of bond polarity, you will learn how to determine if a given molecule is polar or nonpolar.

Chapter Outline:

1. Morphine: A Molecular Imposter

2. Electronegativity and Bond Polarity

    a. Electrons are not usually shared equally between two bonded atoms.

        i. The unequal sharing of electrons results in bond polarity.

    b. A polar covalent bond is one that is between the extremes of an ionic bond and a pure (nonpolar) covalent bond.

    c. Electronegativity (EN) is the ability of an atom to attract electrons to itself in a bond.

        i. Fluorine is the most EN atom on the periodic table. It is arbitrarily assigned an electronegativity value of 4.0 (no units), and all other elements are given values in reference to fluorine.

        ii. Electronegativity increases to the right and decreases down the periodic table.

    d. The greater the electronegativity difference (ΔEN) between two atoms, the more polar the bond.

        i. A small ΔEN (0.0–0.4) results in a nonpolar covalent bond.

        ii. An intermediate ΔEN (0.4–2.0) results in a polar covalent bond.

        iii. A large ΔEN (>2.0) results in an ionic bond.

e. The dipole moment (μ) of a bond quantifies the bond polarity. The units of the dipole moment are debye (D).

$$\mu = q \cdot r$$

where q is the charge of an electron and r is the bond distance.

f. The percent ionic character is the ratio of the actual dipole moment to the dipole moment that would exist if the electrons were transferred to the more electronegative atom (as in an ionic compound).

   i. If the percent ionic character is greater than 50%, the bond is considered ionic.

EXAMPLE:

Use the electronegativity values in Figure 6.3 of your textbook to predict whether the bond that forms between the following atoms will be ionic, covalent, or polar covalent.

a. Li and O

Predict: metal + nonmetal; the bond is ionic. Confirm: Lithium has an EN of 1.0, and oxygen has an EN of 3.5. The ΔEN for this bond is 2.5; this value is greater than 2.0, so the bond is ionic.

b. N and H

Predict: nonmetal + nonmetal; the bond is covalent. Confirm: Nitrogen has an EN of 3.0, and hydrogen has an EN of 2.1. The ΔEN for this bond is 0.9; this value is between 0.4 and 2.0, so the bond is polar covalent.

c. S and Br

Predict: nonmetal + nonmetal; the bond is covalent. Confirm: Sulfur has an EN of 2.5, and bromine has an EN of 2.8. The ΔEN for this bond is 0.3; this value is less than 0.4, so the bond is covalent.

d. Al and F

Predict: metal + nonmetal; the bond is ionic. Confirm: Aluminum has an EN of 1.5, and fluorine has an EN of 4.0. The ΔEN for this bond is 2.5; this value is greater than 2.0, so the bond is ionic.

3. Writing Lewis Structures of Molecular Compounds and Polyatomic Ions

   a. Lewis structures can easily be written by following four steps.

      i. Write the skeletal structure by connecting atoms together with single bonds.

         1. Hydrogen and fluorine atoms are nearly always terminal—they only connect to one other atom and are never the central atom.

         2. The less-EN atoms are used as central atoms, and the more-EN atoms are terminal atoms.

            ▶ Usually, the first nonhydrogen atom in the formula is the central atom.

      ii. Calculate the total number of valence electrons in the molecule (adding an electron for each negative charge and subtracting an electron for each positive charge).

      iii. Distribute remaining electrons as lone pairs around as many atoms as possible.

      iv. Make double and/or triple bonds for electron-deficient (less than eight electrons) atoms by sharing electron pairs from the least-EN adjacent atom.

EXAMPLE:

Draw Lewis structures for each of the following:

a. CH_4

   1. Connect carbon to four hydrogen atoms with bonding pairs:

$$
\begin{array}{c}
H \\
| \\
H - C - H \\
| \\
H
\end{array}
$$

   2. The total number of valence electrons is 1×4 (from carbon) $+ 4 \times 1$ (from hydrogen) $= 8$.

   3. All electrons are used in the four bonds of methane.

   4. All atoms have noble gas configurations: an octet around carbon and duets around each hydrogen.

   Thus, the structure is satisfactory.

b. HCOH

   1. Connect carbon to the oxygen atom and the two hydrogen atoms:

$$
\begin{array}{c}
O \\
| \\
H - C - H
\end{array}
$$

   2. The number of valence electrons is 1×4 (from carbon) $+ 1 \times 6$ (from oxygen) $+ 2 \times 1$ (from hydrogen) $= 12$.

   3. Six of the electrons have been used in forming the bonds. The remaining 6 electrons will be distributed around the carbon and oxygen atoms as lone pairs.

$$
\begin{array}{c}
\ddot{O}: \\
| \\
H - \underset{..}{C} - H
\end{array}
$$

   4. We can see that carbon and hydrogen have full shells, but oxygen is electron-deficient. We will use the electron pair on carbon to form a double bond with oxygen:

   Thus, the structure is satisfactory.

c. SO_2

   1. Sulfur is the less-EN atom, so we will put sulfur in the center and connect it to the oxygen atoms:

$$O - S - O$$

   2. The total number of valence electrons is: 2×6 (from oxygen) $+ 1 \times 6$ (from sulfur) $= 18$.

   3. Four electrons have been used to make the two bonds, so we have fourteen more to distribute:

$$\ddot{\underset{..}{O}} - \ddot{S} - \ddot{\underset{..}{O}}:$$

   4. We can see that one of the oxygen atoms is electron-deficient, so we will use a pair of electrons from sulfur to form a double bond:

$$\ddot{\underset{..}{O}} = \ddot{S} - \ddot{\underset{..}{O}}:$$

   Thus, the structure is satisfactory.

d. NO_3^-

    1. Nitrogen is the central atom since it is less EN than oxygen is:

    2. The total number of valence electrons is: 3×6 (from oxygen) $+ 1 \times 5$ (from nitrogen) $+ 1$ (negatively charged ion) $= 24$.

    3. We have used six of the valence electrons in bonding, so there are 18 left, which we distribute as lone pairs.

    4. We see that the nitrogen atom does not have an octet, so we will form a double bond with one of the oxygen atoms. We will also use brackets around the ion and indicate the total charge:

---

4. Resonance and Formal Charge

    a. Some molecules have multiple equally valid Lewis structures.

        i. In nature, these molecules exist as an average between the possible structures; this average is a resonance hybrid.

    b. Some molecules have more than one valid Lewis structure, but the structures are not equivalent. In this case, the resonance hybrid is a weighted average favoring the more stable structures.

        i. A less stable structure will have a small contribution to the overall structure.

        ii. Formal charge can be used to determine which structure is most reasonable.

            1. Formal charge is the charge an atom would have if all of the bonds to it were perfectly covalent.

$$\text{Formal charge} = (\text{valence electrons}) - (\text{lone pair electrons} + \tfrac{1}{2} \text{ bonding electrons})$$

            2. The sum of all formal charges is equal to zero for a neutral molecule and is equal to the overall charge for an ion.

            3. Small formal charges are preferable to large values (0 is best).

            4. Negative formal charges should be on the more-EN atom.

    c. In both cases, we use a double-headed arrow to show that individual Lewis structures are related by resonance.

EXAMPLE:

Determine the formal charge of each atom in the Lewis structures drawn in the previous example.

a. CH_4: Carbon has four valence electrons and is sharing four bonding pairs, so the formal charge of carbon in CH_4 is $4 - \frac{1}{2}(8) = 0$. Hydrogen has one valence electron, and each is sharing a single pair of bonding electrons, so the hydrogen atoms all have a formal charge of $1 - \frac{1}{2}(2) = 0$. Note that the sum of the formal charges is equal to zero, which is consistent with the neutral charge on this molecule.

b. H_2CO: Carbon has four valence electrons and is sharing four bonding pairs, so the formal charge of carbon is $4 - \frac{1}{2}(8) = 0$. Hydrogen has one valence electron, and each hydrogen atom is sharing a single pair of bonding electrons, so both hydrogen atoms have a formal charge of $1 - \frac{1}{2}(2) = 0$. Oxygen has six valence electrons and two lone pairs, and is sharing two bonding pairs so the formal charge of oxygen is $6 - \{4 + \frac{1}{2}(4)\} = 0$. Note that the sum of the formal charges is equal to zero, which is consistent with the neutral charge on this molecule.

c. SO_2: Sulfur has six valence electrons, is sharing six bonding electrons, and has one lone pair, so the formal charge of sulfur is $6 - \{2 + \frac{1}{2}(6)\} = +1$. The two oxygen atoms in this molecule are different because one is bonded to silicon with a single bond and one is bonded with a double bond. The single bonded oxygen has six valence electrons, three lone pairs, and one bonding pair, so the formal charge is $6 - \{6 + \frac{1}{2}(2)\} = -1$. The double-bonded oxygen has six valence electrons, two lone pairs, and two bonding pairs, so its formal charge is $6 - \{4 + \frac{1}{2}(4)\} = 0$. Note that the sum of the formal charges is equal to zero, which is consistent with the neutral charge on this molecule.

d. NO_3^-: Nitrogen has five valence electrons and is sharing four electron pairs, so the formal charge of nitrogen is $5 - \frac{1}{2}(8) = +1$. There are two different types of oxygen atoms: two with a single bond and one with a double bond. The oxygen atom that is double bonded to nitrogen has six valence electrons, two lone pairs, and two bonding pairs, so the formal charge is $6 - \{4 + \frac{1}{2}(4)\} = 0$. The two oxygen atoms that are bonded to nitrogen with a single bond have six valence electrons, three lone pairs, and one bonding pair, so they have a formal charge of $6 - \{6 + \frac{1}{2}(2)\} = -1$. Note that the sum of the formal charges is equal to -1, which is consistent with the charge of the ion.

EXAMPLE:

Draw resonance structures for NO_3^-:

For NO_3^-, there will be three equivalent resonance structures since there are three identical oxygen atoms:

Notice that the oxygen atoms now have a $-2/3$ formal charge on average.

---

5. Exceptions to the Octet Rule: Odd-Electron Species, Incomplete Octets, and Expanded Octets

    a. Free radicals have an odd number of electrons.

        i. Few radical compounds exist in nature, and most are very reactive.

    b. Some elements exist in compounds with fewer than eight electrons.

        i. Boron tends to have three electron pairs around it.

    c. Some elements can have more than eight electrons around them in compounds.

        i. Only elements in the third row and higher ($n > 2$) can have expanded octets since there are d orbitals available.

EXAMPLE:

Draw Lewis structures for the following molecules:

a. XeF_4

We follow the same four-step procedure as above (total number of valence electrons = 36), with additional electron pairs added to xenon instead of fluorine because xenon is in the fifth row of the periodic table:

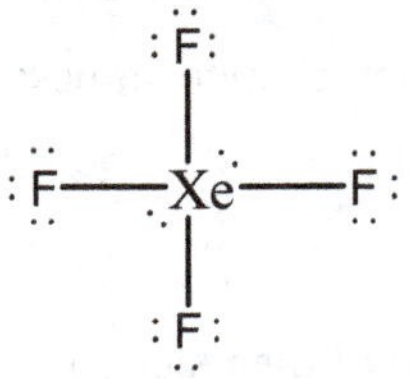

b. BH_3

Recognizing that boron can have fewer than eight electrons around it, we draw the structure (total number of valence electrons = 6):

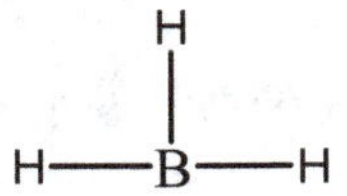

c. SCl_4

Again, we have more valence electrons than can be accommodated using the octet rule; total number of valence electrons = 32. Sulfur is the less-EN atom and is the central atom of this compound; the extra pair of electrons will also be situated on the sulfur atom as halogens do not have more than eight electrons (unless they are the central atom):

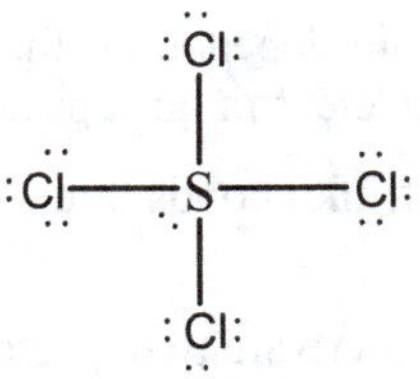

6. Bond Energies and Bond Lengths

    a. The bond energy is the energy required to break 1 mole of chemical bonds.

        i. Bond energy is always positive (endothermic)—energy is always required to break a bond.

        ii. The average bond energy is the average energy of a particular bond in many compounds.

        iii. The greater the strength of the bond, the higher the bond energy. In general, double and triple bonds have energies that are considerably greater than single-bond energies.

    b. The standard enthalpy change of a reaction, ΔH_{rxn}, can be estimated using individual bond energies.

        ΔH_{rxn} = (energy required to break reactant bonds) – (energy released in making product bonds)

        i. A reaction is exothermic when weak bonds are broken and stronger bonds are formed.

        ii. A reaction is endothermic when strong bonds are broken and weaker bonds are formed.

EXAMPLE:

Estimate the enthalpy change for the combustion of methane using average bond energies listed in Table 6.3 of your textbook.

First, we will write the balanced chemical reaction:

$$CH_4(g) + 2O_2(g) \rightarrow CO_2(g) + 2H_2O(g)$$

Now we will make a list of the bonds being broken along with the energy absorbed for each:

4 C–H bonds in each methane molecule: $4 \times +414 \text{ kJ/mol} = +1656 \text{ kJ}$

1 O=O bond in each of two oxygen molecules: $2 \times +498 \text{ kJ/mol} = +996 \text{ kJ}$

The bonds being formed will have negative enthalpy change values:

2 C=O bonds in each carbon dioxide molecule: $2 \times -799 \text{ kJ/mol} = -1598 \text{ kJ}$

2 O–H bonds in each of two water molecules: $4 \times -464 \text{ kJ/mol} = -1856 \text{ kJ}$

The sum of these values gives the enthalpy change of the reaction:

$$\Delta H = (1656 + 996 + -1598 + -1856) \text{ kJ} = -802 \text{ kJ}$$

The negative sign is expected since combustion reactions release heat (i.e., they are exothermic).

---

    c.   The average bond length is the average distance between two atomic nuclei in a large number of compounds.

        i.   For any two bonded atoms, the bond length of a single bond is longer than a double bond, which is longer than a triple bond.

7.   VSEPR Theory: The Five Basic Shapes

    a.   Valence shell electron pair repulsion (VSEPR) theory is a theory for predicting molecular shapes based on the idea that electron groups repel one another.

        i.   The shape of a molecule depends on the total number of electron groups around a central atom and whether those electron groups are lone pairs or bonding pairs.

            1.   Double and triple bonds are similar to single bonds in terms of their spatial requirements

    b.   A linear geometry occurs when there are two electron groups around the central atom.

        i.   The angle between groups is 180°.

    c.   A trigonal planar geometry occurs when there are three electron groups around the central atom.

        i.   The angle between groups is approximately 120°.

            1.   Double bonds repel single bonds more than single bonds repel each other. For example, the angles between the double and single bonds in a particular molecule that is trigonal planar are 121.9°, and the angle between the single bonds is 116.2°.

    d.   A tetrahedral geometry occurs when there are four electron groups around the central atom.

        i.   The angle between the groups is approximately 109.5°.

    e.   A trigonal bipyramidal geometry occurs when there are five electron groups around the central atom.

        i.   There are two different "types" of electron groups in this geometry: the three that define a plane (equatorial); and the two that are perpendicular to that plane and are 180° from one another (axial).

            1.   The angle between the equatorial electron groups is approximately 120°.

            2.   The angle between the equatorial and axial electron groups is approximately 90°.

    f.   An octahedral geometry occurs when there are six electron groups around the central atom.

        i.   The angle between groups is approximately 90°.

EXAMPLE:

Predict the electron and molecular shapes for the following molecules:

a. CF_4

First, we will draw the Lewis structure using the rules from Chapter 6:

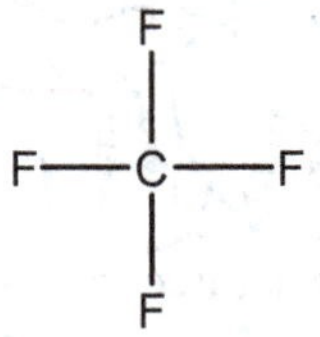

$$
\begin{array}{c}
\quad\;\; F \\
\quad\;\; | \\
F \!-\! C \!-\! F \\
\quad\;\; | \\
\quad\;\; F
\end{array}
$$

We see that the central atom (carbon) has four electron pairs around it, so it has a tetrahedral electron geometry. Since there are no lone pairs, this molecule also has a tetrahedral molecular geometry.

b. IBr_3

The Lewis dot structure is:

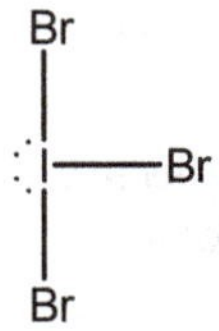

$$
\begin{array}{c}
\quad\; Br \\
\quad\; | \\
\!:\! I \!-\! Br \\
\quad\; | \\
\quad\; Br
\end{array}
$$

We see that the central atom (iodine) has five electron pairs around it, so it has a trigonal bipyramidal electron geometry. Since two of these electron pairs are lone pairs, the molecular geometry is T-shaped.

c. CO_2

The Lewis dot structure is:

$$O = C = O$$

The central atom (carbon) has two electron groups around it, so the electron geometry is linear. Since both electron groups are bonding groups, the molecular geometry is also linear.

d. PH_3

The Lewis structure is:

$$
\begin{array}{c}
\quad\; H \\
\quad\; | \\
\!:\! P \!-\! H \\
\quad\; | \\
\quad\; H
\end{array}
$$

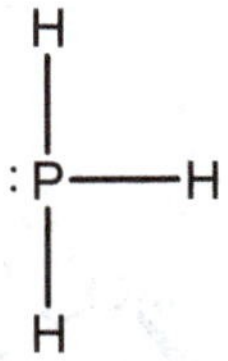

The central atom (phosphorus) has four electron groups around it, so the electron geometry is tetrahedral. Since there is one lone pair, the molecular geometry is trigonal pyramidal.

e. $AlCl_3$

The Lewis structure is:

$$
\begin{array}{c}
Cl \quad\quad Cl \\
\;\backslash \quad / \\
Al \\
| \\
Cl
\end{array}
$$

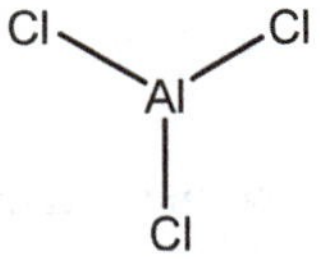

The central atom (aluminum) has three electron groups around it, so the electron geometry is trigonal planar. Since there are no lone pairs on aluminum, the molecular geometry is also trigonal planar.

EXAMPLE:

Predict the molecular shape of each interior atoms of butanoic acid, $CH_2CH_2CH_2COOH$, shown below:

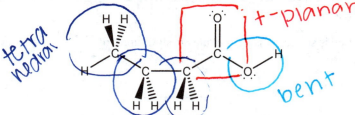

In order from left to right, we can see that the first three carbon atoms are tetrahedral. The fourth carbon atom is trigonal planar. The last atom (oxygen) is bent.

10. Molecular Shape and Polarity

    a. The polarity of a molecule depends on the polarity of its bonds and the molecular shape.

        i. Dipole moments are directional, so they can add together or cancel each other out in a molecule.

        ii. Look for symmetry in a molecule—a perfectly symmetric molecule will not have a dipole even if the bonds in it are polar.

EXAMPLE:

Which of the following molecules is/are polar?

a. SO_2

The Lewis structure is:

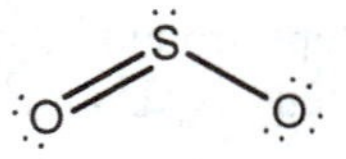

Oxygen is more electronegative than sulfur, so the sulfur-oxygen bonds are polar. Given the bent, asymmetric structure, there will be a net dipole pointing down.

b. ClO_3^-

The Lewis structure is:

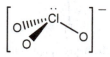

Oxygen is more electronegative than chlorine, so each of the three Cl–O bonds is polar. Since the molecule is asymmetrical (trigonal pyramid), there will be a net dipole pointing down.

Fill in the Blank:

1. Bonds that occur between elements with large electronegativity differences are _____.

2. The _____ of an atom is the charge it would have if all bonding electrons were shared equally between the atoms.

3. The _____ is the energy required to break 1 mole of the bond in the gas phase.

4. The average length of a bond between two particular atoms in a large number of compounds is the _____.

5. A(n) _____ bond is one in which electrons are shared unequally.

6. _____ is the most electronegative element.

7. _____ are species that have an odd number of electrons.

8. Expanded octets occur for elements in the _____ row or below.

9. When three electron pairs are shared between two atoms, a(n) _____ forms.

10. Central atoms that are surrounded by three electron groups have a(n) _____ electron geometry.

11. Electron groups repel each other through _____ forces.

12. A net dipole occurs from the _____ distribution of electrical charge.

13. The angles between atoms in a trigonal pyramidal geometry are _____.

Problems:

1. Each of the following substances contains both ionic and covalent bonds; indicate which bond is which, draw the Lewis structures for each, and assign formal charges to all atoms.

   a. $BaCO_3$

   b. KNO_3

   c. Na_2SO_4

   d. LiIO

   e. $Mg(ClO_3)_2$

2. Calculate the electronegativity difference, dipole moment, and percent ionic character for each of the bonds in the polyatomic ions from Question 1. Use average bond lengths from Table 6.4 in your book.

3. Draw resonance structures, where appropriate, for the polyatomic ions in Question 1.

4. Calculate the enthalpy change for the following reactions using the bond enthalpies given in Table 6.3 in your textbook. State whether the reactions are endothermic or exothermic.

   a. $H_2CO_3(g) \rightarrow H_2O(g) + CO_2(g)$

   b. $CO_2(g) + 4H_2(g) \rightarrow 2H_2O(g) + CH_4(g)$

   c. $2HNO_3(g) \rightarrow H_2O(g) + N_2O_5(g)$

5. Benzene, C_6H_6, is a six-membered ring with alternating double bonds.

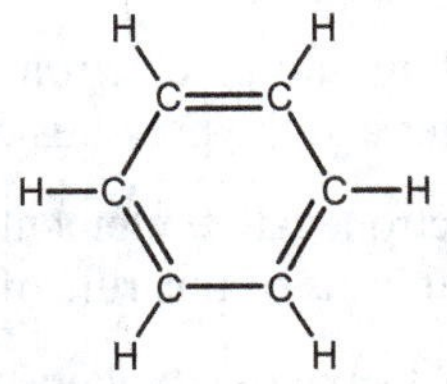

   a. What are the formal charges of all atoms in benzene?

   b. Draw all resonance structures for benzene.

   c. A carbon–carbon single bond has a bond length of 154 pm, and a carbon–carbon double bond has a bond length of 134. Estimate the bond length(s) for all carbon–carbon bonds in benzene.

6. The thiocyanate ion, SCN^-, has three valid Lewis structures.

   a. All three Lewis structures have carbon as the central atom. Explain why this is so.

   b. Which structure makes the largest contribution to the actual overall structure of SCN^-? Explain.

7. Determine the electron geometry and molecular geometry in each:

   a. N_2H_4

   b. HCN

   c. CF_3^-

   d. SiO_4^{4-}

   e. PO_4^{3-}

   f. CS_2

   g. XeF_4

   h. OCN^-

   i. SO_2Cl_2

   j. SCN^-

   k. CH_3OH

   l. CH_3OCH_3

   m. BrI_3

8. Draw the cis/trans isomers for difluoroethene CHFCHF.

   a. Determine the geometry at each carbon.

   b. Determine the hybridization of each carbon atom.

   c. Will either (or both) molecule(s) have a dipole moment? Explain.

Concept Questions:

1. In Chapter 5, we defined ionic substances as bonding between metals and nonmetals. In this chapter, we define an ionic bond as having an electronegativity value difference of greater than 2.0. Explain why both definitions are correct.

2. Carbon is the central atom for life. In fact, the chemistry of carbon-based compounds is the focus of an entire branch of chemistry called organic chemistry. Based on what you have learned about bonding in this chapter, why do you think that carbon is such a good candidate for the formation of so many different compounds?

3. As discussed in this chapter, the size of an atom and its electronegativity value are correlated. Explain this based on the principles that you learned in this chapter and the previous chapter.

4. We learned in this chapter that the least electronegative elements (except hydrogen) are most likely the central atom in a Lewis structure. Explain the physical basis of this rule of thumb.

5. Using the principle that electrons are shared in bonds, explain why boron often has only three bonds in stable compounds.

6. Certain elements can form molecules with an expanded octet.

   a. Which elements can have an expanded octet?

   b. Why do you expect that only certain elements can have an expanded octet? Base your answer on the principles of quantum mechanics that you learned.

7. Water can be formed by reacting hydrogen gas and oxygen gas together. Using the fact that this reaction is exothermic and considering the principles of bond energies described in this chapter, discuss the relative bond strengths of the reactants and products in this reaction.

8. Cl, Br, and I all form oxyanions: chlorate (ClO_3^-), bromate (BrO_3^-), and iodate (IO_3^-). Explain why fluorine does not form an analogous fluorate (FO_3^-) ion.

9. Molecules that have polar covalent bonds can be polar or nonpolar, but molecules with nonpolar covalent bonds are always nonpolar. Explain.

Chapter 6: Chemical Bonding II

Key Learning Outcomes:

- Write Hybridization and Bonding Schemes Using Valence Bond Theory
- Draw Molecular Orbital Energy Diagrams and Predicting Bond Order in a Homonuclear Diatomic Molecule
- Draw Molecular Orbital Energy Diagrams and Predicting Magnetic Properties in a Heteronuclear Diatomic Molecule

Chapter Summary:

In this chapter, you will consider more sophisticated bonding theories. First, valence bond theory will be introduced as a link between the quantum-mechanical atomic orbitals that you learned about in Chapter 3 and the molecular shapes described by VSEPR theory. According to valence bond theory, covalent bonds result from the overlap of atomic orbitals, either pure or hybridized. Through discussion of hybrid orbitals, you will learn about different bonding orientations, sigma and pi, and relate these bonding types to physical properties of molecules. The most complex bonding theory treated here will be molecular orbital theory, which is discussed in the final third of this chapter. Molecular orbital theory is a more rigorous quantum-mechanical description of bonding where electrons are described using (delocalized) bonding and antibonding molecular orbitals instead of localized atomic orbitals. You will see how this theory can be used to determine bond orders and magnetic properties (something that the other theories cannot do). Finally, electron delocalization in polyatomic atoms will be explored briefly.

Chapter Outline:

1. Oxygen: A Magnetic Liquid

    a. A magnetic substance is paramagnetic, meaning it has unpaired electrons.

    b. Oxygen is a paramagnetic liquid. This property is not predicted by Lewis theory alone.

2. Valence Bond Theory: Orbital Overlap as a Chemical Bond

    a. In its simplest form (as discussed in this book), valence bond theory is a qualitative, quantum-mechanical description of orbitals applied to bonding in molecules.

    b. Electrons are treated as if they exist in atomic orbitals that are localized on individual atoms.

        i. The atomic orbitals can be the simple s, p, d, or f orbitals discussed in Chapter 3 or they can be hybrid orbitals, discussed shortly.

    c. When atoms approach each other, the electrons and nuclei interact.

        i. According to valence bond theory, bonds are formed when half-filled orbitals overlap with each other, or when filled orbitals overlap with empty orbitals.

        ii. The molecular geometry is a result of the orbital geometry.

3. Valence Bond Theory: Hybridization of Atomic Orbitals

    a. The overlap of standard atomic orbitals (s, p, d, and f) does not account for the bonding and shapes of all molecules.

    b. Hybridization is a mathematical procedure in which standard atomic orbitals are combined to form hybrid (atomic) orbitals.

        i. Bond energy is minimized by constructing hybrid orbitals that maximally overlap.

            1. It costs energy to make hybrid orbitals; they will form if there is a greater release of energy when bonding occurs.

2. The type of hybrid orbital that forms is the one that gives the lowest energy upon bond formation.

ii. We will assume that only the central atom(s) in a molecule is (are) hybridized.

iii. The total number of hybrid orbitals is equal to the total number of atomic orbitals combined to form them.

iv. The shapes and energies of hybrid orbitals are related to the shapes and energies of the atomic orbitals that comprise them.

c. Four sp^3 orbitals are made by combining one s orbital and three p orbitals of an atom:

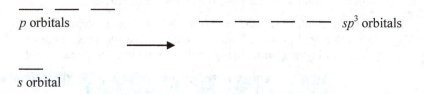

i. The energy of an sp^3 orbital is between the energy of the s and p orbitals, and the energies of all four sp^3 orbitals are the same (i.e., they are degenerate).

ii. The orientation of the four sp^3 orbitals on an atom is tetrahedral; that is, each orbital protrudes out into one of the four corners of a tetrahedron.

d. Three sp^2 orbitals are made by combining one s orbital and two p orbitals of an atom:

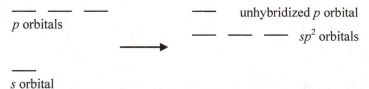

i. The energy of an sp^2 orbital is between the energy of the s and p orbitals, and the energies of all three sp^2 orbitals are the same (i.e., they are degenerate).

ii. One p orbital on the atom remains unhybridized and is perpendicular to the plane defined by the three sp^2 hybridized orbitals, which are oriented so that a lobe points in each corner of an equatorial triangle (trigonal planar arrangement).

iii. The unhybridized p orbital can be used to form a double bond.

1. Double bonds are the result of two bonding interactions: a sigma bond and a pi bond.

a. A pi (π) bond occurs via the side-to-side overlap of atomic orbitals, placing electron density above and below the bonding axis. Thus, pi bonds form from overlapping p orbitals.

b. A sigma (σ) bond occurs via the end-to-end overlap of atomic orbitals, placing electron density along the internuclear bonding axis. Sigma bonds form from overlapping atomic or hybrid orbitals.

2. While there is "free" rotation about single bonds, double bonds do not rotate. This is because, in order to do so, the side-to-side overlap of the p orbitals must be disrupted; that is, a pi bond would need to be broken.

a. This rigidity results in *cis/trans* isomers with substituents on different sides of a double bond.

e. Two *sp* hybrid orbitals are made by combining one *s* and one *p* orbital of an atom:

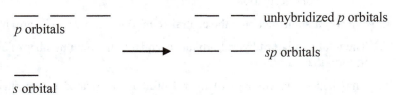

i. The energy of an *sp* orbital is between the energy of the *s* and *p* orbitals, and the energies of both *sp* orbitals are the same (i.e., they are degenerate).

ii. Two *p* orbitals remain unhybridized; they are perpendicular to each other and to the *sp* hybridized orbitals. You can imagine that the two unhybridized orbitals are situated on the *x*- and *y*-axis, while the *sp* orbitals point in either direction of the *z*-axis.

1. The two unhybridized *p* orbitals can be used to form a triple bond, which is composed of one σ bond and two π bonds.

f. Five sp^3d hybrid orbitals are made by combining one *s*, three *p*, and one *d* orbital of an atom:

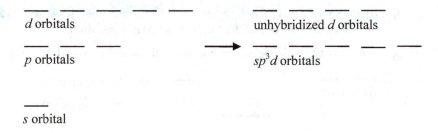

i. The energies of the sp^3d orbitals are the same (i.e., they are degenerate), and four unhybridized *d* orbitals remain.

ii. The geometry of the sp^3d orbitals is in accordance with the trigonal bipyramidal arrangement.

g. Six sp^3d^2 hybrid orbitals are made by combining one *s*, three *p*, and two *d* orbitals:

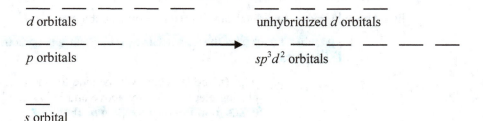

i. The energies of the sp^3d^2 orbitals are the same (i.e., they are degenerate), and three unhybridized *d* orbitals remain.

ii. The geometry of the sp^3d^2 orbitals is in accordance with the octahedral arrangement.

h. The particular hybridization can easily be determined by finding the electron geometry (using VSEPR) and then assigning an appropriate hybridization.

EXAMPLE:

What is the hybridization of each of the central atoms in the first example given in this chapter?

a. CF_4 *tetrahedral, sp^3*

This molecule has a tetrahedral electron geometry, which corresponds to sp^3 hybridization on carbon.

b. IBr_3 *trig bipyramidal, sp^3d*

This molecule has a trigonal bipyramidal electron geometry, which corresponds to sp^3d hybridization on iodine. Note that two of sp^3d hybridized orbitals contain lone pairs.

c. CO_2 *linear, sp*

This molecule is linear with two double bonds, which corresponds to sp hybridization on carbon. Note that the two remaining unhybridized p orbitals are used to form a pi bond with each oxygen atom.

d. PH_3 *H-P-H*

This molecule has a tetrahedral electron geometry, which corresponds to sp^3 hybridization on phosphorus. Note that one of the sp^3 hybrid orbitals contains the lone pair.

e. $AlCl_3$

This molecule has a trigonal planar electron geometry, which corresponds to sp^2 hybridization.

---

4. Molecular Orbital Theory: Electron Delocalization

   a. Valence bond theory is a localized bonding model that treats electrons as if they exist on particular atoms and the selected orbital will be the one with the lowest energy.

   b. Molecular orbital theory is a more rigorous quantum-mechanical description of bonding in which electrons are assigned to orbitals distributed over the entire molecule.

   c. Molecular orbitals (MOs) are constructed using linear combinations of atomic orbitals (LCAOs).

      i. All of the atomic orbitals in a molecule are used to construct its molecular orbitals.

      ii. Atomic orbitals are combined in an additive way (constructive interference of waves) to give bonding orbitals.

      iii. Atomic orbitals are combined in a subtractive way (destructive interference of waves) to give antibonding orbitals.

      iv. The total number of molecular orbitals is equal to the total number of atomic orbitals used to construct them.

   d. The $1s$ orbitals on two atoms combine constructively to give σ_{1s} (bonding) and σ_{1s}^{*} (antibonding) orbitals.

   e. The bond order of a molecule is:

$$\text{Bond order} = \frac{\left(\begin{array}{c}\text{number of electrons} \\ \text{in bonding orbitals}\end{array}\right) - \left(\begin{array}{c}\text{number of electrons} \\ \text{in antibonding orbitals}\end{array}\right)}{2}$$

A bond order greater than zero indicates that it is energetically favorable for a bond to form.

EXAMPLE:

What is the bond order of H_2^+?

We can construct the molecular orbitals from the two $1s$ orbitals:

$$\sigma_{1s}^* \quad \square$$
$$\sigma_{1s} \quad \boxed{\uparrow}$$

We have two valence electrons total, one from each hydrogen atom, and one electron is removed to form the cation. The bond order is thus ½ because only one electron is in the bonding orbital and there are zero electrons in the antibonding orbital.

---

  f.   The $2s$ orbitals of two atoms combine to form σ_{1s} and σ_{1s}^* orbitals.

  g.   The $2p$ orbitals of two atoms combine to form σ_{2p}, σ_{2p}^*, $2\pi_{1p}$, and $2\pi_{1p}^*$ orbitals.

      i.   The energetic ordering of the bonding and antibonding orbitals depends on which atoms are combined to form the molecular orbitals.

         1.   For B_2, C_2, and N_2:

              $$\sigma_{2p}^* \quad \square$$
              $$\pi_{2p}^* \quad \square\square$$
              $$\sigma_{2p} \quad \square$$
              $$\pi_{2p} \quad \square\square$$
              $$\sigma_{2s}^* \quad \square$$
              $$\sigma_{2s} \quad \square$$

         2.   For O_2, F_2, and Ne_2:

              $$\sigma_{2p}^* \quad \square$$
              $$\pi_{2p}^* \quad \square\square$$
              $$\pi_{2p} \quad \square\square$$
              $$\sigma_{2p} \quad \square$$
              $$\sigma_{2s}^* \quad \square$$
              $$\sigma_{2s} \quad \square$$

  h.   Electrons fill molecular orbitals in order of lowest to highest energy, with unpaired, parallel spins in degenerate orbitals (just as they fill atomic orbitals).

      i.   Paramagnetic species have unpaired electrons, while diamagnetic species have all electrons paired.

  i.   When constructing molecular orbital diagrams for heteronuclear diatomic molecules, the energy of the orbitals must be taken into account.

      i.   The more electronegative atom will have a lower energy and will therefore make a greater contribution to the molecular orbital(s).

         1.   Electrons will be more localized on the more electronegative orbital.

      ii.   Nonbonding orbitals are atomic orbitals that are not used in the construction of molecular orbitals.

         1.   Nonbonding orbitals occur because there is poor overlap of the atomic orbitals from the two atoms.

Chapter 7: Chemical Reactions and Chemical Quantities

Key Learning Outcomes:

-   Balance Chemical Equations
-   Make Calculations Involving the Stoichiometry of a Reaction
-   Determine the Limiting Reactant and Calculating Theoretical and Percent Yield
-   Determine the Amount of Reactant in Excess
-   Write Equations for Combustion Reactions
-   Write Reactions for Alkali Metal and Halogen Reactions

Chapter Summary:

In this chapter, you will learn about chemical reactions and chemical quantities. You will begin by learning how to balance chemical reactions. With your understanding of balanced chemical reactions you will be able to construct stoichiometric ratios that will be used as conversion factors in chemical problems. Your understanding of stoichiometry will allow you to calculate reaction yields, the amount of product obtained in a given reaction. Finally, you will be introduced to three common reaction types: combustion, alkali metal, and halogen reactions.

Chapter Outline:

1. Climate Change and the Combustion of Fossil Fuels

    a. The burning of fossil fuels is an example of a chemical reaction.

2. Chemical Changes and Physical Changes

    a. A chemical change is a rearrangement of atoms or alteration of composition.

        i. A chemical property is a property that is displayed when the composition changes.

    b. A physical change is an alteration of state, appearance, or quantity.

        i. A physical property is a property that is displayed by a substance when the composition does not change.

    ▶ Note that physical changes may occur when separating mixtures while chemical changes must occur when separating atoms in a pure substance.

EXAMPLE:

Identify the following as physical or chemical changes.

a. The distillation of saltwater to produce pure water

    This is a physical change in which the liquid water is heated until vaporization and then cooled to condense the vapors to a liquid again.

b. The rusting of an iron nail

    This is a chemical change in which elemental iron reacts with oxygen to form iron oxide.

c. The burning of wood in a fireplace

    This is a chemical change in which combustion (reaction with oxygen) changes the carbon in the wood and oxygen in air into gaseous carbon dioxide and water.

d. Salt precipitating out of a solution

This is a physical change in which the aqueous salt is becoming a solid crystal but the chemical composition remains unchanged.

---

3. Writing and Balancing Chemical Equations

   a. In a chemical reaction, one or more starting substances (termed reactants) are converted into one or more different substances (termed products).

   b. We write chemical reactions in shorthand notation, using an equation that identifies all reactants and products by their chemical formulas and physical states, where the arrow represents the phrase "react(s) to form":

$$\text{reactants} \rightarrow \text{products}$$

   c. Chemical reactions must be balanced in order to adhere to the law of conservation of mass, which states that matter is neither created nor destroyed in a chemical reaction. In other words, the number of atoms of each type of element must be the same on both sides of the equation.

      i. This is achieved by placing appropriate stoichiometric coefficients (whole numbers) in front of reactants and products and NOT by changing the subscripts in a chemical formula (since this changes the identity of the substance).

      ii. Equations can be balanced by inspection. It is simplest to start by balancing an element that appears in only one reactant and product.

EXAMPLE:

Balance the following reactions:

a. $H_2 + Cl_2 \rightarrow HCl$

First, we will set up a table in order to balance the atoms:

Atom	# on Left Side	# on Right Side
Cl	2	1
H	2	1

We will start by putting a "2" in front of HCl so that the Cl atoms are balanced.

$H_2 + Cl_2 \rightarrow 2HCl$

We then recount everything in our table:

Atom	# on Left Side	# on Right Side
Cl	2	2
H	2	2

And we see that the reaction is now balanced.

b. $KClO_3 \rightarrow KCl + O_2$

We will again set up a table:

Atom	# on Left Side	# on Right Side
K	1	1
Cl	1	1
O	3	2

We see that the first two entries, K and Cl, are already balanced, so we look at O. In order to balance with whole numbers, we must multiply the reactant side by 2 and the product side by 3:

$2KClO_3 \rightarrow KCl + 3O_2$

Atom	# on Left Side	# on Right Side
K	2	1
Cl	2	1
O	6	6

We have now unbalanced the first two elements in the table. By multiplying by 2 on the product side to balance the K atoms, we have:

$2KClO_3 \rightarrow 2KCl + 3O_2$

Atom	# on Left Side	# on Right Side
K	2	2
Cl	2	2
O	6	6

And we see that everything is balanced.

c. $CH_3OH + O_2 \rightarrow CO_2 + H_2O$

Again we will start by writing out the table:

Atom	# on Left Side	# on Right Side
C	1	1
H	4	2
O	3	3

For combustion reactions in general (the reaction with oxygen to produce water and carbon dioxide), the easiest way to balance is to consider first C, then H, and finally O. (This strategy can be remembered as alphabetical order: C, H, O). Since carbon is already balanced in this case, we will start with hydrogen by multiplying the water coefficient by 2:

$CH_3OH + O_2 \rightarrow CO_2 + 2H_2O$

Atom	# on Left Side	# on Right Side
C	1	1
H	4	4
O	3	4

Now we need to balance the oxygen by increasing the amount on the reactant side. We have two choices: we can change the number of CH_3OH molecules or change the O_2 molecules. We will balance the O_2 because if we change the number of CH_3OH molecules, we would have to go back and balance C and H again. In this case, in order to balance the oxygen atoms, we will need 3/2 of an O_2 molecule:

$$CH_3OH + 3/2O_2 \rightarrow CO_2 + 2H_2O$$

Atom	# on Left Side	# on Right Side
C	1	1
H	4	4
O	4	4

Now we see that everything balances, but the 3/2 is not conventional. We can multiply all coefficients by 2 in order to eliminate the fraction:

$$2CH_3OH + 3O_2 \rightarrow 2CO_2 + 4H_2O$$

This shows a correctly balanced equation.

---

4. **Reaction Stoichiometry: How Much Carbon Dioxide?**

   a. The coefficients in a chemical reaction specify the relative amounts, in moles, of each substance in the reaction.

   b. Stoichiometry is the numerical relationship between the amounts of each substance in a balanced chemical reaction.

      i. The balanced chemical reaction can be viewed as a recipe.

   c. A chemical reaction can be used to construct conversion factors that relate the numbers of moles of each substance:

   $$H_2(g) + F_2(g) \rightarrow 2HF(g)$$

   Example conversion factors (mole-to-mole ratios) for this reaction are:

   $$\frac{1 \text{ mol } H_2}{1 \text{ mol } F_2}, \frac{1 \text{ mol } H_2}{2 \text{ mol } HF}, \text{ and } \frac{2 \text{ mol } HF}{1 \text{ mol } F_2}.$$

EXAMPLE:

Express the mole-to-mole relationships between each species in the balanced reaction for the combustion of methane:

$$CH_4(g) + 2O_2(g) \rightarrow 2H_2O(g) + CO_2(g)$$

$$\frac{1 \text{ mol } CH_4}{2 \text{ mol } O_2}, \frac{1 \text{ mol } CH_4}{2 \text{ mol } H_2O}, \frac{1 \text{ mol } CH_4}{1 \text{ mol } CO_2}, \frac{1 \text{ mol } O_2}{1 \text{ mol } H_2O}, \frac{2 \text{ mol } O_2}{1 \text{ mol } CO_2}, \text{ and } \frac{2 \text{ mol } H_2O}{1 \text{ mol } CO_2}$$

---

   i. With the mole-to-mole relationships as the central conversion, we can also convert from mass (using molar mass), and molecules, ions, and so on (using Avogadro's number) of one substance to mass, molecules, ions, and the like of another substance in the reaction.

   ▶ Any time you are converting between substances in a chemical reaction, you will use the mole-to-mole ratio; this provides a hint for solving the problem.

5. **Stoichiometric Relationships: Limiting Reactant, Theoretical Yield, Percent Yield, and Reactant in Excess**

   a. The limiting reagent restricts the amount of product that can form in a chemical reaction.

      i. The limiting reactant is the reactant that is completely used up in a chemical reaction.

   b. The excess reagent(s) do not limit the amount of product that forms in a chemical reaction and will be present to some degree after the reaction is complete.

▶ To determine the limiting reagent, calculate the amount of product formed from each reactant. The limiting reagent will be the reacting species that gives the smallest amount of product; all other reacting species will be in excess.

c. The theoretical yield is the amount of product that is formed in a reaction based on the limiting reagent and reaction stoichiometry.

EXAMPLE:

3.04 g of aluminum reacts with 7.42 g of iodine to form aluminum iodide. When the reaction is complete, how many grams of all species are predicted to be present in the reaction flask?

We first need to write a balanced chemical reaction:

$$2Al(s) + 3I_2(s) \rightarrow 2AlI_3(s)$$

Then we can determine the amount of product formed from each of the reactant species:

$$3.04 \text{ g Al} \times \frac{1 \text{ mol Al}}{26.98 \text{ g Al}} \times \frac{2 \text{ mol AlI}_3}{2 \text{ mol Al}} \times \frac{407.68 \text{ g AlI}_3}{1 \text{ mol AlI}_3} = 45.9 \text{ g AlI}_3 \text{ from Al}$$

$$7.42 \text{ g I}_2 \times \frac{1 \text{ mol I}_2}{253.80 \text{ g I}_2} \times \frac{2 \text{ mol AlI}_3}{3 \text{ mol I}_2} \times \frac{407.68 \text{ g AlI}_3}{1 \text{ mol AlI}_3} = 7.95 \text{ g AlI}_3 \text{ from I}_2$$

These are the amounts of AlI_3 that would be produced from each reagent, assuming that there was plenty (at least a stoichiometric amount) of the other reagent. From these values we see that I_2 is the limiting reagent since it results in a smaller amount of product, AlI_3.

Now we can calculate the amount of Al that reacted with the I_2:

$$7.946 \text{ g AlI}_3 \times \frac{1 \text{ mol AlI}_3}{407.68 \text{ g AlI}_3} \times \frac{2 \text{ mol Al}}{2 \text{ mol AlI}_3} \times \frac{26.98 \text{ g Al}}{1 \text{ mol Al}} = 0.526 \text{ g Al reacted}$$

So the mass of Al that remains after the reaction is the amount that we started with minus the amount that reacted: 3.04 g − 0.526 g = 2.51 g.

All of the I_2 reacts so there is none left—this is always the case for the limiting reagent.

This means that after the reaction is complete, there are 2.51 g Al, 0 g of I_2, and 7.95 g of AlI_3.

d. The actual yield is the amount of product that is actually formed in a chemical reaction.

▶ The actual yield is usually lower than the theoretical yield—this is one way to check your work.

e. The percent yield is the actual yield divided by the theoretical yield, multiplied by 100%:

$$\text{percent yield} = \frac{\text{actual yield}}{\text{theoretical yield}} \times 100\%$$

EXAMPLE:

The reaction in the example above is carried out, and 7.04 g of AlI_3 are collected. What is the percent yield of the reaction?

The theoretical yield of AlI_3 is 7.95 g as calculated above, so the percent yield is:

$$\frac{7.04 \text{ g}}{7.95 \text{ g}} \times 100\% = 88.6\%$$

6. Three Examples of Chemical Reactions: Combustion, Alkali Metals, and Halogens

    a. A combustion reaction involves the reaction of a substance with oxygen (O_2) to form one or more oxygen-containing compounds.

        i. Carbon reacts with oxygen to form CO_2.

        ii. Hydrogen reacts with oxygen to form H_2O.

    b. The alkali metals have an ns^1 outer electron configuration, so they react vigorously.

        i. Alkali metals (M) react with halogens (X_2) to form halide salts:

$$2M(s) + X_2(g) \rightarrow 2MX(s)$$

        ii. Alkali metals (M) react with water to form hydroxide salts:

$$2M(s) + 2H_2O(l) \rightarrow 2MOH(aq) + H_2(g)$$

    c. The halogens have an ns^2np^5 outer electron configuration, so they react vigorously.

        i. Halogens (X_2) react with metals (M) to form metal halides:

$$2M(s) + nX_2(g) \rightarrow 2MX_n(s)$$

        ii. Halogens react with hydrogen to form hydrogen halides:

$$H_2(g) + X_2(g) \rightarrow 2HX(g)$$

        iii. Halogens react with each other to form interhalogen compounds:

$$Cl_2(g) + F_2(g) \rightarrow 2ClF(g)$$

Fill in the Blank:

1. A(n) _____ change is one that changes the state or appearance of a substance but does not alter its composition.

2. A(n) _____ property is one that is displayed only when the composition is changed through a chemical change.

3. The species that limits the amount of product that is formed in a chemical reaction is the _____.

4. The _____ is a numerical measure of the relationship between the actual yield and the theoretical yield.

5. The numerical relationship between chemical amounts in a balanced reaction is called reaction _____.

Problems:

1. Classify each of the following as a chemical or physical change:

    a. Removing paint with a solvent

    b. Filtering a precipitate out of a solution

    c. Bread rising in the oven

    d. Ozone depletion

2. Balance the following chemical reactions (note that for problems a–i, physical states have been ignored for simplicity):

    a. $N_2O_4 \rightarrow NO_2$

    b. $Fe + Cl_2 \rightarrow FeCl_3$

Chapter 8: Introduction to Solutions and Aqueous Reactions

Key Learning Outcomes:

- Calculate and Using Molarity as a Conversion Factor

- Determine Solution Dilutions

- Use Solution Stoichiometry to Find Volumes and Amounts

- Predict Compound Solubility

- Write Equations for Precipitation Reactions

- Write Complete Ionic and Net Ionic Equations

- Name Acids

- Write Equations for Acid–Base Reactions

- Solve Calculations Involving Acid–Base Titrations

- Write Equations for Gas-Evolution Reactions

- Assign Oxidation States

- Identify Redox Reactions, Oxidizing Agents, and Reducing Agents Using Oxidation States

- Predict the Sponteneity of Redox Reactions

Chapter Summary:

In this chapter, you will be introduced to solutions. First, you will learn about the most common unit for solution concentration, molarity. Next, you will be introduced to dilution and solution stoichiometry problems. Electrolyte solutions will then be discussed, and solubility rules will be used to predict the product(s) of a precipitation reaction. Various other reaction types, including gas evolution, acid–base, and redox reactions, will then be explored so that you will be able to predict product formation and carry out stoichiometry calculations on the most common reaction types.

Chapter Outline:

1. Molecular Gastronomy

    a. In a precipitation reaction, two solutions are mixed and a solid forms.

    b. A solution in which one component is water is called an aqueous solution.

2. Solution Concentration

    a. A solution is a homogeneous mixture of two or more substances. Particle sizes distinguish solutions from colloids and suspensions.

    b. The solvent is the majority component of a solution, and the solute is the minor component. Either may be a gas, liquid, or solid.

    c. An aqueous solution is a solution in which water is the solvent; it is designated as (aq).

    d. Solution concentrations can be described qualitatively as concentrated or dilute.

    e. Solution concentrations can be described quantitatively using various units. The molar concentration or molarity (abbreviated M) relates moles of solute per liter of solution.

    ▶ You will encounter other units for solution concentrations later on.

EXAMPLE:

Calculate the molarity of a solution that is made from dissolving 5.28 mg of sodium chloride in 48.5 mL of water. Assume that the solution volume is the same as the solvent volume.

First, we need to calculate the number of moles of NaCl:

$$5.28 \text{ mg NaCl} \times \frac{1 \text{ g}}{1000 \text{ mg}} \times \frac{1 \text{ mol NaCl}}{58.44 \text{ g NaCl}} = 0.0000903 \text{ mol NaCl}$$

Now we convert the volume into liters of water:

$$48.5 \text{ mL} \times \frac{1 \text{ L}}{1000 \text{ mL}} = 0.0485 \text{ L}$$

The molarity is then the number of moles of solute divided by the volume (in L) of solution:

$$\frac{0.0000903 \text{ mol}}{0.0485 \text{ L}} = 0.00186 \text{ M}$$

---

    f.   Molarity can be used as a conversion factor between chemical amount (mol) and volume (L).

EXAMPLE:

How many grams of sodium are in 12.5 L of a 5.82 M aqueous solution of sodium carbonate?

This is a multiple-step conversion factor problem using the molarity to convert from liters to moles:

$$12.5 \text{ L solution} \times \frac{5.82 \text{ mol Na}_2\text{CO}_3}{1 \text{ L solution}} \times \frac{2 \text{ mol Na}^+}{1 \text{ mol Na}_2\text{CO}_3} \times \frac{22.99 \text{ g}}{1 \text{ mol Na}^+} = 3350 \text{ g Na}^+$$

---

    g.   Stock solutions (concentrated solutions) can be diluted; the concentration of the dilute solution can be calculated using the concentration of the stock solution and the volumes according to $M_1V_1 = M_2V_2$.

    ▶   Note that the units on each side of this equation are moles. In a dilution, the moles of solute remain unchanged (are equal in both solutions).

EXAMPLE:

How many liters of a 4.95 M aqueous solution of HCl are required to prepare 10.0 L of a 0.510 M solution?

This is a simple dilution problem where the unknown is the volume of the concentrated solution:

$$(4.95 \text{ M})V_1 = (0.510 \text{ M})(10.0 \text{ L})$$

$$V_1 = 1.03 \text{ L}$$

---

3.   Solution Stoichiometry

    a.   Many chemical reactions occur in solution, and so it is necessary to use molarity in a stoichiometry problem.

EXAMPLE:

When aqueous solutions of lead(II) nitrate and sodium chloride are combined, a lead(II) chloride precipitate forms:

$$Pb(NO_3)_2(aq) + 2NaCl(aq) \rightarrow PbCl_2(s) + 2NaNO_3(aq)$$

If 30.0 mL of a 0.541 M $Pb(NO_3)_2$ solution is added to 24.5 mL of a 1.452 M NaCl solution, how much solid $PbCl_2$ (in grams) is formed?

This is a conversion factor problem using molarity and the reaction stoichiometry. To determine the product yield, we need to identify the limiting reagent.

We can calculate the mass of product formed from $Pb(NO_3)_2$:

$$30.0 \text{ mL} \times \frac{1 \text{ L}}{1000 \text{ mL}} \times \frac{0.541 \text{ mol } Pb(NO_3)_2}{1 \text{ L}} \times \frac{1 \text{ mol } PbCl_2}{1 \text{ mol } Pb(NO_3)_2} \times \frac{278.11 \text{ g } PbCl_2}{1 \text{ mol } PbCl_2} = 4.51 \text{ g } PbCl_2$$

Now we will do the same calculation using NaCl:

$$24.5 \text{ mL} \times \frac{1 \text{ L}}{1000 \text{ mL}} \times \frac{1.452 \text{ mol NaCl}}{1 \text{ L}} \times \frac{1 \text{ mol } PbCl_2}{2 \text{ mol NaCl}} \times \frac{278.11 \text{ g } PbCl_2}{1 \text{ mol } PbCl_2} = 4.95 \text{ g } PbCl_2$$

We can see from these calculations that $Pb(NO_3)_2$ is the limiting reagent, so 4.51 g of $PbCl_2$ will form.

---

4. Types of Aqueous Solutions and Solubility

   a. Water is a polar molecule, meaning that one end of it is partially positive and one end is partially negative.

      i. Some ionic compounds dissolve in water because the anion is attracted to the positive end of the water molecule and the cation is attracted to the negative end.

   b. Substances that, when dissolved in water, form solutions that conduct electricity are called electrolytes.

      i. An electrolyte must contain mobile charge carriers such as ions.

      ii. Strong electrolytes completely dissociate, and weak electrolytes dissociate only partially when dissolved in water.

   c. Nonelectrolytes do not dissociate into ions when dissolved in water, and their solutions therefore do not conduct electricity.

   d. Acids are molecular compounds that dissociate into ions when dissolved in water.

      i. A strong acid completely ionizes, and a weak acid only partially dissociates in water.

      ▸ Acids are an example of an electrolyte. A solution of a strong acid will be a better electrical conductor than a weak acid.

   e. A soluble compound is, essentially, one that dissolves in water (or another solvent), while an insoluble compound does not. The solubility of a solute in a given solvent is reported in units of concentration and depends on temperature and pressure.

      ▸ We will learn how to predict the solubility of molecular compounds in water in later chapters.

      ▸ If we consider ionic compounds in water, the solubility rules can be simplified in two ways. First, the solubility rules trump the insolubility rules, so you don't need to memorize the exceptions. Second, the exceptions always include the same six metal cations: Ca^{2+}, Ba^{2+}, Sr^{2+}, Ag^+, Pb^{2+}, and Hg_2^{2+}.

1. Calcium, strontium, and barium are in the same group in the periodic table.

2. Silver, mercury, and lead are one away from each other on the periodic table: start at silver, skip one to the right (cadmium), below is mercury; then skip one to the right (thallium) and end at lead.

EXAMPLE:

Predict which of the following substances will form an aqueous solution that will conduct an electrical current.

a. HNO_3

Nitric acid is a strong acid, which means that it will dissociate completely in water. A solution of nitric acid will therefore conduct electricity.

b. NaCl

Sodium salts are always soluble in water. The presence of sodium cations and chloride anions in water means that the solution will conduct electricity.

c. CaS

Calcium sulfide is not soluble in water, so a solution of it will not conduct electricity.

d. $Fe(NO_3)_3$

All salts containing the nitrate ion are soluble in water. A solution of iron(III) nitrate will conduct electricity since iron(III) cations and nitrate anions will be present in solution.

e. $C_6H_{12}O_6$

Glucose is a molecular compound that dissolves in water, but glucose does not dissociate into ions. So a solution of glucose will not conduct electricity.

f. AgI

Silver halides are insoluble in water, so a solution of it will not conduct electricity.

---

5. Precipitation Reactions

    a. Precipitation reactions are those in which a solid (precipitate) forms upon the mixing of two aqueous solutions.

        i. Only insoluble compounds form precipitates.

    b. When two aqueous solutions are mixed, the anions can pair with the cations of the other species (termed a double displacement reaction); if an insoluble species is formed, the reaction is a precipitation reaction.

6. Representing Aqueous Reactions: Molecular, Complete Ionic, and Net Ionic Equations

    a. A molecular equation gives the complete neutral formulas for all species as though they existed as molecular compounds in solution.

    b. A complete ionic equation shows the individual ions present in the solution.

    ▶ Any species that is in the aqueous phase can potentially dissociate into ions. Gases, liquids, and solids do not dissociate.

    c. The net ionic equation shows only species that change during a chemical reaction and eliminates spectator ions, which appear in the same form on both sides of the reaction.

    ▶ Note that species that appear on both sides of the equation do not change and therefore do not participate in the chemistry. Canceling out these spectator ions is a way to show only the chemistry that is occurring.

► When predicting products in a precipitation reaction, follow these steps:

1. Write all ions present in the solution as reactants.

2. Change partners—the cations from one salt combine with anions from the other in whatever ratio will give a neutral salt.

3. Balance the chemical reaction equation.

4. Determine the phase of each species using solubility rules.

EXAMPLE:

When an aqueous solution of barium acetate is combined with an aqueous solution of sodium sulfate, a precipitate forms. Predict the products formed in the reaction, and write the molecular, complete ionic, and net ionic equations.

$$Ba(C_2H_5O_2)_2(aq) + Na_2SO_4(aq) \rightarrow$$

We will first determine the products of the reaction by exchanging the anions of the reactants and being sure to form charge-neutral species. The phases of all species are determined using the solubility rules.

$$Ba(C_2H_5O_2)_2(aq) + Na_2SO_4(aq) \rightarrow BaSO_4(s) + NaC_2H_3O_2(aq)$$

Now we will balance the reaction to give the "molecular" equation:

$$Ba(C_2H_5O_2)_2(aq) + Na_2SO_4(aq) \rightarrow BaSO_4(s) + 2NaC_2H_3O_2(aq)$$

The full ionic equation shows all aqueous species (i.e., dissolved ionic compounds) separated into their ions. This equation shows all species that are present in the container when these solutions are mixed.

$$Ba^{2+}(aq) + 2C_2H_5O_2^-(aq) + 2Na^+(aq) + SO_4^{2-}(aq) \rightarrow BaSO_4(s) + 2Na^+(aq) + 2C_2H_3O_2^-(aq)$$

Now we write the net ionic equation by eliminating the spectator ions (Na^+ and $C_2H_3O_2^-$):

$$Ba^{2+}(aq) + SO_4^{2-}(aq) \rightarrow BaSO_4(s)$$

7. Acid–Base Reactions

a. Acid–base reactions are often called neutralization reactions and produce water and a salt.

    i. An acid is a substance that produces H^+ (or H_3O^+) in water.

        1. H_3O^+ is a proton that is associated with a water molecule due to electrostatic interactions. H_3O^+ is called the hydronium ion.

    ii. A base is a substance that produces OH^- in water.

    ► You will be given a more complete set of definitions for acids and bases later on.

b. A polyprotic acid is an acid in which more than one ionizable proton can be sequentially released. A diprotic acid is one in which two protons are sequentially released.

c. Acids are molecular compounds that produce hydrogen ions (H^+) in water. Acids are named according to which type of acid it is:

    i. Binary acids contain only two elements: hydrogen and one other element.

        1. Binary acids are named using "hydro-(base name of the nonmetal)-ic acid."

EXAMPLE:

Provide names for the following acids.

a. $H_2S(aq)$

   Hydrosulfuric acid

b. $HCl(aq)$

   Hydrochloric acid

c. $H_2Te(aq)$

   Hydrotelluric acid

d. $HBr(aq)$

   Hydrobromic acid

Note that all substances are molecular (and named according to the rules for molecular compounds) when in the gas phase.

---

   ii. Oxyacids contain one of the oxyanions and enough hydrogen atoms to make the species charge-neutral.

     1. Oxyacids are named based on the relevant oxyanion.

       a. If the oxyanion ends in -ate, then the acid is named "(base name of oxyanion)-ic acid"

       b. If the oxyanion ends in -ite, then the acid is named "(base name of the oxyanion)-ous acid"

EXAMPLE:

Provide names for the following acids.

a. H_2CO_3

   Carbonic acid

b. H_2SO_3

   Sulfurous acid

c. $HClO_4$

   Perchloric acid

d. H_2SeO_4

   Selenic acid

---

  d. In a general neutralization reaction, the acid donates a proton to the base to produce water and a salt:

$$ACID + BASE \rightarrow WATER + SALT$$

$$HA + BOH \rightarrow H_2O + AB$$

   ▶ Neutralization reactions are really just another example of ions trading partners (i.e., double displacement reactions), as we have seen already with precipitation reactions.

  e. An acid–base titration is an analytical tool in which the concentration of an acid is determined by neutralizing it with a base of known concentration (or vice versa); a titration problem is a stoichiometry problem involving acids and bases.

i. The equivalence point is the point in a titration where the moles of acid and base are stoichiometrically equal.

ii. An acid–base indicator is a dye that changes color at a particular acidity and is used to visually identify the equivalence point.

▶ Note that the point where the indicator changes color is called the endpoint of the titration. It is desirable for the endpoint to be as close as possible to the equivalence point, but this is not always the case. In this chapter, we will assume that they are the same.

EXAMPLE:

A 0.4512 M solution of nitric acid is titrated with 25.4 mL of a 0.5420 M solution of NaOH. What volume of acid was utilized in the experiment?

First, we need to write a balanced chemical reaction:

$$HNO_3(aq) + NaOH(aq) \rightarrow NaNO_3(aq) + H_2O(l)$$

We can calculate the moles of sodium hydroxide used:

$$25.4 \text{ mL NaOH} \times \frac{1 \text{ L}}{1000 \text{ mL}} \times \frac{0.5420 \text{ mol NaOH}}{\text{L}} = 0.01377 \text{ mol NaOH}$$

We can see from the balanced chemical reaction that we need 1 mole of nitric acid for every mole of sodium hydroxide. We can use this to determine the volume of acid:

$$0.01377 \text{ mol NaOH} \times \frac{1 \text{ mol HNO}_3}{1 \text{ mol NaOH}} \times \frac{1 \text{ L}}{0.4512 \text{ mol HNO}_3} \times \frac{1000 \text{ mL}}{1 \text{ L}} = 30.5 \text{ mL HNO}_3$$

Check: The two solutions have very similar molarity values, so it makes sense that the volume required for an equimolar amount of each solution is similar.

8. Gas-Evolution Reactions

a. Gas-evolution reactions are those where a gas is produced.

b. Gases can form directly in a reaction or indirectly from the decomposition of a product.

c. The formations of sulfides, carbonates, sulfites, and ammonia are all indicative of a gas-evolution reaction.

▶ H_2SO_3, H_2CO_3, and NH_4OH all decompose to give $H_2O(l)$ and a gas.

EXAMPLE:

How many grams of carbon dioxide gas are produced when a stoichiometric amount of hydrochloric acid is added to 3.452 g of sodium bicarbonate dissolved in water?

We will first write the balanced chemical reaction:

$$NaHCO_3(aq) + HCl(aq) \rightarrow H_2CO_3(aq) + NaCl(aq) \rightarrow H_2O(l) + CO_2(g) + NaCl(aq)$$

In this reaction, we have indicated that the carbonic acid that is produced decomposes into carbon dioxide and water.

We can use the reaction stoichiometry to calculate the theoretical yield of carbon dioxide:

$$3.452 \text{ g NaHCO}_3 \times \frac{1 \text{ mol NaHCO}_3}{84.01 \text{ g}} \times \frac{1 \text{ mol CO}_2}{1 \text{ mol NaHCO}_3} \times \frac{44.01 \text{ g CO}_2}{1 \text{ mol CO}_2} = 1.808 \text{ g CO}_2$$

9. Oxidation–Reduction Reactions

   a. An oxidation–reduction reaction (also called a redox reaction) involves the transfer of electrons between reacting species.

      i. Many, but not all, redox reactions involve oxygen.

      ii. Oxidation is the loss of electrons, and reduction is the gain of electrons.

         1. A mnemonic device for redox reactions is "leo the lion goes ger" where "leo" stands for "<u>l</u>oss of <u>e</u>lectrons is <u>o</u>xidation" and "ger" stands for "<u>g</u>ain of <u>e</u>lectrons is <u>r</u>eduction."

      iii. Oxidation and reduction occur simultaneously.

   b. The oxidation state (or oxidation number) is the hypothetical charge an atom would acquire if all the electrons in its bonds were assigned to the more electronegative atom sharing them.

   c. The rules for assigning oxidation state are listed below and go in order of priority (meaning that the first item takes priority over any item below it):

      i. Free (uncombined) elements have an oxidation state of zero.

      ii. For a simple monatomic ion, the oxidation state is equal to the net charge of the ion. For a polyatomic ion or (neutral) compound, the sum of the oxidation states of individual atoms must add up to the overall charge of the species.

      iii. Group IA elements have an oxidation state of +1.

      iv. Group IIA elements have an oxidation state of +2.

      v. Fluorine has an oxidation state of −1.

      vi. Hydrogen has an oxidation state of +1 when bonded to a nonmetal and an oxidation state of −1 when bonded to a metal.

      vii. Oxygen has an oxidation state of −2 (except when bonded to itself in peroxides or to fluorine).

      viii. Group VIIA elements have an oxidation state of −1.

      ix. Group VIA elements have an oxidation state of −2.

      x. Group VA elements have an oxidation state of −3.

EXAMPLE:

What is the oxidation state of the underlined atom in each of the following compounds?

a. H$\underline{F}$

According to rule v from above, fluorine has an oxidation state of −1.

b. H$\underline{N}O_3$

According to rule vi from above, hydrogen has an oxidation state of +1. According to rule vii, oxygen has an oxidation state of −2. Since the sum of the oxidation states must equal the overall charge of the compound (zero), nitrogen has an oxidation state of $0 - (+1 + 3(2)) = +5$.

c. $\underline{Na}_2SO_4$

Sodium is a Group IA element, which means that it has an oxidation state of +1 according to rule iii.

d. $C_2\underline{H}_6$

According to rule vi, hydrogen has an oxidation state of +1.

---

d. Redox reactions always involve a change in the oxidation numbers of the elements in the reaction.

    i. An atom is oxidized when its oxidation number increases.

    ii. An atom is reduced when its oxidation number decreases.

EXAMPLE:

Determine the species that is oxidized and the species that is reduced when solid magnesium is burned:

$$2Mg(s) + O_2(g) \rightarrow 2MgO(s)$$

Using the rules for oxidation states given above, we can determine the oxidation states of all species.

On the reactant side:

    Mg: 0

    O: 0

On the product side:

    Mg: +2

    O: −2

Thus, we see that magnesium is oxidized (increase in oxidation number) and oxygen is reduced (decrease in oxidation number).

e. The reducing agent is the species that is oxidized; it is the agent of reduction.

f. The oxidizing agent is the species that is reduced; it is the agent of oxidation.

EXAMPLE:

Which species is the oxidizing agent, and which species is the reducing agent in the reaction of magnesium and oxygen given in the example above?

Since magnesium is oxidized, it is the reducing agent, and since oxygen is reduced, it is the oxidizing agent.

EXAMPLE:

Write the combustion reaction for methane gas, CH_4, and determine which species is oxidized and which species is reduced.

The combustion reaction of a hydrocarbon results in the production of carbon dioxide and water:

$$CH_4(g) + O_2(g) \rightarrow CO_2(g) + 2H_2O(g)$$

Using the rules for assigning oxidation states, we can find the oxidation states of all species:

On the reactant side:

    C: −4

    H: +1

    O: 0

On the product side:

C: +4

H: +1

O: −2

And we see that carbon is oxidized (thus, CH_4 is the reducing agent), while oxygen is reduced (making O_2 the oxidizing agent).

Fill in the Blank:

1. _____ reactions are those in which electrons are transferred from one species to another.

2. When a species is dissolved in water but does not form a solution that conducts electricity, it is called a(n) _____.

3. Acids that contain more than one ionizable proton are called _____.

4. We can qualitatively describe the concentration of a solution as _____ or _____.

5. A reaction in which two solutions combine to form a solid is called a(n) _____.

6. The oxidation number of the oxidizing agent _____ in a redox reaction.

7. The equivalence point in a titration is visualized using a(n) _____.

8. The sum of oxidation states of all atoms in an ion is equal to _____.

9. _____ are species that are not involved in the chemistry of a solution phase reaction.

10. A(n) _____ is a species that dissociates completely in solution to form ions.

11. A(n) _____ is a solution stored in concentrated form.

Problems:

1. Provide formulas for the following compounds:

   a. Carbonic acid

   b. Hydroiodic acid

   c. Perchloric acid

2. Provide names for the following compounds:

   a. H_3PO_3

   b. $HClO_2$

3. When aqueous solutions of sodium carbonate and hydrochloric acid react, the production of a gas is observed.

   a. Write the balanced molecular equation for the reaction of sodium carbonate and hydrochloric acid.

   b. Write the net ionic equation for the reaction.

   c. What volume of a 1.54 M HCl solution will be required to neutralize 60.5 g of sodium carbonate?

   d. How much gas is produced (in grams) when 45.7 mL of a 1.54 M HCl solution is combined with 52.4 mL of a 1.38 M Na_2CO_3 solution?

4. A 10.0 mL sample of a 0.154 M aqueous solution of lead(II) nitrate is combined with a 14.0 mL sample of 0.201 M aqueous solution of ammonium sulfide. The reaction results in a precipitate.

   a.   Write the balanced molecular equation for the reaction of lead(II) nitrate and ammonium sulfide.

   b.   Write the net ionic equation for the reaction.

   c.   Determine the amount of solid (in g) that is expected to form in this reaction.

   d.   Determine the concentration of ammonium ions present after the reaction is complete.

   e.   Do you expect the conductivity of the final solution (after mixing and reaction) to be greater than, less than, or the same as the conductivity of the original 10.0 mL $Pb(NO_3)_2$ solution? Explain.

5.   10.0 g of NaOH are dissolved in water to prepare 150.0 mL of solution.

   a.   What is the concentration of the initial NaOH solution?

   b.   The NaOH stock solution is then diluted by taking 10.0 mL of the original solution and adding 90.0 mL of water to it. What is the concentration of the newly prepared solution?

   c.   The diluted solution (from part b) is used to titrate 53.0 mL of a H_2SO_4 solution. Write the molecular, complete ionic, and net ionic equations for this reaction.

   d.   If 32.4 mL of the NaOH solution are required to completely neutralize the acid, what is the concentration of the H_2SO_4 solution?

6.   Consider a solution made from combining 10.0 mL of a 1.0 M $Mg(C_2H_3O_2)_2$ solution and 40.0 mL of a 2.0 M Na_2CO_3 solution.

   a.   When these solutions are combined, but before any reaction occurs, what will be the concentration of sodium carbonate?

   b.   What chemical reaction will occur between these two solutions?

   c.   What will be the concentration of sodium ions before the reaction?

   d.   What will be the concentration of sodium ions after the reaction?

   e.   What mass of solid can be collected after the reaction occurs?

   f.   Imagine that we could test the conductivity of the solution (using a light bulb) before the reaction took place (but after the solutions had been combined) and again after the reaction took place. Would the bulb be brighter before or after the reaction occurs? Explain your answer.

Concept Questions:

1.   A solution is a homogeneous mixture of two or more substances. Explain why this means that the concentration of the solute can be expressed as moles of solute per liter of solution even when the sample size is on the milliliter scale.

2.   There are situations when the qualitative description of a solution as dilute or concentrated is just as valuable as a quantitative statement of molarity. Describe a situation where this would be appropriate.

3.   When solid NaCl and solid $AgNO_3$ are crushed and combined, no reaction takes place. When the two solids are dissolved in water and mixed, however, a precipitate forms. Explain why there is such a dramatic difference between these two scenarios.

4.   The net ionic equation provides a complete description of the chemistry of a reaction, while the complete ionic equation shows the actual nature of the reaction. Explain this statement.

5.   The solubility of ionic substances is explained using the attraction of polar water molecules to the charged species that comprise ionic substances. Some ionic substances, however, do not dissolve in water; for example, many sulfates that do not contain Group IA cations are insoluble. Suggest one possible reason for this.

Chapter 9: Thermochemistry

Key Learning Outcomes:

- Calculate Changes in Internal Energy from Heat and Work

- Find Heat from Temperature Changes

- Determine Quantities in Thermal Energy Transfer

- Calculate Work from Volume Changes

- Use Bomb Calorimetry to Calculate ΔE_{rxn}

- Distinguish between Endothermic and Exothermic Processes

- Determine Heat from ΔH and Stoichiometry

- Find ΔH_{rxn} Using Calorimetry

- Find ΔH_{rxn} Using Hess's Law

- Find ΔH_{rxn} Using Bond Energies

- Find ΔH_{rxn}^{o} Using Standard Enthalpies of Formation

- Predict Relative Lattice Energies

Chapter Summary:

     This chapter introduces you to the subject of thermochemistry, which is the study of energy in chemical reactions. You will begin this exploration by gaining an understanding of the two main forms of energy and the manner in which it can be transferred. The first law of thermodynamics will then be explained to provide context for understanding energy transfer. Calorimetry will be explored as a method for determining the heat transferred to and from a substance as well as the heat transfers that occur in chemical reactions. Pressure-volume work will then be explained so that you can calculate the energy change of any system using heat and work. The relationship between internal energy and enthalpy will be introduced, and the conditions under which they are equal will be explored. Experimental and theoretical methods for determining the enthalpy of a chemical reaction will be explained alongside their physical meanings. We will use our understanding of bond energies to estimate enthalpies of reactions, and then we will define the zero of enthalpy and the standard state so that reaction enthalpies can be calculated from tabulated values. Finally, we will discuss lattice energies for ionic compounds including a discussion of the periodic trend associated with this value.

Chapter Outline:

1. Fire and Ice

    a. Thermochemistry is the study of the relationship between chemical reactions and energy.

2. The Nature of Energy: Key Definitions

    a. Energy is a physical quantity that corresponds to the capacity to do work.

        i. Energy is possessed by an object.

    b. Energy is transferred or exchanged via heat or work.

        i. Work is the result of a force acting through a distance.

        ii. Heat is the flow of energy that results from temperature differences.

   c. Kinetic energy is the energy associated with motion.

      i. Thermal energy is associated with temperature; it is a type of kinetic energy—it results from the motion of atoms or molecules.

   d. Potential energy is the energy associated with position.

      i. Chemical energy (or chemical potential) is associated with the relative positions of electrons and nuclei; it is a type of potential energy.

   e. The law of conservation of energy states that energy cannot be created or destroyed, although it can be transferred between objects and it can change form.

   f. A thermodynamic universe is divided into the system and the surroundings.

      i. The system is the part of the universe that is of interest; in a chemical reaction, the system is the reactants and products of the reaction.

      ii. The surroundings are everything in the universe other than the system.

         1. There are certain situations where the universe does not need to be considered. As we will see, there are times when a portion of the universe can be isolated for study. In this case, the surroundings are defined as anything that can exchange energy with the system.

   g. The joule (J) is the SI unit of energy. Calories (cal), nutritional calories (Cal), and kilowatt-hours (kWh) are also commonly used.

      i. $1 \text{ J} = 1 \dfrac{\text{kg} \cdot \text{m}^2}{\text{s}^2}$

      ▶ We can remember the units of energy by remembering that KE = ½mv^2, where the SI unit of mass is kg and velocity is m/s.

      ii. 1 cal = 4.184 J exactly

         1. The calorie is defined as the amount of energy required to raise the temperature of 1 g of a substance by 1 °C (at a pressure of 1 atm).

      iii. 1 Cal = 1000 cal (as used in the food industry)

      iv. $1 \text{ kWh} = 3.60 \times 10^6 \text{ J}$ (as used for electricity)

EXAMPLE:

The energy content of a typical candy bar is about 250 Cal. How long (in seconds) could you light up a 100.-W light bulb using the candy bar? (One watt is defined as 1 joule per second.)

We need to convert the energy from the candy bar from Cal to joules to find the amount of energy available to light the bulb:

$$250 \text{ Cal} \times \frac{1000 \text{ cal}}{1 \text{ Cal}} \times \frac{4.184 \text{ J}}{1 \text{ cal}} = 1.046 \times 10^6 \text{ J}$$

Now we can convert to time using the conversion factor given and the fact that a 100.-W light bulb will use 100 joules per second:

$$1.046 \times 10^6 \text{ J} \times \frac{1 \text{ s}}{100 \text{ J}} = 1.046 \times 10^4 \text{ s}$$

Our answer should have two significant figures, so the energy from a candy bar can light up a 100.-W light bulb for 1.0×10^4 s.

---

3. The First Law of Thermodynamics: There Is No Free Lunch

    a. The first law of thermodynamics states that the total energy of the universe is constant.

        i. This is another statement of the law of conservation of energy.

    b. The internal energy of any system is the sum of the kinetic and potential energies of all of the particles in the system.

        i. The internal energy of a system is a state function, given the symbol E.

            1. A state function is any property of a system that depends only on the state of a system (temperature, pressure, and amount and type of substance present) and not on how the system was prepared.

            2. The change in a state function is defined as the difference between the final and initial states:

$$\Delta E = E_{final} - E_{initial}$$

        ii. For a chemical reaction, the change in energy is defined by the difference in internal energy of the products and the reactants:

$$\Delta E_{reaction} = E_{products} - E_{reactants}$$

            1. We can depict changes in energy using an energy diagram, a plot of energy (y-axis) versus reaction progress (x-axis).

        iii. The sum of the energies of the system and the surroundings is the energy of the universe. Since the total energy change of the universe is zero, the energy change of the system (ΔE_{sys}) is equal and opposite to the energy change of the surroundings (ΔE_{surr}):

$$\Delta E_{universe} = \Delta E_{sys} + \Delta E_{surr} = 0$$

$$\Delta E_{sys} = -\Delta E_{surr}$$

EXAMPLE:

What is the energy change of a system if the surroundings absorb 843 kJ of energy during some process?

The energy change of the system is equal and opposite to the energy change of the surroundings. Since the surroundings are absorbing 843 kJ of energy, ΔE_{surr} is equal to +843 kJ of energy. This means that the system has lost 843 kJ of energy, so $\Delta E_{sys} = -843$ kJ.

---

        iv. During a chemical reaction or physical transformation, reactants and products define the system.

            1. If $\Delta E_{sys} > 0$, the products must have a higher internal energy than the reactants because energy is absorbed from the surroundings.

            2. If $\Delta E_{sys} < 0$, the reactants must have a higher internal energy than the products because energy is released to the surroundings.

        v. Energy can be exchanged through heat (q) and work (w):

$$\Delta E = q + w$$

        vi. Unlike energy, heat and work are path functions (as opposed to state functions); their values, therefore, depend on how a change is carried out.

EXAMPLE:

When a 5.0-g ball is dropped from a building, work is done as the ball moves over the distance that it falls. If the building is 100. m tall, what is the energy change that results from the work and where does the energy go when the ball stops moving?

The work done is the force times distance. In this example, the force is the acceleration due to gravity (9.81 m/s^2) times the mass of the ball (in SI units of kg):

$$w = 5.00 \text{ g} \times \frac{1 \text{ kg}}{1000 \text{ g}} \times 9.81 \frac{\text{m}}{\text{s}^2} \times 100 \text{ m} = 0.04905 \frac{\text{kg} \cdot \text{m}^2}{\text{s}^2} = 4.905 \text{ J}$$

This is the work that is done by the ball. Since the ball is doing work, it is losing energy, so the sign of the energy change is negative. Our answer should have two significant figures, so the final answer is that the energy change is –4.9 J.

The energy must go somewhere after the work is done because $\Delta E_{univ} = 0$ for any process. The energy is being transferred to the molecules of the ground as the ball hits it. The molecules will gain kinetic energy, which we do not see but could feel as heat.

---

4. Quantifying Heat and Work

   a. Temperature is a measure of thermal energy; heat is thermal energy in transit.

      i. Thermal energy always flows from high to low temperature.

   b. Thermal equilibrium is the condition at which no further heat transfer takes place. When the temperature of two systems in contact is the same throughout, thermal equilibrium has been established.

   c. The heat capacity (C) of a system is the quantity of heat required to raise its temperature by 1 °C.

   $$q = C \times \Delta T$$

   $$C = \frac{q}{\Delta T}$$

      i. Heat capacity has units of J/°C and is an extensive property, meaning that the heat transfer depends on the quantity of material.

      ii. The specific heat capacity (C_s) is the quantity of heat required to raise the temperature of 1 g of a substance by 1 °C. The specific heat capacity is an intensive property: it depends only on the identity of the substance (as well as its physical state), not on the size of the sample.

   $$q = C_s \times m \times \Delta T$$

      iii. The molar heat capacity (C_m) is the quantity of heat required to raise the temperature of 1 mole of a substance by 1 °C. Molar heat capacity is an intensive property.

EXAMPLE:

How much heat is transferred to the air when 1.54 L of water at 100 °C is allowed to sit in a room that is at 25 °C?

First, we need to convert the volume of water to a mass. The density of water is 0.958 g/mL.

$$1.54 \text{ L} \times \frac{1000 \text{ mL}}{\text{L}} \times \frac{0.958 \text{ g}}{\text{mL}} = 1475.32 \text{ g}$$

Recalling that the definition of a calorie is the amount of energy required to change the temperature of 1 g of water by 1 °C, we know that the specific heat capacity of water must be 4.184 J/g °C. We also need to consider the final temperature of the water. Because the room is open to the universe, the temperature change of the air will be negligible, so we can assume that the final temperature of the water will be the same as the initial temperature of the room. We can then calculate the heat lost by the water:

$$q = 4.184 \frac{J}{g \cdot {}^\circ C} \times 1475.32 \, g \times (100 \, {}^\circ C - 25 \, {}^\circ C) = 4.6296 \times 10^5 \, J$$

We need three significant figures in our answer, so we report that 4.63×10^5 J was transferred to the air in the room when the water cooled.

---

iv. When heat is transferred from the system to the surroundings, energy is conserved and in the absence of work:

$$q_1 = -q_2$$

$$m_1 \times C_1 \times \Delta T_1 = -m_2 \times C_2 \times \Delta T_2$$

▶ The negative sign applies to the product of the three terms on the right.

EXAMPLE:

A 5.4-g sample of an unknown metal is heated to 100.0 °C and is placed in a beaker containing 142 g of water at 24.2 °C. The final temperature of the water is 25.1 °C. What is the heat capacity of the metal?

Heat is being transferred from the metal to the water, so:

$$q_{water} = -q_{metal}$$

$$m_{metal} \times C_{s,metal} \times \Delta T_{metal} = -m_{water} \times C_{s,water} \times \Delta T_{water}$$

Using the information given in the problem, we have:

$$5.4 \, g \times C_{s,metal} \times (25.1 \, {}^\circ C - 100.0 \, {}^\circ C) = -142 \, g \times 4.184 \, \frac{J}{g \, {}^\circ C} \times (25.1 \, {}^\circ C - 24.2 \, {}^\circ C)$$

Rearranging to solve for s_{metal} gives:

$$C_{s,metal} = \frac{-142 \, g \times 4.184 \, \dfrac{J}{g \, {}^\circ C} \times (25.1 \, {}^\circ C - 24.2 \, {}^\circ C)}{5.4 \, g \times (25.1 \, {}^\circ C - 100.0 \, {}^\circ C)}$$

And we find that the specific heat capacity of the metal is 1 J/g °C.

---

d. Pressure–volume work involves gases and occurs when an applied force is the result of a volume change against a constant pressure.

i. Work is force times a distance (D), and pressure (P) is force over area (A). Combining these:

$$w = F \times D \text{ and } P = F/A \text{ so } w = P \times A \times D$$

ii. We can change the $A \times D$ term to ΔV since the area times distance will be equal to the change in volume of a reaction vessel:

$$w = -P \times \Delta V$$

The negative sign is included by the convention of defining work of the system: when work is done *by* the system on the surroundings (pushes against the atmosphere, meaning $\Delta V > 0$), the system loses energy, and so the work is negative. Conversely, when work is done *on* the system, work is positive.

▶ It is wise to get a handle on the physical meaning of the signs of thermodynamic values as they are easily mixed up in calculations.

EXAMPLE:

A large-displacement motorcycle can have gas cylinders that are as large as 0.500 L. What is the energy change due to the work done by a motorcycle piston when it travels its maximum distance at sea level?

The change in volume will be +0.500 L since the maximum volume will be completely closed (0.000 L) to completely open (0.500 L). The pressure will be 1.0 atm since the change in volume is carried out at sea level against the atmosphere. The work is therefore:

$$w = -1\,\text{atm} \times 0.500\ \text{L} = -0.500\ \text{atm} \cdot \text{L}$$

The work is negative because the motorcycle is doing work on the surroundings. Now, we need to convert to more conventional units:

$$-0.500\,\text{atm} \cdot \text{L} \times \frac{101.3\ \text{J}}{\text{atm} \cdot \text{L}} = -50.65\ \text{J}$$

Our answer should have three significant figures and should be negative, since the problem suggests that the motorcycle engine is doing work. The change in energy from the work is therefore -50.7 J.

5. Measuring ΔE for Chemical Reactions: Constant-Volume Calorimetry

    a. In order to calculate the energy change of a system, we need to calculate the heat absorbed by/released by and the work done by/done on the system.

        i. The heat is calculated by measuring the temperature change.

        ii. The work is calculated by measuring the volume change

    b. When a reaction is carried out at a constant volume, the change in volume is zero and therefore the pressure–volume work is zero.

        i. The energy change will therefore result only from heat transfer:

        $$\Delta E_{rxn} = q_v$$

    c. We can measure the constant-volume heat transfer using an apparatus called a bomb calorimeter that maintains a constant volume.

        i. If the calorimeter is designed so that no heat can escape it, then the heat lost/gained by the reaction (system) will be gained/lost by the calorimeter (surroundings):

        $$q_{cal} = -q_{rxn}$$

        ii. By measuring the change in temperature of the calorimeter, we can determine the constant-volume heat transfer and the energy change of the reaction:

        $$\Delta E_{rxn} = q_{rxn} = C_{cal} \times \Delta T$$

EXAMPLE:

Bomb calorimeters are often used to measure the heat released in a combustion reaction, as mentioned in the text. 1.76 g of methane gas [$CH_4(g)$] is burned in a bomb calorimeter with excess molecular oxygen at 25 °C. What is the final temperature of the calorimeter and its contents given that 1 mole of methane gas releases 882 kJ of heat when 1.0 mol of it is combusted? The heat capacity of the calorimeter is 4.319 kJ/°C.

We first need to find out how much heat is released when 1.76 g of methane is combusted; this is a conversion factor problem:

$$1.76\ \text{g CH}_4 \times \frac{1\ \text{mol CH}_4}{16.042\ \text{g CH}_4} \times \frac{882\ \text{kJ released}}{1\ \text{mol CH}_4} = 96.7660\ \text{kJ released}$$

$$\text{Therefore, } q_{rxn} = -96.7660 \text{ kJ}$$

In a calorimetry experiment, the heat of the reaction is equal to the heat capacity of the calorimeter multiplied by the temperature change.

$$q_{rxn} = C_{cal} \times \Delta T$$

$$96.7660 \text{ kJ} = 4.319 \frac{\text{kJ}}{°C} \times \Delta T$$

$$\Delta T = 22.405 \text{ °C}$$

Since the change in temperature is the final temperature minus the initial temperature, the final temperature will be:

$$T_f - 25.0 \text{ °C} = 22.405 \text{ °C}$$

$$T_f = 47.405 \text{ °C}$$

Our final answer is reported to the tenths place, so the final temperature of the calorimeter is 47.4 °C. Note that the final temperature exceeds the initial temperature since the reaction is exothermic.

6. Enthalpy: The Heat Evolved in a Chemical Reaction at Constant Pressure

   a. The enthalpy (H) of a system is defined as the internal energy plus the product of gas pressure times volume:

   $$H = E + PV$$

      i. Enthalpy is a state function.

   b. At constant pressure, the change in enthalpy is:

   $$\Delta H = \Delta E + P\Delta V$$

      i. The change in internal energy is equal to the heat at constant pressure (q_p) and the pressure–volume work:

      $$\Delta H = (q_p + w) + P\Delta V$$

      $$\Delta H = (q_p + w) + (-w)$$

      $$\Delta H = q_p$$

      ii. The change in internal energy is a measure of all of the energy exchanged, while the change in enthalpy is only a measure of the heat exchanged at constant pressure.

   c. At constant volume and a given pressure, the change in enthalpy is equal to the change in internal energy since the PV work is zero.

   d. When the change in enthalpy of a chemical reaction or physical transformation is positive, heat is flowing into the system; this is an endothermic process.

      i. In an endothermic process, the potential energy of the reactants is lower than the potential energy of the products.

   e. When the change in enthalpy of a chemical reaction or physical transformation is negative, heat is flowing out of the system; this is an exothermic process.

      i. In an exothermic process, the potential energy of the reactants is higher than the potential energy of the products.

   f. The change in enthalpy of a reaction, ΔH_{rxn} is called the heat of a reaction and is an extensive property.

g. The ΔH_{rxn} value can be used in stoichiometry problems.

    i. A thermochemical equation gives the enthalpy change associated with a balanced stoichiometric equation.

EXAMPLE:

State whether each of the following examples is endothermic or exothermic:

a. Water freezing

Water freezes when heat is removed (the converse, ice melting, requires heat) from the system, meaning that the potential energy of the reactants is higher than the potential energy of the reactants. This is therefore an exothermic reaction.

b. Wood burning

Burning/combustion is a chemical change; heat is released from the system (which is why a fireplace keeps us warm). Thus, the process is exothermic.

c. Alcohol evaporating from your skin

Evaporation is a physical change; as liquid is converted to gas, heat must be absorbed to break molecular contacts. (When alcohol evaporates from our skin, it feels cold!). Thus, this is an endothermic process.

d. Combustion of propane

Again, burning/combustion is a chemical change in which heat is released (and we can use it to cook our food). Heat is released from the system because the products are more stable (at a lower potential energy) than the reactants, meaning that the reaction is exothermic.

e. Dry ice subliming

Sublimation occurs when a solid is transformed directly to a gas. Gas molecules have greater kinetic energy and are more widely spaced than molecules in the solid, which vibrate about fixed positions. Since energy must be provided to the system in order for this to occur, it is an endothermic process.

---

7. Measuring ΔH_{rxn} for Chemical Reactions: Constant-Pressure Calorimetry

    a. The heat of a reaction can be measured with a "coffee-cup" calorimeter, an apparatus that maintains constant-pressure conditions. The heat of the solution, q_{soln}, is:

$$q_{soln} = m_{soln} \times C_{s,soln} \times \Delta T$$

    b. The solution will absorb the heat released or will provide the heat absorbed by the reaction.

$$q_{rxn} = -q_{soln} = \Delta H_{rxn}$$

EXAMPLE:

A common experiment that is carried out in a coffee-cup calorimeter is the dissolution of a salt in water. When 5.45 g of ammonium nitrate is dissolved in 154.2 g of water, the temperature of the water changes from 25.5 °C to 25.0 °C. What is the enthalpy change (in kJ/mol) for the dissolution of ammonium nitrate in water? Assume that the specific heat of the solution is equal to the heat capacity of the water.

Note that the temperature of the water decreases upon addition of NH_4NO_3; the dissolution process must be endothermic.

We will use the expression for the q_{soln} given above with the information provided in the problem:

$$q_{soln} = (5.45 \text{ g} + 154.2 \text{ g}) \times 4.184 \text{ J/g °C} \times (25.0 \text{ °C} - 25.5 \text{ °C})$$

$$q_{soln} = -333.99 \text{ J}$$

We know that the enthalpy change of the reaction is equal and opposite to the heat of the solution. ΔH is therefore 333.99 J.

This is the heat that is absorbed from the solution when 5.45 g of ammonium nitrate are dissolved in the water. We can normalize the calculated value of ΔH to find ΔH_{rxn} by dividing the calculated value by the number of moles of ammonium nitrate used.

$$5.45 \text{ g NH}_4\text{NO}_3 \times \frac{1 \text{ mol NH}_4\text{NO}_3}{80.04 \text{ g NH}_4\text{NO}_3} = 0.06809 \text{ mol NH}_4\text{NO}_3$$

$$\frac{333.99 \text{ J}}{0.06809 \text{ mol NH}_4\text{NO}_3} \times \frac{1 \text{ kJ}}{1000 \text{ J}} = 4.905 \text{ kJ/mol}$$

Our answer should have one significant figure (resulting from the temperature change term), so the ΔH_{rxn} for the dissolution of ammonium nitrate in water is 5 kJ/mol.

---

8.   Relationships Involving ΔH_{rxn}

    a.   If a chemical reaction is multiplied by a factor, then the ΔH_{rxn} value must be multiplied by that factor; this reflects the fact that ΔH is extensive.

    b.   If a chemical reaction is reversed, the ΔH_{rxn} value changes sign.

    c.   If a reaction can be expressed as the sum of a series of steps, ΔH_{rxn} for the overall reaction is the sum of the heats of reactions for each step.

        i.   This is called Hess's law and is valid because ΔH_{rxn} is a state function.

EXAMPLE:

Calculate the enthalpy change for the combustion of propane, C_3H_8, using the reactions given below:

1. $3C(s) + 4H_2(g) \rightarrow C_3H_8(g)$ $\Delta H = -103.8 \text{ kJ}$

2. $2H_2(g) + O_2(g) \rightarrow 2H_2O(g)$ $\Delta H = -484.0 \text{ kJ}$

3. $C(s) + O_2(g) \rightarrow CO_2(g)$ $\Delta H = -393.5 \text{ kJ}$

We first need to write the balanced chemical reaction for the combustion of propane:

$$C_3H_8(g) + 5O_2(g) \rightarrow 3CO_2(g) + 4H_2O(g) \qquad \Delta H_{rxn} = ?$$

Now we can manipulate the given reactions so that they add up to give the overall reaction. We will start by figuring out whether any of the reactions need to be reversed. We will do this by looking for species in the given reactions that also appear in the overall reaction.

Reaction(1) has propane as a product; in our reaction we want propane to be a reactant, so we will flip the reaction and reverse the sign of the enthalpy change. Reaction(2) has oxygen as a reactant and water as a product; in our reaction this is also what we want, so we will leave reaction(2) alone. Reaction(3) has oxygen as a reactant and carbon dioxide as a product; again the positions of these species are the same in our overall reaction, so we will leave them alone. We now have:

1. $C_3H_8(g) \rightarrow 3C(s) + 4H_2(g)$ $\Delta H = +103.8 \text{ kJ}$

2. $2H_2(g) + O_2(g) \rightarrow 2H_2O(g)$ $\Delta H = -484.0 \text{ kJ}$

3. $C(s) + O_2(g) \rightarrow CO_2(g)$ $\Delta H = -393.5 \text{ kJ}$

Now we need to multiply each reaction by a factor so that the stoichiometry of the overall reaction is obtained and all species that are not in the overall reaction cancel out. Reaction(1) has one propane molecule, which is the same amount of propane in our reaction, so we will not change this reaction. In reaction(2) we see that two molecules of water are formed; because we need four molecules of water, we will multiply the entire reaction by two (note that we are not worried about oxygen yet because it appears as a reactant in both reactions 2 and 3). Reaction(3) forms 1 mole of carbon dioxide, while the overall reaction produces 3 moles of carbon dioxide; we will multiply the third reaction by three.

1. $C_3H_8(g) \rightarrow 3C(s) + 4H_2(g)$ $\qquad \Delta H = +103.8 \text{ kJ}$

2. $2 \times \{2H_2(g) + O_2(g) \rightarrow 2H_2O(g)\}$ $\qquad \Delta H = 2 \times (-484.0 \text{ kJ})$

   $4H_2(g) + 2O_2(g) \rightarrow 4H_2O(g))$ $\qquad \Delta H = -968.0 \text{ kJ}$

3. $3 \times \{C(s) + O_2(g) \rightarrow CO_2(g)\}$ $\qquad \Delta H = 3 \times (-393.5 \text{ kJ})$

   $3C(s) + 3O_2(g) \rightarrow 3CO_2(g)$ $\qquad \Delta H = -1180.5 \text{ kJ}$

Now we will add up the reactions $(1 + 2 + 3)$ and the corresponding enthalpy changes to give the following thermochemical equation:

$$C_3H_8(g) + 4H_2(g) + 2O_2(g) + 3C(s) + 3O_2(g) \rightarrow 3C(s) + 4H_2(g) + 4H_2O(g) + 3CO_2(g) \qquad \Delta H = -2044.7 \text{ kJ}$$

Finally, we will collect like species and cancel out any substances that appear on both sides of the equation:

$$C_3H_8(g) + 5O_2(g) \rightarrow 4H_2O(g) + 3CO_2(g) \qquad \Delta H = -2044.7 \text{ kJ}$$

Check: The reaction is the same as the overall reaction that we want, so ΔH_{rxn} is -2044.7 kJ.

---

9. Determining Enthalpies of Reaction from Bond Energies

   a. Bond energies correspond to the energy required to break a particular bond, but they also correspond to the energy given off upon bond formation (the sign is changed).

   b. The standard enthalpy change of a reaction, ΔH_{rxn}, can be estimated using individual bond energies.

   ΔH_{rxn} = (energy required to break reactant bonds) – (energy released in making product bonds)

   i. A reaction is exothermic when weak bonds are broken and stronger bonds are formed.

   ii. A reaction is endothermic when strong bonds are broken and weaker bonds are formed.

EXAMPLE:

Estimate the enthalpy change for the combustion of methane using average bond energies listed in Table 10.3 of your textbook.

First we will write the balanced chemical reaction:

$$CH_4(g) + 2O_2(g) \rightarrow CO_2(g) + 2H_2O(g)$$

Now we will make a list of the bonds being broken along with the energy absorbed for each:

4 C–H bonds in each methane molecule: $\qquad\qquad 4 \times +414 \text{ kJ/mol} = +1656 \text{ kJ}$

1 O=O bond in each of two oxygen molecules: $\qquad\qquad 2 \times +498 \text{ kJ/mol} = +996 \text{ kJ}$

The bonds being formed will have negative enthalpy change values:

2 C=O bonds in each carbon dioxide molecule: $\qquad\qquad 2 \times -799 \text{ kJ/mol} = -1598 \text{ kJ}$

2 O–H bonds in each of two water molecules: $\qquad\qquad 4 \times -464 \text{ kJ/mol} = -1856 \text{ kJ}$

The sum of these values gives the enthalpy change of the reaction:

$$\Delta H = (1656 + 996 + -1598 + -1856) \text{ kJ} = -802 \text{ kJ}$$

The negative sign is expected since combustion reactions release heat (i.e., are exothermic).

---

10. Determining Enthalpies of Reaction from Standard Enthalpies of Formation

  a. The standard state (designated by the symbol °) is defined depending on the phase of the substance.

     i. The standard state of a gas is 1 atm.

     ii. The standard state for a liquid or solid is the pure substance in its most stable form at a pressure of 1 atm and a given temperature (usually 25 °C).

     iii. The standard state for a substance in solution is a concentration of exactly 1 M.

  b. The standard enthalpy change ($\Delta H°$) is the change in enthalpy when all reactants and products are in their standard states.

  c. The standard enthalpy of formation (ΔH_f^o) is defined using the zero of enthalpy.

     i. Absolute enthalpies cannot be measured. By definition, the zero of enthalpy is the ΔH_f^o of a pure element in its standard state.

     ii. The ΔH_f^o for a pure compound is the change in enthalpy that results when 1 mole of the compound is produced from its constituent elements in their standard states, that is, their most stable form at 1 atm and a specified temperature.

  d. The $\Delta H°$ for a reaction can be determined by breaking the reactants into their constituent elements [$-\Delta H_f^o$ (reactants)] and then forming products from the elements [ΔH_f^o (products)]:

$$\Delta H_{rxn}^o = \sum n_p \Delta H_f^o (\text{products}) - \sum n_r \Delta H_f^o (\text{reactants})$$

  where the ΔH_f^o values are weighted by the stoichiometry of the reaction.

  ▶ This is an application of Hess's law.

EXAMPLE:

Using the enthalpies of formation in your book, calculate the standard enthalpy of formation for the dissolution of solid silver nitrate in a 1 M aqueous solution of sodium chloride to form solid silver chloride and aqueous sodium nitrate. Will the solution get hotter or colder when the silver nitrate is added?

First, we need to write the balanced chemical reaction for the process:

$$AgNO_3(s) + NaCl(aq) \rightarrow AgCl(s) + NaNO_3(aq)$$

We can write an expression for the standard enthalpy change of the reaction using the enthalpies of formation of the reactants and products:

$$\Delta H_{rxn}^o = \left[1 \times \Delta H_f^o (AgCl(s)) + 1 \times \Delta H_f^o (NaNO_3(aq)) \right] - \left[1 \times \Delta H_f^o (AgNO_3(s)) + 1 \times \Delta H_f^o (NaCl(aq)) \right]$$

The values are given in Appendix II in your book:

$$\Delta H^o_{rxn} = \left[-127.0 \text{ kJ/mol} + -447.5 \text{ kJ/mol}\right] - \left[-124.4 \text{ kJ/mol} + -407.2 \text{ kJ/mol}\right]$$

$$\Delta H^o_{rxn} = -42.9 \text{ kJ}$$

Our answer has the correct number of significant figures since this is an addition/subtraction problem.

Since the change in enthalpy is negative, the system loses energy in the course of the reaction, meaning that the solution will absorb energy and get warmer.

---

11. Lattice Energies for Ionic Compounds

    a.   Lattice energy is the energy associated with the formation of a crystalline lattice of alternating anions and cations from the gaseous ions.

    b.   The Born–Haber cycle is a hypothetical series of steps that together describe the formation of an ionic compound from its constituent elements. For example:

$$Na(s) + \tfrac{1}{2}Cl_2(g) \rightarrow NaCl(s) \qquad \Delta H^o_f$$

       i.   The first step is the formation of gaseous Na from the solid (sublimation energy).

$$Na(s) \rightarrow Na(g) \qquad \Delta H^o_1$$

       ii.   The second step is the formation of chlorine atoms from molecular chlorine (bond energy).

$$\tfrac{1}{2}Cl_2(g) \rightarrow Cl(g) \qquad \Delta H^o_2$$

       iii.   The third step is the ionization of sodium to form Na^+ ions (ionization energy).

$$Na(g) \rightarrow Na^+(g) \qquad \Delta H^o_3$$

       iv.   The fourth step is the ionization of chlorine atoms to form Cl^- ions (electron affinity).

$$Cl(g) + e^- \rightarrow Cl^-(g) \qquad \Delta H^o_4$$

       v.   The fifth step is the formation of the ionic solid (lattice energy).

$$Na^+(g) + Cl^-(g) \rightarrow NaCl(s) \qquad \Delta H^o_5$$

       Thus $\Delta H^o_f = \Delta H^o_1 + \Delta H^o_2 + \Delta H^o_3 + \Delta H^o_4 + \Delta H^o_5$

    c.   The lattice energy changes with size and charge due to Coulomb's law.

       i.   The lattice energy decreases with increasing cation and anion size because the distance between the ions increases (r in Coulomb's law).

       ii.   The lattice energy increases with increasing ion charge because of an increase in attraction (q_1 and q_2 in Coulomb's law).

EXAMPLE:

Determine the lattice energy for the formation of Li_2O using information provided in Appendix II in your textbook. The ionization energy of lithium is 520.2 kJ/mol, and the energy required for oxygen to gain two electrons is +650. kJ/mol.

ΔH_1 is equal to the ΔH^o_f of Li(g). From Appendix II, ΔH^o_f (Li(g)) = +159.3 kJ/mol. Note that we need two lithium ions because of the formula of lithium oxide, so double the enthalpy: $\Delta H^o_f = +318.6$ kJ/mol.

ΔH_2 is equal to the ΔH^o_f of O(g). From Appendix II, ΔH^o_f (O(g)) = +249.2 kJ/mol.

ΔH_3 is equal to the ionization energy of lithium, $\Delta H_{IE}^o = +520$ kJ/mol. Again, we need to multiply this by two in order to account for the formation of two lithium ions: $\Delta H_3^o = +1040.4$ kJ/mol.

ΔH_4^o is the energy required for oxygen to gain two electrons (the sum of the first and second electron affinity values for oxygen), $\Delta H_4^o = +650.$ kJ/mol.

ΔH_f^o is the enthalpy of formation for Li_2O. From Appendix II, $\Delta H_f^o = -597.9$ kJ/mol.

Rearranging the expression for the Born–Haber cycle to solve for the lattice energy gives:

$$\Delta H_5^o = \Delta H_5^o - (\Delta H_1^o + \Delta H_2^o + \Delta H_3^o + \Delta H_4^o)$$

Using the values found above, we have:

$$\Delta H_5^o = (597.9 - (318.6 + 249.2 + 1040.4 + 650.)) \text{ kJ/mol}$$

$$\Delta H_5^o = -1660.3 \text{ kJ/mol}$$

According to significant figure rules, we should only report our answer to the ones digit. The lattice energy is -1660 kJ/mol.

Fill in the Blank:

1. _____ is energy associated with motion, while _____ is energy associated with position.

2. The SI units of energy are the _____.

3. A(n) _____ is defined as the amount of energy required to raise the temperature of 1 g of water by exactly 1 °C.

4. The first law of thermodynamics states that the total energy of the universe is _____.

5. A(n) _____ function is one that depends only on the initial and final values, but not on the manner in which the system was prepared.

6. When heat is absorbed by a system, the sign of ΔH is _____.

7. A positive value of work is obtained when work is done _____ the system.

8. When two objects are at the same temperature, they are said to be at _____ equilibrium.

9. The measure of a substance's intrinsic capacity to absorb heat is called its _____.

10. When a system expands, the sign of the work is _____.

11. At constant pressure, the enthalpy is equal to the _____.

12. A(n) _____ reaction is one that absorbs heats from its surroundings.

13. According to _____, the heat of an overall reaction is equal to the sum of the heats of reaction for each step.

14. The standard state of a gas is defined as the pure gas at _____.

15. The zero of enthalpy is defined as the enthalpy of formation for _____.

16. The energy associated with forming a crystalline lattice of alternating positively and negatively charged ions is called the _____.

17. The _____ is a hypothetical series of steps that represent the formation of an ionic compound from its constituent elements.

18. The lattice energy _____ when the size of the cation increases.

Problems:

1. Determine the sign of heat, work, and change in energy for the systems (italicized) described.

   a. A *candle* is burned in an open room.

   b. A *car* rolls down a hill.

   c. A *piano* is lowered down the side of a building.

   d. *Water* is heated on a stovetop.

2. When gaseous ammonia is oxidized, it produces gaseous nitrogen monoxide and water, as well as heat:

$$4NH_3(g) + 5O_2(g) \rightarrow 4NO(g) + 6H_2O(g) \qquad \Delta H° = -906 \text{ kJ}$$

   a. Is the reaction endothermic or exothermic? Explain your answer.

   b. If the reaction is carried out in a bomb calorimeter, will the temperature of the calorimeter increase or decrease? Explain your answer.

   c. Is the internal energy of the products or reactants higher? Explain your answer.

3. Butane, C_4H_{10}, is a fuel for lighters. When 1.0 mol of butane burns at a constant pressure, it produces 2658 kJ of heat and does 3 kJ of work.

   a. What is the enthalpy change of this reaction?

   b. What is the internal energy change of this reaction?

   c. Explain what kind of work is done in this process.

   d. In this reaction, is it reasonable to use the values of energy and enthalpy interchangeably? Explain your answer.

4. Gas grills use the combustion of propane (C_3H_8) gas to cook food.

   a. Write the balanced chemical reaction for the combustion of propane. (All species should be written in the gas phase.)

   b. Calculate the standard enthalpy change for this reaction using the standard enthalpy of formation values given.

   c. Does the sign of ΔH that you calculated in part b make physical sense? Explain.

   d. Is the reaction endothermic or exothermic? Explain.

   e. Was the internal energy of the reactants greater than, less than, or equal to the internal energy of the products? Explain.

   f. Do you expect the internal energy change to be greater than or less than the enthalpy change of the reaction? Explain.

5. One method for rating fuels with respect to global warming is to determine the heat that they release relative to the amount of CO_2 produced (kJ/mol CO_2). The greater the heat produced relative to the amount of CO_2 released, the more efficient the fuel.

   a. Using the standard enthalpies of formation, calculate the enthalpy change for the combustion of solid carbon, gaseous methane (CH_4), and liquid octane (C_8H_{18}). Assume that each substance reacts with a stoichiometric amount of oxygen to produce carbon dioxide gas and water vapor. The standard enthalpy of formation for octane is −252.1 kJ/mol.

   b. Calculate the amount of heat produced per mol of CO_2 released for each fuel.

   c. Which gas is most environmentally friendly based on your results in part b?

6. Calculate the molar heat capacity (in cal/J °C) given the specific heat capacities for the following substances:

   a. aluminum, $C_s = 0.897$ J/g °C

   b. copper, $C_s = 0.385$ J/g °C

    c.   gold, $C_s = 0.129$ J/g °C

    d.   magnesium, $C_s = 1.02$ J/g °C

    e.   zinc, $C_s = 0.387$ J/g °C

    f.   iron, $C_s = 0.450$ J/g °C

7. A sample of iron at 80.5 °C is placed in a 125 mL sample of water at 24.2 °C. The temperature of the water rises by 2.7 °C. What is the mass of the metal sample?

8. A typical experiment used to measure the specific heat capacity of a metal is to heat the metal in a boiling water bath for a period of time and then place the hot metal sample into a coffee-cup calorimeter. The temperature change of the water is measured, and the specific heat capacity of the metal is calculated.

    a.   Calculate the specific heat capacity of an unknown 5.423-g metal sample if the temperature of 30.68 g of water in the calorimeter changes by 2.72 °C to a final temperature of 28.22 °C.

    b.   The metal is one of the substances listed in Problem 4. Which metal is it?

    c.   In this type of experiment, it is important that the metal sample is placed in a clean dry test tube that is then submerged in a boiling water bath until thermal equilibrium is established. Explain how your results would be different if you instead placed the metal sample directly into the boiling water.

9. Molecular hydrogen and oxygen gas are converted to water in a hydrogen fuel cell.

    a.   Given that hydrogen fuel cells are being made to replace combustion engines, what do you expect the sign of the ΔH_{rxn} to be?

    b.   Calculate the ΔH_{rxn} for this reaction using the standard enthalpies of formation.

    c.   How many moles of oxygen and hydrogen would you need to convert to water in order to accelerate a large family car (~3500 lb) from 0 to 60 mph over a 10-second time period?

10. Hydrogen and iodide gases react together to form hydrogen iodide gas in a bomb calorimeter in an endothermic reaction.

    a.   Using the enthalpies of formation, determine the enthalpy of the reaction.

    b.   Equimolar amounts of the reactants are placed in a 5.0-L bomb calorimeter. The initial pressure inside the calorimeter is 2.5 atm. What is the change in internal energy of the reactants and products when the reaction is carried out in the bomb?

    c.   Calculate the temperature change of the calorimeter when the reactant gases are initially at 25.4 °C. The heat capacity of the calorimeter is determined to be 4.230 kJ/°C.

11. Calculate the standard heat of combustion of methane using the following equations:

    i. $2O(g) \rightarrow O_2(g)$ $\Delta H = -249$ kJ/mol

    ii. $H_2O(l) \rightarrow H_2O(g)$ $\Delta H = +44$ kJ/mol

    iii. $2H(g) \rightarrow H_2(g)$ $\Delta H = -803$ kJ/mol

    iv. $C(s) + 2O(g) \rightarrow CO_2(g)$ $\Delta H = -643$ kJ/mol

    v. $C(s) + O_2(g) \rightarrow CO_2(g)$ $\Delta H = -394$ kJ/mol

    vi. $C(s) + 2H_2(g) \rightarrow CH_4(g)$ $\Delta H = -75$ kJ/mol

    vii. $2H_2(g) + O_2(g) \rightarrow 2H_2O(g)$ $\Delta H = -484$ kJ/mol

12. Use the equations from Problem 9 to determine the heat of the reaction between methane and oxygen to produce liquid water instead of gaseous water.

13. Draw the potential energy diagrams for the processes carried out in Problems 9 and 10.

14. Calculate the enthalpy change for the following reactions using the bond enthalpies given in Table 10.3 in your textbook. State whether the reactions are endothermic or exothermic.

    a. $H_2CO_3(g) \rightarrow H_2O(g) + CO_2(g)$

    b. $CO_2(g) + 4H_2(g) \rightarrow 2H_2O(g) + CH_4(g)$

    c. $2HNO_3(g) \rightarrow H_2O(g) + N_2O_5(g)$

15. Calculate the lattice energy for the formation of the following ionic compounds (pertinent enthalpy change values can be found in your book).

    a. MgO

    b. RbCl

    c. CsI

Concept Questions:

1. Most chefs prefer gas stove tops because they allow greater control over the heat that is delivered to food. Gas stoves, however, are the least efficient type of stove (in terms of heat loss). Electric stoves are about 15% more efficient. Even better are induction cooktops, which utilize electromagnetic radiation to heat up the cooking vessel itself instead of transferring heat to the cooking vessel. Explain, using the principles used in this chapter, why induction cooktops are more efficient than electric cooktops and why electric cooktops are more efficient than gas cooktops.

2. Using the specific heat capacity values of various metals given in Problem 6 above, which substance would be most ideal for use as a cooking pan? Explain your answer.

3. Molar heat capacities are related to the microscopic complexity of the substance: the greater the complexity, the higher the heat capacity.

    a. Considering the values that you calculated in Problem 6 above, what general statement can you make about the microscopic complexities of metals?

    b. Specific heat capacity is often used to confirm the identity of a metal. As you saw in Problem 6 above, molar heat capacity is not as useful. Explain how these two statements can be rectified.

    c. The molar heat capacity of liquid water is approximately 18.0 cal/mol °C. What conclusion can you draw about the molecular complexity of liquid water?

4. When an ice cube is placed in water, the water molecules comprising the ice cube change temperature more than the water molecules of the liquid.

    a. Which has a higher molar heat capacity, water or ice?

    b. State which water molecules (those from ice or those from liquid) will have a larger change in their molecular motions.

    c. We will learn later that heat transferred to a substance results in an increase in the random motion of the components of the substance. We will also learn that random motion increases for all spontaneous processes. Explain why this means that when the ice cube is placed in the glass of liquid water, the liquid will never freeze.

5. The enthalpy of a reaction is also called the heat of a reaction. This seems a bit troubling since the enthalpy of a reaction is actually the heat of a reaction at constant pressure. No one, however, makes this distinction because of the conditions in which most reactions are carried out in a laboratory. Explain why this is so.

6. The enthalpy of formation of a pure element in its most stable form at standard conditions is zero. Explain why this is so.

7. Internal energy is a physically important quantity since it describes all of the kinetic and potential energies of all components that make up the system. Enthalpy, on the other hand, does not have such a physically tidy definition but is the more often measured/discussed quantity. Explain why enthalpy is an easier quantity to measure and discuss.

8. In this chapter, you learned that ionic size and ion charge both have a large impact on the magnitude of the lattice energy. Consider the magnitude of the change for each factor and explain why the charge has a larger influence than size alone. You need to consider the equation for the Coulombic potential and any other factors that might be important.

Chapter 10: Gases

<u>Key Learning Outcomes:</u>

- Convert between Pressure Units

- Relate Volume and Pressure: Boyle's Law

- Relate Volume and Temperature: Charles's Law

- Relate Volume and Moles: Avogadro's Law

- Determine P, V, n, or T Using the Ideal Gas Law

- Relate the Density of a Gas to Its Molar Mass

- Calculate the Molar Mass of a Gas with the Ideal Gas Law

- Calculate Total Pressure, Partial Pressures, and Mole Fractions of Gases in a Mixture

- Calculate the Root Mean Square Velocity of a Gas

- Calculate the Effusion Rate of the Ratio of Effusion Rates of Two Gases

- Relate the Amounts of Reactants and Products in Gaseous Reactions: Stoichiometry

<u>Chapter Summary:</u>

In this chapter, you will be introduced to the physical properties and behaviors of gases. You will begin by exploring kinetic molecular theory, which will provide a physical basis for the gas laws. Next you will consider the simple gas laws, each of which relates two of the four variables that determine gas behaviors: temperature, pressure, volume, and sample size (number of moles). You will then see how the simple gas laws are combined to give the ideal gas law as well as consider a derivation of the law using kinetic molecular theory; the idea gas law will then be used in density, molar mass, and molar volume calculations. Once you are comfortable with such calculations, you will explore mixtures of gases and will learn how to calculate the partial pressures of components in a mixture. Next, you will be introduced to the concepts of mean free path, diffusion, and effusion. You will then return to stoichiometry, focusing on chemical reactions that deal with gaseous reactants and/or products. With a firm grasp of ideal gases in hand, real gases and their deviations from ideal behavior will be explored and quantified in the van der Waals equation.

<u>Chapter Outline:</u>

1. Supersonic Skydiving and the Risk of Decompression

    a. Pressure is the force exerted per unit area by gas particles as they strike the surfaces around them.

2. A Particulate Model for Gases: Kinetic Molecular Theory

    a. Kinetic molecular theory (KMT) assumes that gases are in constant random motion and that:

        i. The size of the gas particles is negligible as compared to the empty space in the container they occupy.

        ii. The kinetic energy of the particles is proportional to the temperature in kelvins.

        iii. The collisions between gas particles and the walls of a container are elastic.

            1. An elastic collision is one in which no energy is lost, although energy can be exchanged between colliding particles.

3. Pressure: The Result of Particle Collisions

    a. The pressure of a gas (P) is the force (F) that results from the collisions of the gas particles divided by the area (A) of the surface with which they collide:

$$P = \frac{F}{A}$$

    b. Common units of pressure are mmHg, torr, atm, psi, bar, and the SI unit of pressure, Pa (= N/m^2).

        1 atm = 1.01325×10^5 Pa = 1.01325 bar = 14.696 psi = 760 mmHg (0 °C) = 760 torr

    c. Atmospheric pressure can be measured using a barometer, which is an evacuated tube inverted in a pool of mercury or other liquid. The pressure from the atmosphere pushes the mercury up the tube until the weight of the mercury column is equal and opposite to the force from the atmosphere.

EXAMPLE:

The barometric pressure is often given in millibars. The barometric pressure in northern California today is 1018 millibars. Given that 1 atm is equal to 1013.25 millibars, determine the barometric pressure of northern California in the following units:

a. mmHg

$$1018 \text{ mbar} \times \frac{1 \text{ atm}}{1013.25 \text{ mbar}} \times \frac{760 \text{ mmHg}}{1 \text{ atm}} = 763.6 \text{ mmHg}$$

b. torr

$$1018 \text{ mbar} \times \frac{1 \text{ atm}}{1013.25 \text{ mbar}} \times \frac{760 \text{ torr}}{1 \text{ atm}} = 763.6 \text{ torr}$$

c. atm

$$1018 \text{ mbar} \times \frac{1 \text{ atm}}{1013.25 \text{ mbar}} = 1.005 \text{ atm}$$

d. psi

$$1018 \text{ mbar} \times \frac{1 \text{ atm}}{1013.25 \text{ mbar}} \times \frac{14.7 \text{ psi}}{1 \text{ atm}} = 14.8 \text{ psi}$$

e. kPa

$$1018 \text{ mbar} \times \frac{1 \text{ atm}}{1013.25 \text{ mbar}} \times \frac{101,325 \text{ Pa}}{1 \text{ atm}} \times \frac{1 \text{ kPa}}{1000 \text{ Pa}} = 101.8 \text{ kPa}$$

---

    d. A manometer is used to measure pressures of samples in a laboratory. It consists of a "U-tube" that contains mercury and is open to the atmosphere on one end and the sample vessel on the other. The difference in mercury height between the two sides of the "U-tube" is equal to the difference in pressure between the vessel and the atmosphere.

4. The Simple Gas Laws: Boyle's Law, Charles's Law, and Avogadro's Law

    a. The properties of gases can be described completely using the temperature, pressure, volume, and chemical amount (moles). We can look at the relationships between any two of these variables by holding the other two constant.

    b. Boyle's law: Pressure and volume are indirectly proportional; if a gas is compressed (at constant temperature and number of particles), the average speed of the gas particles will not change but the frequency of collisions with the container walls will increase, and thus, so will the gas pressure.

Volume and pressure are inversely related when the temperature and number of moles of gas are held constant.

$$V \propto \frac{1}{P} \text{ or } VP = \text{constant}$$

We can compare two different sets of conditions (that have the same temperature and sample size) with the following equation:

$$V_1 P_1 = V_2 P_2$$

▶ Note that for this (and all of the simple gas laws) any units of pressure or volume can be used as long as the units are the same on both sides of the equation.

EXAMPLE:
_____

A 5.0 L plastic container is opened to the atmosphere and then sealed. The container is crushed so that the volume is 3.2 L. What is the pressure (in atm) in the crushed container?

We first need to realize that the initial pressure in the container will be equal to atmospheric pressure (i.e., $P_1 = 1$ atm). Then we can use Boyle's law to solve for the final pressure, P_2:

$$V_1 P_1 = V_2 P_2$$

$$(5.0 \text{ L}) \times (1 \text{ atm}) = (3.2 \text{ L}) \times P_2$$

$$P_2 = 1.6 \text{ atm}$$

Check: Our answer is reasonable since the pressure is expected to increase when the volume decreases.

---

c. Charles's law: volume and temperature are directly related (constant pressure and number of moles); when the volume increases, the number of collisions with the walls of the container can remain constant only if the particles are moving faster. Particles move faster (i.e., have a greater kinetic energy) when the temperature increases.

Volume and temperature (in kelvins) are directly related when the number of moles and pressure are held constant:

$$V \propto T \text{ or } \frac{V}{T} = \text{constant}$$

We can compare two different sets of conditions with an equation:

$$\frac{V_1}{T_1} = \frac{V_2}{T_2}$$

EXAMPLE:
_____

A 2.73 dm^3 sample of gas at 21.0 °C is warmed to 100.0 °C. What is the final volume of the sample?

We will use Charles's law to solve this problem, but we first need to convert the temperatures to kelvins.

$$T_1 = 21.0 + 273.15 = 294.2 \text{ K}$$

$$T_2 = 100.0 + 273.15 = 373.2 \text{ K}$$

$$\frac{2.73 \text{ dm}^3}{294.2 \text{ K}} = \frac{V_2}{373.2 \text{ K}}$$

$$V_2 = 3.46 \text{ dm}^3$$

Check: Our answer is reasonable since we expect the volume to increase when the temperature increases.

---

d.   Avogadro's law: volume and number of moles are directly proportional; when the volume increases, the number of collisions with the walls of the container can remain constant (constant pressure and temperature) only if the number of moles of the gas increases.

Number of moles and volume of gas are directly related when the temperature (in kelvins) and pressure are held constant:

$$V \propto n \text{ or } \frac{V}{n} = \text{constant}$$

We can compare two different sets of conditions with an equation:

$$\frac{V_1}{n_1} = \frac{V_2}{n_2}$$

EXAMPLE:

10.0 mol of H_2 and 5.0 mol of O_2 react to completion according to the equation:

$$2H_2(g) + O_2(g) \rightarrow 2H_2O(g)$$

If the reactants are placed in a sealed 12.0 L container, what will be the volume of the container after the reaction takes place (assuming that temperature and pressure remain constant)?

The number of moles before the reaction takes place is 15.0 mol, and the number of moles after the reaction takes place is 10.0 moles (using the reaction stoichiometry). So we can use Avogadro's law to solve for V_2:

$$\frac{12.0 \text{ L}}{15.0 \text{ mol}} = \frac{V_2}{10.0 \text{ mol}}$$

$$V_2 = 8.00 \text{ L}$$

Check: Our answer is reasonable since a decrease in the number of moles of gas should result in a decreased volume.

---

5.   The Ideal Gas Law

a.   We can combine the simple gas laws into a single expression:

$$V \propto \frac{nT}{P}$$

This can be rearranged slightly to eliminate the fractional form:

$$PV \propto nT$$

And the proportionality can be converted into an equality with the introduction of a constant, R:

$$PV = nRT$$

$$R = 0.08206 \frac{\text{L} \cdot \text{atm}}{\text{mol} \cdot \text{K}}$$

This equation is called the ideal gas law, and R is called the gas constant or ideal gas constant. Note that R can be expressed in other units of pressure.

▶ The simple gas laws can always be derived from the ideal gas law by considering a ratio at two different conditions:

$$\frac{P_1 V_1}{P_2 V_2} = \frac{n_1 R T_1}{n_2 R T_2}$$

The equation can simply be reduced to one of the simple gas laws by canceling out any terms that remain constant in the problem.

EXAMPLE:

What is the volume of a container that holds 46.3 g of $CO_2(g)$ at 283 K and 1.4 atm?

In order to use the ideal gas law, we need to convert the mass of CO_2 into the number of moles:

$$46.3\,g \times \frac{1\,mol}{44.01\,g} = 1.05\,mol\,CO_2$$

Rearranging the ideal gas law and solving for volume gives:

$$V = \frac{nRT}{P} = \frac{(1.05\,mol) \times \left(0.08206 \dfrac{L \cdot atm}{mol \cdot K}\right) \times (283\,K)}{1.4\,atm}$$

$$V = 17\,L$$

Check: We have used the appropriate physical parameters, units cancel, and our answer is reported to the correct sig. figs.

EXAMPLE:

What is the temperature of 0.587 mol of a gas that is held in a sealed 4.73 L container at 34.3 psi?

In order to use the ideal gas equation, we need to convert the units of pressure to the same pressure units in the gas constant (atm):

$$34.3\,psi \times \frac{1\,atm}{14.7\,psi} = 2.33\,atm$$

Now we can rearrange the ideal gas law to solve for temperature:

$$T = \frac{PV}{nR} = \frac{(2.33\,atm) \times (4.73\,L)}{(0.587\,mol) \times \left(0.08206 \dfrac{L \cdot atm}{mol \cdot K}\right)}$$

$$T = 229\,K$$

Check: We have used the appropriate physical parameters, units cancel, and our answer is reported to the correct sig. figs.

b. The ideal gas law can be derived directly using the postulates of kinetic molecular theory.

  i. The pressure of a sample of gas(es) is equal to the total force divided by the area:

$$P = \frac{F_{total}}{A}$$

  ii. The total force is the sum of the force from collisions between gas particles and the wall multiplied by the number of collisions that occur in a given time period.

1. The force of a collision is equal to the mass of a gas particle times its acceleration:

$$F_{\text{collision}} = ma$$

2. The acceleration of a gas particle is equal to its change in velocity over the time period:

$$F_{\text{collision}} = m\frac{\Delta v}{\Delta t}$$

3. The change in velocity that results from an elastic collision with the wall will be twice the velocity. This is because the velocity changes direction:

$$F_{\text{collision}} = m\frac{2v}{\Delta t}$$

4. The total number of collisions will be proportional to the number of particles that are within a distance in order to hit the wall ($v\Delta t$) in a given time (t), the area of the wall (A), and the density of the particles (moles per volume):

$$\text{number of collisions} \propto v\Delta t \times A \times \frac{n}{V}$$

5. The total force is then:

$$F_{\text{total}} \propto F_{\text{collision}} \times \text{number of collisions}$$

$$F_{\text{total}} \propto m\frac{2v}{\Delta t} \times v\Delta t \times A \times \frac{n}{V}$$

$$F_{\text{total}} \propto 2mv^2 \times A \times \frac{n}{V}$$

iii. The mass times velocity squared (kinetic energy) is proportional to the temperature, the pressure is equal to the total force over area, and the constants can be removed because it is a proportionality:

$$P = \frac{F_{\text{total}}}{A} = \frac{2mv^2 \times A \times \dfrac{n}{V}}{A}$$

$$P \propto \frac{T \times n}{V}$$

1. We can introduce a constant and rearrange to get the ideal gas law:

$$PV = nRT$$

6. Applications of the Ideal Gas Law: Molar Volume, Density, and the Molar Mass of a Gas

   a. The molar volume of a gas at standard temperature and pressure (STP) is the volume that an ideal gas occupies at 0 °C and 1 atm.

   i. Standard temperature and pressure is also known as standard conditions.

   ii. We can calculate the molar volume using the ideal gas law:

$$V = \frac{nRT}{P}$$

iii. The molar volume of a gas at STP is 22.4 L.

$$V = \frac{(1 \text{ mol}) \times \left(0.08206 \ \frac{\text{L} \cdot \text{atm}}{\text{mol} \cdot \text{K}} \right) \times (273.15 \text{ K})}{1 \text{ atm}} = 22.41 \text{ L}$$

▶ We can use this as a conversion factor between volume (L) and chemical amount (mol).

b. The density of 1 mole of a gas at standard conditions can be calculated by dividing the molar mass by the molar volume:

$$\text{density} = \frac{\text{molar mass (MM)}}{\text{molar volume } (V_m)} = \frac{m/n}{V/n} = \frac{m}{V}$$

i. Gas density can also be calculated using the ideal gas law and the relationship between number of moles, mass (m), and molar mass (MM):

$$V = \frac{\dfrac{m}{\text{MM}} RT}{P}$$

$$d = \frac{m}{V} = \frac{MMP}{RT}$$

EXAMPLE:

A sample of gaseous ammonia is held in a closed container at 1.58 atm and 456 K. What is the density of ammonia in the container?

The molar mass of ammonia, NH_3, is 17.04 g/mol; using the density form of the ideal gas law gives:

$$d = \frac{m}{V} = \frac{(17.04 \text{ g/mol}) \times (1.58 \text{ atm})}{\left(0.08206 \ \dfrac{\text{L} \cdot \text{atm}}{\text{mol} \cdot \text{K}} \right) \times (456 \text{ K})}$$

$$d = \frac{m}{V} = 0.719 \text{ g/L}$$

Check: We have used the appropriate physical parameters, units cancel, and our answer is reported to the correct sig. figs.

c. The molar mass of a gas can be calculated using a rearranged form of the above expression for density:

$$\text{MM} = \frac{mRT}{PV}$$

EXAMPLE:

A 5.6 L container holds 39 g of an unknown gas at a temperature of 100 °C and a pressure of 4.8 atm. Is the identity of the gas O_2, CO, CO_2, or CH_4?

We will use the information provided and compare the experimental molar mass to the molar masses of the given species.

$$\text{MM} = \frac{mRT}{PV}$$

$$MM = \frac{(39 \text{ g}) \times \left(0.08206 \dfrac{\text{L} \cdot \text{atm}}{\text{K} \cdot \text{mol}}\right) \times (373.2 \text{ K})}{(4.8 \text{ atm}) \times (5.6 \text{ L})}$$

$$MM = 44 \text{ g/mol}$$

This is the same molar mass as CO_2, so we can conclude that the unknown gas is CO_2.

7. Mixtures of Gases and Partial Pressures

    a.   Ideal gases do not interact (i.e., do not experience attractive or repulsive forces), so the pressure exerted by a single component of a mixture of ideal gases is the partial pressure of that component.

    b.   The partial pressure of an ideal gas, at a given temperature and volume, is determined by the number of moles of that gas:

$$P_a = \frac{n_a RT}{V}$$

    c.   Dalton's law of partial pressures states that the total pressure exerted by a mixture of ideal gases is equal to the sum of the partial pressures of each component of the mixture:

$$P_{total} = P_a + P_b + P_c + \cdots$$

Kinetic molecular theory states that pressure is a result of collisions between the gas particles and containers of the wall; because gases do not interact and have no size differences, the identity of the gas particles is not relevant to their behavior in this context.

    d.   The partial pressure of an ideal gas is related to the mole fraction (X_a) of the gas:

$$P_a = X_a P_{total}$$

$$X_a = \frac{n_a}{n_{total}}$$

EXAMPLE:

A mixture of neon, argon, and krypton gases has a total mass of 10.45 g and a total pressure of 1.25 atm. If there are 0.278 mol of neon and 0.0589 mol of argon in the mixture, what is the partial pressure of krypton?

We first need to find out how many moles of krypton are present in the mixture. In order to do this, we need to calculate the mass of neon and grams of argon:

$$0.278 \text{ mol Ne} \times \frac{20.18 \text{ g Ne}}{1 \text{ mol Ne}} = 5.61 \text{ g Ne}$$

$$0.0589 \text{ mol Ar} \times \frac{39.95 \text{ g Ar}}{1 \text{ mol Ar}} = 2.35 \text{ g Ar}$$

Next, we can determine by difference the mass of Kr in the gas mixture:

$$10.45 \text{ g gas} - 5.61 \text{ g Ne} - 2.35 \text{ g Ar} = 2.49 \text{ g Kr}$$

We can now calculate the moles of Kr:

$$2.49 \text{ g Kr} \times \frac{1 \text{ mol Kr}}{83.80 \text{ g Kr}} = 0.0297 \text{ mol Kr}$$

The mole fraction of Kr is:

$$\frac{0.0297 \text{ mol Kr}}{0.0297 \text{ mol Kr} + 0.278 \text{ mol Ne} + 0.0589 \text{ mol Ar}} = 0.0811$$

The partial pressure of Kr is the mole fraction of Kr times the total pressure:

$$0.081 \times 1.25 \text{ atm} = 0.101 \text{ atm}$$

Check: The mole fraction of Ar atoms present in the gas sample is small (~8%), so its contribution to the overall pressure is low.

EXAMPLE:

Equimolar amounts of CO, CO_2, and O_2 are placed in a sealed 1.24 L container at 250 K. The total pressure in the container is 5.4 kPa. What is the mass of O_2 in the container?

Since there is an equimolar amount of each gas, the mole fraction of O_2 in the container is 1/3.

Given that the partial pressure of O_2 is the mole fraction times the total pressure, the partial pressure of O_2 is 1/3 of the total pressure, or 1.8 kPa.

We can calculate the number of moles of O_2 using the ideal gas equation (with pressure converted to atm):

$$n = \frac{\left(1.8 \text{ kPa} \times \dfrac{1 \text{ atm}}{101.325 \text{ kPa}}\right) \times (1.24 \text{ L})}{\left(0.08206 \dfrac{\text{L} \cdot \text{atm}}{\text{mol} \cdot \text{K}}\right) \times (250 \text{ K})}$$

$$n = 0.00107 \text{ mol}$$

Now we can convert the number of moles to mass to solve the problem:

$$0.00107 \text{ mol } O_2 \times \frac{32.00 \text{ g } O_2}{1 \text{ mol } O_2} = 0.034 \text{ g } O_2$$

---

    e.   The vapor pressure of a liquid is the pressure exerted by the vapor in equilibrium with the liquid in a closed container; the vapor pressure is temperature-dependent.

        i.   When gases are collected over a liquid, the vapor pressure of the liquid must be taken into account.

        ▶   You must take the vapor pressure of water into account whenever reactions are done over liquid water.

8.   Temperature and Molecular Velocities

    a.   Since the temperature determines the average kinetic energy of a sample of gas particles, gases with a lower mass will travel, on average, faster at a given temperature:

$$KE = \frac{1}{2}mv^2$$

        i.   The average kinetic energy of one mole of an ideal gas will be related to the average molecular speed:

$$KE_{ave} = \frac{1}{2}N_A m\bar{u}^2$$

where $\bar{u}$ is the average speed and N_A is Avogadro's number.

    ii.   Kinetic molecular theory states that the kinetic energy is proportional to the temperature:

$$KE_{ave} = \frac{3}{2}RT$$

where 3/2 R is the proportionality constant.

    iii.   We can set the two equations for the average kinetic energy equal to one another:

$$\frac{3}{2}RT = \frac{1}{2}N_A m \overline{u^2}$$

    iv.   Solving for the speed gives:

$$\overline{u^2} = \frac{3RT}{N_A m}$$

Since $N_A m$ is equal to the molar mass (MM):

$$\sqrt{\overline{u^2}} = \sqrt{\frac{3RT}{MM}}$$

    v.   The expression $\sqrt{\overline{u^2}}$ is called the root mean square velocity (u_{rms}) and is a measure of the average speed of the gas particles:

$$u_{rms} = \sqrt{\frac{3RT}{MM}}$$

   ▸   Note the value and units of the ideal gas constant, $R = 8.3145$ J/mol·K

  b. When the temperature of a gas increases, the distribution of gas particle speeds gets broader and has a higher average value. When the temperature of a gas decreases, the distribution of speeds gets narrower and has a lower average value.

  c. If we compare two different gas samples at the same temperature, the gas that has a higher molar mass will have a narrower distribution and a lower average value of gas particle speeds.

  ▸ Temperature and mass are independent of each other, so it's important to read problems carefully so that you are certain of which variables can change in a specific scenario.

EXAMPLE:

The root mean square velocity of a diatomic gas is measured at 543 K and found to be 21.99 m/s. What is the identity of the gas?

We can rearrange the expression above to solve for the molar mass:

$$MM = \frac{3RT}{\overline{u^2}} = \frac{3 \times \left(8.3145 \dfrac{J}{mol \cdot K}\right) \times (543\ K)}{(21.99\ m/s)^2}$$

$$MM = 28.0\ g/mol$$

Since we are told that the gas is diatomic, it must be N_2 gas.

Check: We have used the appropriate physical parameters and constants, units cancel, and our answer is reported to the correct sig. figs.

---

9. Mean Free Path, Diffusion, and Effusion of Gases

   a. The mean free path of a gas is the average distance that a gas particle can travel before colliding with another gas particle in a sample.

   b. Diffusion is the process by which gas particles spread out. The rate of diffusion is dependent on the u_{rms} of the gas and, thus, temperature and molar mass.

   c. Effusion is the process of a gas escaping a container with a small hole in it. The rate of effusion is dependent on the u_{rms} of the gas and, thus, temperature and molar mass.

      i. Graham's law of effusion relates the rate of effusion of two gas samples to the molar masses of each gas:

$$\frac{rate_A}{rate_B} = \sqrt{\frac{MM_B}{MM_A}}$$

EXAMPLE:

How much longer will methane gas take to effuse out of a container than ammonia gas?

We can calculate the rate at which ammonia (NH_3, MM = 17.0 g/mol) will effuse as compared to methane (CH_4, MM = 16.0 g/mol):

$$\frac{rate_{CH_4}}{rate_{NH_3}} = \sqrt{\frac{17.0 \text{ g/mol}}{16.0 \text{ g/mol}}} = 1.03$$

CH_4 effuses out of the container 1.03 times faster than NH_3 does; since rate is proportional to distance/time, it will take 1.03 times as long for ammonia to effuse out of the container.

Check: Since NH_3 is heavier than CH_4, its gas molecules will move, on average, more slowly (at a given temperature) and thus will take longer to effuse.

---

10. Gases in Chemical Reactions: Stoichiometry Revisited

    a. Stoichiometry problems utilize mol–mol ratios, and since moles are one variable in the ideal gas equation, it is often possible to measure the volume, temperature, and/or pressure of gaseous reactants or products (instead of mass, moles, or particles). We can use the ideal gas law to convert other variables into moles at the beginning or end of the problem.

EXAMPLE:

Upon heating, solid potassium chlorate decomposes to solid potassium chloride and oxygen gas according to the equation:

$$2KClO_3(s) \rightarrow 2KCl(s) + 3O_2(g)$$

When potassium chlorate is allowed to decompose at 50.0 °C, the oxygen gas is collected over water in a 2.45-L container. If the total pressure in the container is 0.875 atm, what mass of the $KClO_3$ decomposed?

Since the gas is collected over water, we first need to find the partial pressure of water at 50.0 °C. In Table 11.3 of the text we see that the vapor pressure of water at 50.0 °C is 92.6 mmHg. We can find the partial pressure of oxygen gas by subtracting the vapor pressure of water from the total pressure:

$$92.6 \text{ mmHg} \times \frac{1 \text{ atm}}{760 \text{ mmHg}} = 0.1218 \text{ atm}$$

$$0.875 \text{ atm} - 0.122 \text{ atm} = 0.753 \text{ atm}$$

We can now use the ideal gas law to find the number of moles of O_2 that were produced, remembering to convert the temperature to kelvins:

$$n = \frac{(0.753 \text{ atm}) \times (2.45 \text{ L})}{\left(0.08206 \dfrac{\text{L} \cdot \text{atm}}{\text{mol} \cdot \text{K}}\right) \times (323.15 \text{ K})}$$

$$n = 0.06957 \text{ mol}$$

We can use the reaction stoichiometry and the molar mass of potassium chlorate to find the mass of $KClO_3$ reacted:

$$0.06957 \text{ mol } O_2 \times \frac{2 \text{ mol } KClO_3}{3 \text{ mol } O_2} \times \frac{122.55 \text{ g } KClO_3}{1 \text{ mol } KClO_3} = 5.68 \text{ g } KClO_3$$

Check: We have used the appropriate physical parameters, units cancel, and our answer is reported to the correct sig. figs.

EXAMPLE:

What volume of gaseous CO_2 (in mL) will be produced from the decomposition of 10.0 g of carbonic acid (H_2CO_3) at STP?

$$H_2CO_3(aq) \rightarrow H_2O(l) + CO_2(g)$$

This is a stoichiometry problem involving the molar volume of a gas at STP:

$$10.0 \text{ g } H_2CO_3 \times \frac{1 \text{ mol } H_2CO_3}{62.03 \text{ g } H_2CO_3} \times \frac{1 \text{ mol } CO_2}{1 \text{ mol } H_2CO_3} \times \frac{22.4 \text{ L } CO_2}{1 \text{ mol } CO_2} = 3.61 \text{ L } CO_2$$

Check: We have used the appropriate physical parameters, units cancel, and our answer is reported to the correct sig. figs.

---

11. Real Gases: The Effects of Size and Intermolecular Forces

    a. Real gases behave ideally when the size of the gas particle is negligible compared to the distance between the gas particles and when forces between particles are insignificant.

        i. Ideal behavior is best approximated at low pressures and high temperatures.

    b. The effect of finite volume of gas particles becomes important at high pressures. A correction to the ideal gas law for volume is:

$$V = \frac{nRT}{P} + nb$$

The value of "b" (in units of L/mol) depends on the identity of the gas and is determined experimentally. At high pressure the molar volume of a gas increases.

    c. Intermolecular forces are attractions that exist between atoms or molecules; at low pressures particles are too far away from each other to exert large forces, and at high temperatures particles are moving too fast to be affected by these forces.

        i. Intermolecular forces become important at high pressures and low temperatures. The correction to the ideal gas law for intermolecular forces is:

$$P = \frac{nRT}{V} - a\left(\frac{n}{V}\right)^2$$

The value of "a" (in units of $\text{atm} \cdot \text{L}^2/\text{mol}^2$) depends on the identity of the gas, because different gases have different intermolecular forces and is determined experimentally.

At low temperatures and high pressures, the pressure decreases because the gas particles are "sticking" to each other.

d. We can combine the corrections for volume and intermolecular forces into a single expression called the van der Waals equation:

$$\left[P + a\left(\frac{n}{V}\right)^2\right] \times (V - nb) = nRT$$

▶ Gas behavior approaches ideality when $a, b \sim 0$.

e. Different molecules and atoms will have different deviations from ideal behavior:

   i. Negative deviations ($nT/PV < R$) occur when intermolecular forces are great.

   ii. Positive deviations ($nT/PV > R$) occur when particle volume has a large effect.

EXAMPLE:

The van der Waals constants for CO are $a = 1.485$ atm·L^2/mol^2 and $b = 0.03985$ L/mol. What will be the temperature of a sample of CO gas when 5.4 moles are placed in a 10.4-L container at 1.5 atm?

We can use the van der Waals equation and solve for temperature:

$$T = \frac{\left[P + a\left(\frac{n}{V}\right)^2\right] \times (V - nb)}{nR}$$

$$T = \frac{\left[1.5\ \text{atm} + \left(1.485\frac{\text{L}^2 \cdot \text{atm}}{\text{mol}^2}\right)\left(\frac{5.4\ \text{mol}}{10.4\ \text{L}}\right)^2\right] \times \left[10.4\ \text{L} - (5.4\ \text{mol})\left(0.03985\frac{\text{L}}{\text{mol}}\right)\right]}{(5.4\ \text{mol})(0.08206\frac{\text{L} \cdot \text{atm}}{\text{mol} \cdot \text{K}})}$$

$$T = 44\ \text{K}$$

Fill in the Blank:

1. _____ states that the total pressure of a gas sample is equal to the sum of the partial pressures of its components.

2. The average distance that a particle can travel in between collisions is called its _____.

3. The average kinetic energy of a particle is proportional to its _____ in units of _____.

4. _____ is the force exerted per unit area by gas particles as they strike the walls of a container.

5. Pressure and volume are _____ related.

6. _____ law relates the temperature of a gas to its volume.

7. A positive deviation from ideal behavior occurs primarily because of the particle's _____.

8. In a gas at a given temperature, lighter particles travel _____ than heavier particles.

9. A mercury barometer is a device that measures pressure in units of _____.

10. The SI unit of pressure is the _____.

11. The molar volume of an ideal gas at STP is always _____.

12. The partial pressure of a gas in equilibrium with its liquid phase is its _____.

13. The constant of proportionality in the ideal gas law is equal to _____.

Problems:

1. The density of saltwater is 1.025 g/mL. Calculate the conversion factor between mmHg and millimeters of saltwater.

2. The pressure on the top of Mount Everest is about 30. kPa.

    a. What is the pressure in units of atm, psi, torr, and mmHg?

    b. If the percentage of oxygen in air on top of Mount Everest is about 21.0%, what is the partial pressure of oxygen on top of Mount Everest?

    c. If the content of oxygen in the air on top of Mount Everest is only 2/3 of that at sea level, what is the partial pressure of oxygen at sea level?

    d. A sample of air from on top of Mount Everest is collected in a 1.0-L jar and brought back down. If the sample was collected on a "warm" sunny day, the temperature of the sample would be about −15 °F on the mountain. If the temperature is 50 °F in Oakland when the jar is brought home, what will be the new pressure of the gas in the jar?

3. A 12.0 g sample of nitrogen gas is placed into a sealed 5.0 L container at 25 °C. What will the pressure be inside the container?

4. Assume the gas sample in problem 3 is heated up to 100 °C. What will be the pressure?

5. A cylinder of helium gas is advertised to fill 50 9" balloons.

    a. If a 9" balloon can hold a volume of 0.27 cubic feet and the temperature of the gas is 25 °C, what is the mass of helium contained in the full cylinder? You can assume that the pressure inside a balloon is equal to 1.01 atm.

    b. What is the density of helium in one balloon?

    c. What would be the density of a balloon filled with nitrogen if all other factors remained constant?

    d. Air leaks out of the balloon so that the pressure decreases by a rate of 0.0012 atm per day. How long will it take (in hours) for the balloon to be half of its original size?

6. A mixture of gases contains 5.0 g of He, 2.0 g of Ne, and 5.0 g of Ar at STP.

    a. What is the partial pressure (in atm) of helium in the mixture?

    b. What volume will the gaseous mixture occupy?

7. When 10.0 g of hydrogen gas and 10.0 g of chlorine gas react together, they form hydrogen chloride gas.

    a. What will the partial pressure of $H_2(g)$ be at the end of the reaction if it is carried out in a sealed 10.0-gallon container at 100 °C?

    b. What will be the total pressure (in mmHg) in the container?

8. Solid calcium phosphide is reacted with an excess of water to produce aqueous calcium hydroxide and gaseous phosphine:

$$Ca_3P_2(s) + 6H_2O(l) \rightarrow 3Ca(OH)_2(aq) + 2PH_3(g)$$

What will be the total pressure within the container when 3.0 g of Ca_3P_2 are reacted in a closed 5.5-mL container at 20 °C?

9. 10.5-mL of a 0.115 M HCl solution and 0.0425 g of solid Na_2CO_3 react together.

   a. Write the balanced chemical equation for the reaction that occurs.

   b. What volume (in mL) will be produced if this reaction is carried out at STP?

   c. If the gas formed in this reaction was collected in a 1.0-L flask over water at 25.0 °C, what will be the total pressure of the gas in the flask?

10. What volume of carbon dioxide gas will be produced when 2.0 g of methane gas is combusted in a sealed container with a stoichiometric amount of oxygen gas at STP?

11. Pennies are currently made by coating zinc with copper. The mass of a penny is determined to be 2.482 g, and then the penny is scratched to expose the zinc. The scratched penny is placed in an aqueous solution of HCl, where it decomposes in a single displacement reaction.

    a. Write the balanced chemical reaction that occurs between solid zinc and aqueous HCl.

    b. The product gas is collected over water at 25 °C. The total pressure after the reaction is 791 mmHg. What is the partial pressure of the product gas?

    c. The gas collected occupies a volume of 0.899 L. What is the percent zinc in the penny if all of the zinc reacts with the acid?

12. State the three main postulates of the kinetic molecular theory.

13. What is the average molecular speed of a collection of nitrogen molecules at 100 °C?

14. How much more time will it take for a sample of nitrogen gas to effuse out of a container as compared to a sample of hydrogen gas?

15. A sample of methane (CH_4) gas is held in a 4.0-L container at standard conditions.

    a. What mass of methane is in the container if its behavior is ideal?

    b. Use the values of a and b listed in Table 11.4 of your textbook to determine the mass of methane in the container if its behavior is better described using the van der Waals equation.

    c. Do you think that methane is appropriately described as an ideal gas at STP based on your calculations in parts a and b? Explain your answer.

Concept Questions:

1. You car has tires that each contain gas at 25–35 psi.

   a. Without doing any calculations, is the pressure in your tires greater than, less than, or equal to the atmospheric pressure? How can you tell? (*Hint:* What happens when you put a tire gauge on your valve stem?)

   b. If the tire pressure is different from atmospheric pressure, why don't your tires collapse or explode as you are driving down the street?

   c. Most people do not check their tire pressure often enough and drive around with tires that are underinflated. Explain why this is very dangerous.

2. Equal molar amounts of gas A and gas B are combined in a 1.0-L container at room temperature. Gas B has a molar mass that is twice that of gas A. Answer the following questions about this system with a brief explanation of your reasoning.

   a. Which gas, A or B, has a higher kinetic energy?

   b. Which gas, A or B, has a higher partial pressure?

   c. Which gas, A or B, has a higher root mean square velocity?

   d. Which gas, A or B, contributes more to the average density of the mixture?

3. Consider two 1.0-L flasks of gas at room temperature. One flask contains 10 g of F_2, and the other contains 10 g of H_2.

    a. Is the kinetic energy of the F_2 gas greater than, less than, or equal to the kinetic energy of the H_2 gas? Explain.

    b. Is the average velocity of the F_2 gas greater than, less than, or equal to the average velocity of the H_2 gas? Explain.

    c. Is the pressure in the flask containing the F_2 gas greater than, less than, or equal to the pressure in the flask containing the H_2 gas? Explain.

    d. Is the volume of the F_2 gas greater than, less than, or equal to the volume of the H_2 gas? Explain.

4. Use kinetic molecular theory to explain the relationship between temperature and pressure.

5. Use kinetic molecular theory to explain the relationship between volume and pressure.

6. Compare and contrast the deviation from ideal behavior for carbon dioxide and carbon tetrachloride using the values of the van der Waals constants listed in Table 11.4 of your textbook.

Chapter 11: Liquids, Solids, and Intermolecular Forces

Key Learning Outcomes:

- Determine Whether a Molecule Has Dipole–Dipole Forces

- Determine Whether a Molecule Displays Hydrogen Bonding

- Use the Heat of Vaporization in Calculations

- Use the Clausius–Clapeyron Equation

- Navigate Within a Phase Diagram

Chapter Summary:

 This chapter focuses on the interactions between ions, atoms, and molecules. You will learn how these interactions lead to the phases of matter and the energy associated with changing from one phase to another. First, you will learn about the various types of intermolecular forces such as dispersion forces, dipole–dipole forces, hydrogen bonding, and ion–dipole forces. Your understanding of intermolecular forces will be used to explore liquid properties such as surface tension, viscosity, and capillary action. Vapor pressure will then be explored as an indicator of the strength of intermolecular forces, and the connection between temperature and vapor pressure will be quantified using the Clausius–Clapeyron equation. After a discussion of boiling and condensation, you will learn about sublimation and fusion. With your knowledge of phase transitions, you will learn how to interpret and sketch the heating (or cooling) curve for water. The relationship between temperature and pressure will be discussed in the context of a phase diagram. Finally, you will explore the unique qualities of water.

Chapter Outline:

1. Water, No Gravity

   a. Intermolecular forces are the attractive forces that exist between all molecules and atoms.

2. Solids, Liquids, and Gases: A Molecular Comparison

   a. Gases, liquids, and solids differ in the freedom of movement of their constituent building blocks (atoms, ions, or molecules).

      i. Gases assume the shape and volume of their containers, are low in density, and are compressible.

      ii. Liquids have high densities compared to gases, assume the shape of their containers, and are not easily compressed.

      iii. Solids have high densities compared to gases, have definite volume and shape, are not compressible, and may be crystalline (ordered) or amorphous (disordered).

   b. The phase of matter changes when the temperature and/or pressure changes.

      i. The stronger the intermolecular forces between the particles comprising the macroscopic sample, the higher the temperature needs to be in order to overcome them.

3. Intermolecular Forces: The Forces That Hold Condensed States Together

   a. Intermolecular forces originate from interactions between the charges, partial charges, and temporary charges associated with atoms, ions, and molecules.

      i.   Just as intramolecular bonds form in order to lower the potential energy of atoms, molecules interact in order to lower their collective potential energy according to Coulomb's law:

$$E = \frac{1}{4\pi\varepsilon_o} \frac{q_1 q_2}{r}$$

      ii.   In general, intermolecular interactions are weaker than intramolecular bonds because the magnitudes of the charges are smaller and the distance between the particles is larger.

  b. Dispersion forces (London forces) result from fluctuations in the electron distribution within molecules, ions, or atoms.

      i.   All chemical species have electrons, and therefore, all species experience dispersion forces.

      ii.   Instantaneous or temporary dipoles that result from fluctuations will induce similar dipoles in surrounding species.

      iii.   The magnitude of dispersion forces depends on the polarizability (ease with which electron clouds can be distorted) of an atom, ion, or molecule.

          1.   The larger an atom, molecule, or other species, the more polarizable it is.

          2.   As polarizability increases, dispersion forces become stronger. Thus, dispersion forces increase with increasing molar mass.

      iv.   Dispersion forces also depend on molecular shape.

          1.   The more surface area a molecule has, the greater its dispersion forces as electrons can interact over a greater distance.

      v.   The stronger the dispersion forces are between adjacent chemical species, the higher will be the melting and boiling points of that substance.

  c. Dipole–dipole forces exist for polar molecules with permanent dipoles.

      i.   The positive end of one molecule is attracted to the negative end of an adjacent molecule.

      ii.   Dipole–dipole forces are stronger than dispersion forces when comparing species with the same molar mass; dispersion forces can become more important than dipole–dipole interactions when very large molecules are considered.

      iii.   The larger the dipole moment of a molecule, the stronger will be the intermolecular forces and the higher will be the melting and boiling points.

          Polarity influences the ability of two substances to form a homogeneous mixture: polar substances are able to combine with other polar substances, and nonpolar substances are able to combine with other nonpolar substances.

  d. Hydrogen bonding is the attractive force between the hydrogen atom attached to a small, electronegative atom in one molecule and a small, electronegative atom in another molecule.

      i.   Hydrogen bonding is an extreme example of dipole–dipole interactions.

          1.   Hydrogen bonding occurs between molecules that have HF, HO, and HN bonds.

            ▶   The hydrogen bond is NOT the HF, HO, or HN covalent bond but is the interaction of the H in one of these bonds on one molecule with the F, O, or N of the bond in another molecule. A hydrogen bond is not actually a bond!

      ii.   Hydrogen bonds are stronger than dispersion forces and dipole–dipole forces.

  e. Ion–dipole forces occur when an ionic compound is mixed with a polar covalent compound.

      i.   These are the strongest of the intermolecular forces discussed so far.

      ii.   Ion–dipole forces are important for the solubility of ionic compounds in water.

1. When ionic compounds dissolve in water, the anions are surrounded by the positive end of water and the cations are surrounded by the negative end of water. This process is called solvation (or hydration in the case of water).

EXAMPLE:

Identify the type of intermolecular forces that <u>dominate</u> in the following examples.

a. Dichloromethane, $CH_2Cl_2(l)$

In dichloromethane, there are two polar C–Cl bonds. Since the structure of the molecule has two chlorine atoms pointed in the same direction, the molecule will have a net dipole and the individual molecules will orient themselves so that the dipoles align. In a macroscopic sample of dichloromethane, dipole–dipole interactions dominate.

b. Octane, $CH_3CH_2CH_2CH_2CH_2CH_2CH_2CH_3(l)$

Octane has C–H bonds that are slightly polar, but because the molecule is symmetrical, these tiny dipoles cancel and the molecule is nonpolar. Octane will have strong dispersion forces because it is so large.

c. Octanol, $CH_3CH_2CH_2CH_2CH_2CH_2CH_2CH_2OH(l)$

Octanol has a polar O–H group that can potentially form hydrogen bonds with other octanol molecules. Since it is so large, however, dispersion forces will be the dominant intermolecular interactions.

d. Sodium chloride in water, $NaCl(aq)$

Sodium chloride dissociates into ions when dissolved in water. The sodium and chloride ions will interact strongly with the water through ion–dipole interactions.

e. Hydrogen fluoride, $HF(l)$

Hydrogen fluoride forms a hydrogen bonding network since the H–F bond, and thus the molecule, is very polar.

---

4. Intermolecular Forces in Action: Surface Tension, Viscosity, and Capillary Action

   a. Surface tension is the energy required to increase the surface area of a liquid by a unit amount; it measures the tendency of a liquid to minimize its surface area.

      i. Surface tension decreases as the intermolecular forces decrease.

   b. Viscosity is the resistance of a liquid to flow.

      The units of viscosity are poise (1 P = 1 g/cm·s). Units of centipoise (cP) are often used because the viscosity of water at room temperature is about 1 cP.

      i. The stronger the intermolecular forces, the higher the viscosity will be.

      ii. Viscosity also depends on molecular shape since long or branched molecules can become tangled in each other.

      iii. Viscosity is temperature-dependent and generally decreases with increasing temperature.

   c. Capillary action is the ability of a liquid to flow against gravity up a narrow tube.

      i. Capillary action depends on the balance between adhesive forces (forces of attraction between the tube and the substance) and cohesive forces (intermolecular forces).

         1. If the adhesion is greater than the cohesion, the liquid will spontaneously move against gravity up a tube.

         2. If the cohesion is greater than the adhesion, the liquid will not fill the tube (or will fill it to a level lower than the surface of the bulk sample) when it is placed in a beaker of the liquid.

      ii. The meniscus of a liquid in a tube is an indicator of its capillary action (i.e., concave versus convex).

5. Vaporization and Vapor Pressure

    a. Vaporization is the escape of molecules from the liquid phase into the gas phase.

    b. Higher temperatures lead to a higher average kinetic energy of a collection of molecules.

        i. In vaporization, faster moving molecules escape from the liquid surface, where there are fewer intermolecular forces.

        ii. In condensation, slower moving molecules in the gas phase are captured by the intermolecular forces of the liquid.

    c. Increasing the temperature increases the number of surface molecules with enough energy to overcome the intermolecular forces and therefore increases the rate of vaporization.

    d. At a given temperature, a substance with weaker intermolecular forces will have relatively more molecules escape from the liquid phase.

        i. A volatile substance has weak intermolecular forces and therefore is vaporized at lower temperatures and pressures compared to a nonvolatile substance.

    e. The heat of vaporization (ΔH_{vap}) is the amount of heat required to vaporize (i.e., convert from liquid to gas) 1 mole of a substance.

        i. For a given substance, $\Delta H_{vap} = -\Delta H_{cond}$.

        ii. Vaporization is always endothermic, and condensation is always exothermic.

EXAMPLE:

Calculate the quantity of methane gas (in grams) that you would need to burn in order to boil 100.0 g of water. The heat of combustion for methane is −802.5 kJ/mol, and the heat of vaporization for water is 40.7 kJ/mol.

Given $\Delta H_{vap}(H_2O)$, we will first calculate the heat needed to boil 100.0 g of $H_2O(l)$:

$$100.0 \text{ g } H_2O \times \frac{1 \text{ mol } H_2O}{18.02 \text{ g } H_2O} \times \frac{40.7 \text{ kJ}}{1 \text{ mol } H_2O} = 225.9 \text{ kJ}$$

Next, we can calculate the quantity of methane that must be combusted in order to produce this heat:

$$225.9 \text{ kJ} \times \frac{1 \text{ mol } CH_4}{802.5 \text{ kJ}} \times \frac{16.04 \text{ g } CH_4}{1 \text{ mol } CH_4} = 4.51 \text{ g } CH_4$$

So 4.51 g of methane needs to be burned in order to boil 100.0 g of water.

---

    f. Vapor pressure is the pressure of a gas in dynamic equilibrium with its liquid (or solid).

        i. Dynamic equilibrium occurs when the rate of condensation is equal to the rate of evaporation. This can be achieved in a closed container.

        ii. The vapor pressure of a substance increases with increasing temperature since a larger fraction of molecules are able to escape the liquid phase into the gas phase.

        iii. Increasing the pressure above a liquid disturbs the equilibrium and results in condensation in order to reduce the pressure and reestablish equilibrium.

    g. Boiling occurs when the vapor pressure of a substance is equal to the external pressure.

        i. The normal boiling point is the temperature at which the vapor pressure is equal to 1 atm.

        ii. The temperature of a liquid cannot be higher than the boiling point since all of the added energy is used to effect the phase transition.

h. The vapor pressure of a substance is temperature dependent:

$$P_{vap} = \beta \exp\left(-\frac{\Delta H_{vap}}{RT}\right)$$

where β is a constant that depends on the identity of the gas, R is the gas constant, and T is the absolute temperature.

i. This expression can be rearranged to give the Clausius–Clapeyron equation, which demonstrates a linear relationship between $\ln(P_{vap})$ and $1/T$:

$$\ln P_{vap} = -\frac{\Delta H_{vap}}{R}\frac{1}{T} + \ln \beta$$

where the slope is $-\Delta H_{vap}R$.

ii. A two-point form of the Clausius–Clapeyron equation allows one to determine the vapor pressure of a substance at any temperature if the normal boiling point and the heat of vaporization are known:

$$\ln \frac{P_2}{P_1} = -\frac{\Delta H_{vap}}{R}\left(\frac{1}{T_2} - \frac{1}{T_1}\right)$$

EXAMPLE:

What is the normal boiling point of acetone if the vapor pressure at 20.0 °C is 181.7 mmHg and the heat of vaporization is 29.1 kJ/mol?

Given ΔH_{vap}(acetone), T_1, P_1, and P_2, we use the two-point Clausius–Clapeyron equation, rearranged to solve for the temperature (T_2):

$$\frac{1}{T_1} - \frac{R}{\Delta H_{vap}}\ln \frac{P_2}{P_1} = \frac{1}{T_2}$$

$$T_2 = \frac{1}{\dfrac{1}{T_1} - \dfrac{R}{\Delta H_{vap}}\ln \dfrac{P_2}{P_1}}$$

Now we can include the values given in the problem with the temperature converted to kelvin units, the pressure at the normal boiling point of 760 mmHg, and the heat of vaporization in J/mol:

$$T_2 = \frac{1}{\dfrac{1}{293\ K} - \dfrac{8.314\ \dfrac{J}{mol \cdot K}}{29100\ J/mol}\ln \dfrac{760\ mmHg}{181.7\ mmHg}} = 332.9\ K$$

So the normal boiling point is 60 °C. Check: We expect $T_2 > T_1$ because $P_2 > P_1$.

iii. In a closed container, the heating of a liquid will cause a gas to form; as the gas forms, the pressure increases, which tends to cause the gas to condense. Eventually, a new phase, a supercritical fluid, will form.

1. Supercritical fluids have properties of both liquids and gases.

2. The critical temperature is the temperature above which a liquid cannot exist (i.e., the vapor cannot be liquefied), no matter what pressure is applied.

3. The critical pressure is the pressure required to bring about a transition to the liquid phase at the critical temperature.

6. Sublimation and Fusion

   a. Sublimation is the phase transition from solid to gas, and deposition is the phase transition from gas to solid.

   b. Melting is the phase transition from solid to liquid, and freezing (or fusion) is the phase transition from liquid to solid.

      i. At the melting point, the molecules, ions, or atoms have sufficient thermal energy to overcome the intermolecular forces that hold them at stationary points.

      ii. Solids cannot be at temperatures higher than the melting point, just as liquids cannot exist above the boiling temperature.

   c. The heat of fusion (ΔH_{fus}) of a substance is the amount of heat required to melt 1 mole of its solid.

      i. For a given substance, $\Delta H_{fus} = -\Delta H_{melt}$

      ii. Freezing is always exothermic, and melting is always endothermic.

   d. Usually, $\Delta H_{vap} > \Delta H_{fus}$ because intermolecular forces must be totally overcome in order to transition from the liquid to the gas phase (compared to the solid to liquid transition).

7. Heating Curve for Water

   a. The heating curve for water is a graphical representation of how the temperature of water changes as heat is added:

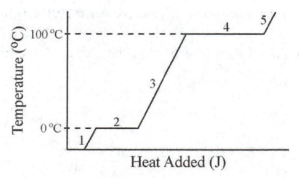

   b. The heat added in each region of the curve can be calculated systematically.

      i. In region 1, $H_2O(s)$ is being heated; the added heat is dependent on the mass of ice, the specific heat capacity of ice, and the temperature change:

$$q = m \cdot C_{s(solid)} \cdot \Delta T$$

      ii. In region 2, the $H_2O(s)$ is melting to form liquid water; the added heat is related to the quantity (in mol) and the heat of fusion:

$$q = n \cdot \Delta H_{fus}$$

      iii. In region 3, $H_2O(l)$ is being heated; the added heat is dependent on the mass of the liquid water, the specific heat capacity of liquid water, and the temperature change:

$$q = m \cdot C_{s(liquid)} \cdot \Delta T$$

      iv. In region 4, $H_2O(l)$ is boiling to form water vapor; the added heat is related to the quantity (in mol) and the heat of vaporization:

$$q = n \cdot \Delta H_{vap}$$

v. In region 5, the water vapor is being heated; the added heat is dependent on the mass of water vapor, the specific heat capacity of water vapor, and the temperature change:

$$q = m \cdot C_{s(gas)} \cdot \Delta T$$

EXAMPLE:

Calculate the amount of heat required to convert 10.0 g of ice at −10 °C to steam at 110 °C. The heat capacity of ice is 2.09 J/g· °C, the heat capacity of liquid water is 4.184 J/g· °C, and the heat capacity of steam is 2.09 J/g· °C. For H_2O, the heat of fusion is 6.02 kJ/mol and the heat of vaporization is 40.7 kJ/mol.

We will calculate the heat required for each of the five regions of the heating curve.

First we calculate the heat needed to warm the ice from −10 °C to the melting point of 0 °C:

$$q = m \cdot C_{s,ice} \cdot \Delta T = 10.0 \text{ g} \cdot 2.09 \text{ J/g} \cdot °C \cdot [0 - (-10)] \, °C = 209 \text{ J}$$

Next we calculate the heat required to convert the ice into liquid water:

$$q = n \cdot \Delta H_{fus} = 10.0 \text{ g} \cdot \frac{1 \text{ mol } H_2O}{18.02 \text{ g } H_2O} \cdot 6.02 \text{ kJ/mol} = 3.34 \text{ kJ} \cdot \frac{1000 \text{ J}}{1 \text{ kJ}} = 3340 \text{ J}$$

Now we calculate the heat needed to warm the liquid water at 0 °C to the boiling point at 100 °C:

$$q = m \cdot C_{s,liquid} \cdot \Delta T = 10.0 \text{ g} \cdot 4.184 \text{ J/g} \, °C \cdot (100 - 0) \, °C = 4184 \text{ J}$$

Now we calculate the heat required to vaporize the liquid water:

$$q = n \cdot \Delta H_{vap} = 10.0 \text{ g} \cdot \frac{1 \text{ mol } H_2O}{18.02 \text{ g } H_2O} \cdot 40.7 \text{ kJ/mol} = 22.59 \text{ kJ} \cdot \frac{1000 \text{ J}}{1 \text{ kJ}} = 22590 \text{ J}$$

Finally, we calculate the heat needed to warm the steam from 100 °C to 110 °C:

$$q = m \cdot C_{s,steam} \cdot \Delta T = 10.0 \text{ g} \cdot 2.09 \text{ J/g} \, °C \cdot (110 - 100) \, °C = 209 \text{ J}$$

The total heat required is the sum of these five steps:

$$q_{total} = 209 \text{ J} + 3340 \text{ J} + 4184 \text{ J} + 22590 \text{ J} + 209 \text{ J} = 30532 \text{ J}$$

So, 3.05×10^4 J are required to convert 10.0 g of ice at −10 °C to steam at 110 °C.

c. Notice that the temperature does not change in regions 2 and 4, meaning that the kinetic energy of the molecules is constant and all of the added heat goes into the formation of a new phase (i.e., breaking intermolecular forces of attraction).

d. The kinetic energy of a given phase is related to the heat capacity of that phase.

8. Phase Diagrams

a. A phase diagram is a map of the phases of a substance as a function of temperature (*x*-axis) and pressure (*y*-axis):

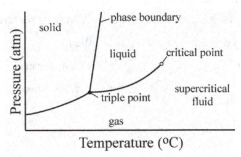

        i.   There are three main regions in the phase diagram: solid, liquid, and gas.

            1.   The phase diagram shows the conditions where a particular state is stable.

        ii.   Each line or curve in a phase diagram is a set of temperatures and pressures at which the substance is in equilibrium between the two states on either side of the line (i.e., gas-solid, liquid-gas, liquid-solid equilibrium curves).

        iii.   The triple point is the set of conditions at which the three states are in equilibrium.

        iv.   The critical point shows the temperature and pressure above which the substance is a supercritical fluid.

        v.   A phase diagram can feature only one critical point but one or more triple points.

    b.   Linear movement, left/right or up/down, in the phase diagram represents a change in the temperature and/or pressure respectively.

9.   Water: An Extraordinary Substance

    a.   Water has a permanent dipole due to its bent geometry. The large dipole moment results in strong hydrogen bonds between water molecules.

    b.   Water has a high boiling point for its molar mass.

    c.   Water has a high heat capacity.

    d.   Water expands upon freezing, which means that ice is less dense than liquid water.

        i.   Most organisms cannot survive freezing due to the expansion of water upon freezing.

<u>Fill in the Blank:</u>

1. The point in the phase diagram called the _____ is the temperature and pressure above which a supercritical fluid exists.

2. When the pressure of a gas in dynamic equilibrium with its liquid phase is _____, the gas will condense to form more liquid.

3. The amount of heat required to melt 1 mole of a solid is called the _____.

4. According to Coulomb's law, the potential energy between like charges _____ with increasing distance between them.

5. Hydrogen bonds occur when hydrogen is bonded to one of three atoms: _____.

6. The energy required to increase a liquid's surface area by a unit amount is its _____.

7. The ability of a liquid to flow against gravity up a narrow tube is _____.

8. Atomic solids that are held together by covalent bonds are called _____.

9. Polar molecules have _____ that interact with neighboring molecules.

10. The heat change associated with vaporization is always _____, and the heat change associated with condensation is always _____.

11. The _____ is the temperature at which its vapor pressure is equal to the external pressure.

12. The phase transition from a solid directly to a gas is called _____.

13. The density of liquid water is _____ than solid water (ice).

14. Dispersion forces _____ with increasing molar mass.

Problems:

1. Identify the dominant intermolecular force in macroscopic samples of each of the following molecules, atoms, or ions:

   a. Br_2

   b. Butane, $CH_3CH_2CH_2CH_3$

   c. Sulfuric acid solution, $H_2SO_4(aq)$

   d. Hydrogen sulfide, H_2S

   e. Methyl amine, CH_3NH_2

   f. Xenon, Xe

   g. Dimethyl ether, CH_3OCH_3

   h. Magnesium chloride, $MgCl_2(aq)$

2. Rank the following in order of increasing vapor pressure:

   a. He, NH_3, NF_3, NaCl

   b. NF_3, NCl_3, NBr_3, NH_3

   c. HF, F_2, ClF

   d. $CH_3CH_2CH_3$, $CH_3CH_2CH_2CH_2CH_2CH_2CH_2CH_3$, $CH_3CH_2CH_2CH_2CH_3$

3. Consider taking a sample of dry ice, CO_2, at −35.0 °C and heating it until it is a gas at 123 °C. The pressure is held at 1 atm for all questions.

   a. At what temperature and phase during this process will the sample have the largest amount of kinetic energy? Explain your answer.

   b. At what temperature and phase during this process will the sample have the largest internal energy? Explain your answer.

4. Three 50.0-mL containers, each containing an equimolar sample of one of the following gases, are held at constant temperature. Use the data below and the ideal gas law's assumption that molecules have no intermolecular forces to answer the following questions.

	Pressure of CO_2 (mmHg)	Pressure of NH_3 (mmHg)	Pressure of H_2 (mmHg)
Ideal gas prediction	489	489	489
Real gas measurement	1881	202	489

   a. Explain how the intermolecular forces cause NH_3 to deviate from ideal behavior.

   b. Which gas do you expect to have the highest boiling point?

   c. A supercritical fluid is often modeled using the ideal gas law. The idea is that the constituents have enough kinetic energy to overcome all intermolecular forces. Use this idea to determine which gas has the lowest critical temperature.

5. A cooling curve indicates how the temperature of a substance changes as heat is released.

   a. The vapor pressure of ammonia is 0.409 bar at −50 °C. Given that ammonia's normal boiling point is −33.3 °C, calculate the heat of vaporization for ammonia.

   b. The melting point of ammonia is −77.3 °C at 1 atm. The specific heat of liquid ammonia is 4.52 J/g·K, and the specific heat of solid and gaseous ammonia is 0.028 J/g·K. Sketch the heating curve for ammonia starting with a temperature of −100.0 °C and ending at a temperature of 0.0 °C.

   c. Calculate the amount of heat that must be supplied to a 10.0 g sample of solid ammonia at −100.0 °C to convert it to gaseous ammonia at 0.0 °C. The heat of fusion is 5.63 kJ/mol.

6. The average bond energy of an oxygen-hydrogen bond is 467 kJ/mol, and the average strength of a hydrogen bond is 20 kJ/mol.

   a. How much energy is required to break all of the hydrogen bonds in 15 g of liquid water, assuming that each water molecule forms two hydrogen bonds?

   b. How much energy will be required to break all of the oxygen-hydrogen bonds in 15 g of water?

   c. The heat of vaporization of water is 2.26 kJ/g. Compare this value to the values that you calculated in parts a and b. Explain any similarities and/or differences.

7. Consider taking a sample of dry ice, CO_2, at -35.0 °C and heating it until it is a gas at 123 °C. The pressure is held at 1 atm for all questions.

   a. At what temperature and phase during this process will the sample have the largest amount of kinetic energy? Explain your answer.

   b. At what temperature and phase during this process will the sample have the largest internal energy? Explain your answer.

   c. The normal (1 atm) sublimation point for CO_2 is -78.5 °C, the triple point is -56.6 °C and 5.11 atm, and the critical point is 31.0 °C and 73.0 atm. The solid phase is denser than the liquid phase. Sketch the phase diagram of CO_2.

   d. Is it possible for the liquid phase of CO_2 to be stable at -60 °C? Explain why or why not.

   e. What is the vapor pressure of solid CO_2 at -78.5 °C? Explain how you arrived at your answer.

   f. A sample of solid CO_2 is originally at a pressure of 1 atm and a temperature of -80.0 °C. The temperature is increased to 0.0 °C, while the pressure is held constant. Is the sample still a solid?

   g. A sample of solid CO_2 is originally at a pressure of 1 atm and a temperature of -80.0 °C. The temperature is increased to 0.0 °C, while the pressure is increased to 50 atm. Will the phase changes (if they occur) be the same as in part f? Why or why not?

8. Consider the phase diagram for water shown below.

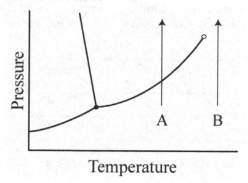

   a. Indicate the liquid, solid, gas, and supercritical fluid regions on the diagram.

   b. An isotherm is a curve showing a change that occurs while maintaining a constant temperature. For example, A and B are isotherms for water. Sketch a graph of the pressure versus volume for each of these isothermal changes. *Hint:* You need to consider how the pressure and volume are related at a phase change.

   c. Using your plots of pressure versus volume, give another definition for an ideal gas.

Concept Questions:

1. Consider the two structural isomers with the molecular formula $C_4H_{10}O$: diethyl ether ($CH_3CH_2OCH_2CH_3$) and butanol ($CH_3CH_2CH_2CH_2OH$).

   a. Which compound has larger dispersion forces? Explain.

    b.   Which substance has a higher boiling point? Explain.

    c.   Which substance do you expect to have a higher heat of vaporization? Explain.

2.   Two beakers are placed side by side in a 25 °C room. One beaker contains 2.5 L of pentane, $CH_3CH_2CH_2CH_2CH_3(l)$, and one beaker contains 2.5 L of pentanol, $CH_3CH_2CH_2CH_2CH_2OH(l)$.

    a.   If left out for long enough, both beakers will eventually empty due to evaporation. Which beaker will be empty first? Explain your answer.

    b.   Which beaker will have a larger concentration of its vapor above the solution? Explain your answer.

    c.   Both beakers are placed on a hot plate and heated up. Which beaker will begin boiling at a lower temperature? Explain your answer.

3.   In DNA, the two strands are held together by hydrogen bonding. Explain why you think that nature utilizes hydrogen bonds instead of covalent bonds. Why do you think that dispersion forces aren't as well suited to this job?

4.   Table 12.7 in your textbook shows the heats of vaporization for some common substances. Explain the trend in the heats of vaporization as they relate to the intermolecular forces that dominate in each substance.

5.   When water evaporates, the temperature of the remaining liquid decreases. Explain this decrease using the Boltzmann distribution of molecular speeds.

6.   Based on your understanding of intermolecular forces, do you think that an aqueous sodium chloride solution will have a higher or lower boiling point than pure water? Explain your answer.

7.   The heating curve of a substance can be related to a slice of the phase diagram at a given temperature. Each region of the heating curve therefore corresponds to a region of the phase diagram. Identify the connections between the five regions of the heating curve and the phase diagram of water.

Chapter 12: Phase Diagrams and Crystalline Solids

Key Learning Outcomes:

- Use Bragg's Law in X-Ray Diffraction Calculations
- Calculate the Packing Efficiency of a Unit Cell
- Relate Unit Cell Volume, Edge Length, and Atomic Radius
- Relate Density to Crystal Structure
- Classify Crystalline Solids

Chapter Summary:

You will begin this chapter by exploring how the relationship between temperature and pressure affects the phase of a substance. Next you will be introduced to X-ray crystallography, a technique that allows scientists to probe the structure of crystalline solids. Simple crystalline solids will be described, followed by discussions of ionic and network covalent solids. Carbon and silicates, both network covalent solids, followed by ceramics, cement, and glass, will all be explored, with a focus on their structure and general properties. After discussing these common substances, we will turn to semiconductors, with a special focus on their conductivity, which will be explained using band theory. Finally, polymer structure, function, and synthesis will be explored, especially as it relates to the formation of common plastics.

Chapter Outline:

1. Friday Night Experiments: The Discovery of Graphene

   a. Graphene is the thinnest material known, one atom thick and one of the strongest. It conducts heat and electricity, is transparent, and is completely impermeable.

2. Crystalline Solids: Determining Their Structure by X-Ray Crystallography

   a. The long-range order (arrangement) of atoms in solids can be studied using an experimental technique called X-ray crystallography.

   b. X-rays directed at a single crystal scatter when interacting with electrons in atoms; the result is a pattern of constructive and destructive interference that is captured on a detection screen.

      i. The pattern of constructive and destructive interference is called a diffraction pattern.

      ii. This pattern can be used to determine the location of atoms in the crystal using Bragg's law:

$$n\lambda = 2d \cdot \sin \theta$$

      where d is the distance between atomic layers, λ is the wavelength of the incident light, n is the integer number of wavelengths, and θ is the angle of reflection.

EXAMPLE:

Find the distance between atomic layers of an unknown crystal given that the angle of maximum reflection is 42.1° when 184 pm light is shone upon it. Assume that $n = 1$.

We rearrange the Bragg's law expression to solve for *d*:

$$d = \frac{n\lambda}{2 \cdot \sin \theta}$$

$$d = \frac{1 \cdot 184 \text{ pm}}{2 \cdot \sin(42.1)} = 137 \text{ pm}$$

---

3. Crystalline Solids: Unit Cells and Basic Structures

   a. A crystalline lattice is the regular arrangement of atoms or molecules in a crystalline solid.

   b. A unit cell is a small collection of atoms, ions, or molecules; it is the simplest repeating unit of the crystal.

   c. A lattice point is a point in space that is occupied by an atom, ion, or molecule.

   d. The coordination number is the number of atoms or molecules with which each atom or molecule is in direct contact.

   e. The packing efficiency is the percentage of the volume of a cubic cell that is occupied by spheres (representing atoms or ions).

   f. There are three main cubic unit cells:

      i. The simple cubic unit cell is a cube that has an atom at each corner.

         1. One complete atom is found in each simple cubic unit cell.

         2. The coordination number is 6, and the packing efficiency is 52%.

      ii. The body-centered cubic unit cell is a cube with an atom at each corner and one in the center.

         1. Two complete atoms are found in each body-centered cubic unit cell.

         2. The coordination number is 8, and the packing efficiency is 68%.

      iii. The face-centered cubic unit cell is a cube with an atom at each corner and one in the center of each face of the cube.

         1. Four complete atoms are found in each face-centered cubic unit cell.

         2. The coordination number is 12, and the packing efficiency is 74%.

EXAMPLE:

The density of iron solid is 7.87 g/mL. Calculate the radius of an iron atom (in pm) given that it has a body-centered cubic closest-packing structure.

Given density (m/V), we can calculate the atomic radius of Fe once we have determined the volume of the unit cell.

First, we can calculate the mass of a single unit cell, taking into account the number of atoms in a unit cell (2 in a body-centered cubic arrangement) and the mass of a single atom:

$$2 \text{ Fe atoms} \times \frac{1 \text{ mol Fe atoms}}{6.022 \times 10^{23} \text{ Fe atoms}} \times \frac{55.85 \text{ g Fe}}{1 \text{ mol Fe atoms}} = 1.855 \times 10^{-22} \text{ g Fe}$$

Next, we can determine the volume of the unit cell using the mass and density:

$$1.855 \times 10^{-22} \text{ g Fe} \times \frac{1 \text{ ml}}{7.87 \text{ g Fe}} \times \frac{1 \text{ cm}^3}{1 \text{ mL}} \times \left(\frac{1 \text{ m}}{100 \text{ cm}}\right)^3 = 2.357 \times 10^{-29} \text{ m}^3$$

The edge length for a body-centered cubic unit cell is given in terms of the atomic radius:

$$l = \frac{4r}{\sqrt{3}}$$

Since the length is the cube root of the volume of a unit cell, we can solve for the radius:

$$\sqrt[3]{V} = \frac{4r}{\sqrt{3}}$$

$$r = \frac{\sqrt{3}}{4}\sqrt[3]{V} = \frac{\sqrt{3}}{4}\sqrt[3]{2.357 \times 10^{-29}} = 1.24 \times 10^{-10} \text{ m}$$

Converting to a more standard atomic radius unit of pm gives our solution:

$$1.24 \times 10^{-10} \text{ m} \times \frac{10^{12} \text{ pm}}{1 \text{ m}} = 124 \text{ pm}$$

---

    g.   Closest-packed structures are the ways in which atoms or ions form layers stacked on one another. We can represent the layer types as A, B, and C:

        i.   Simple cubic packing involves one layer directly on top of another: AAAAA…

        ii.   Hexagonal closest packing involves shifting one layer by ½ an atom from the layer below it. The third layer is the same as the first: ABABAB…

        iii.   Cubic closest packing involves shifting one layer by ½ of an atom from the layer below it and the layer below that. The fourth layer is the same as the first: ABCABC…

4.   Crystalline Solids: The Fundamental Types

    a.   Molecular solids are composed of molecules held together by intermolecular forces and have low melting points.

    b.   Ionic solids are composed of ions held together by coulombic interactions.

        i.   The melting points of ionic solids are higher than molecular solids because of the stronger forces.

    c.   Atomic solids are species in which atoms are the composite unit; there are three types of atomic solids:

        i.   Nonbonding atomic solids are held together by dispersion forces and have low melting points.

        ii.   Metallic solids are held together by metallic bonds (as discussed in Chapter 9) and generally have high melting points.

        iii.   Network covalent solids are held together by covalent bonds and have very high melting points.

5.   The Structure of Ionic Solids

    a.   The coordination number of an ionic solid depends on the relative size of the cations and anions.

        i.   A larger size difference results in a cubic cell with a smaller coordination number.

6.   Network Covalent Atomic Solids: Carbon and Silicates

    a.   Carbon forms four common covalent solids: diamond, graphite, fullerenes, and nanotubes.

        i.   Diamond has carbon atoms that are bonded to four other carbon atoms in a tetrahedral arrangement.

  ii. Graphite has carbon atoms that are bonded to three other carbon atoms in a trigonal planar arrangement.

  iii. Fullerenes are clusters of carbon atoms (36 to over 100) that are held together with dispersion forces.

  iv. Nanotubes are sheets of C_6 rings that assume a cylindrical shape.

 b. Silicates such as quartz form network covalent bonds.

7. Ceramics, Cement, and Glass

 a. Ceramics are inorganic nonmetallic solids prepared from powders that are mixed with water, formed into a desired shape, and then heated.

  i. Ceramics are usually hard, strong, and nonconductive.

  ii. Ceramics can be classified as one of three types:

   1. Silicate ceramics have the silicate structure with some of the Si atoms replaced with Al (aluminosilicate). These produce clay when weathered.

   2. Oxide ceramics have high melting points so they can be used in high heat applications, but they tend to be brittle and crack easily with sudden temperature changes. The most common are Al_2O_3 and MgO.

   3. Nonoxide ceramics include Si_3N_2, BN, and SiC.

    a. BN is isolectronic with C_2 and therefore has similar properties and structures.

 b. Portland cement is a mixture of limestone ($CaCO_3$), silica (SiO_2), and smaller amounts of Al_2O_3, Fe_2O_3, and gypsum ($CaSO_4 \cdot 2H_2O$).

  i. Cement reacts with water to harden by forming Si-O-Si bridges that produce fibrous structures that bond strongly to each other.

  ii. Concrete is portland cement with sand and pebbles added to it.

 c. Glass is the amorphous structure of SiO_2 and is called vitreous silica or fused silica.

  i. Soda-lime glass is 70% SiO_2 with Na_2O and CaO. Soda-lime glass is common glass.

  ii. Adding boric oxide instead of CaO gives borosilicate glass (Pyrex) which does not crack under thermal shock (abrupt temperature changes).

  iii. Leaded glass (crystal) is composed of PbO and SiO_2. It looks brilliant since it has a higher index of refraction.

8. Semiconductors and Band Theory

 a. When metals bond together, their atomic orbitals interact in such a way as to form delocalized molecular orbitals over the entire crystal.

  i. The bonding orbitals are collectively referred to as the valence band.

  ii. The antibonding orbitals are collectively referred to as the conduction band.

 b. In general, metals can conduct electricity when electrons from the valence band can be promoted into the conduction band.

 c. Species in which there is a large difference in energy between the valence band and the conduction band – termed the band gap – will be insulators (they will not conduct electricity).

 d. Species in which there is a small difference in energy between the valence band and the conduction band will be semiconductors.

  i. Semiconductors act like conductors when energy is supplied to them and act like insulators when there is not sufficient energy for an electron to be promoted to the conduction band.

e. Doping involves the replacement of atoms in a metal with other atoms (termed dopants) in order to alter the conductivity.

    i. N-type doping increases conductivity by introducing extra electrons into the semiconductor.

    ii. P-type doping introduces holes (missing electrons) into the semiconductor, which also increases conductivity.

f. A p-n junction is a region in a semiconductor that is p-type on one side and n-type on the other side.

    i. Junctions like these are used in diodes: when the extra electron combines with the hole, energy is released as light.

9. Polymers

a. Polymers are long, chainlike molecules that are composed of repeating units called monomers.

b. An addition polymer is a polymer formed by an addition reaction, where the many monomers (typically alkenes) link without loss of atoms.

    i. Reaction of two monomers produces a dimer; as more monomer units are added, a long chain forms.

    ii. Polymers are named using the name of the monomer with the prefix "poly-."

c. Condensation polymers are formed by a condensation reaction, resulting in the elimination of an atom or small group of atoms during the process.

d. Copolymers are those that form from more than one type of monomer.

Fill in the Blank:

1. A polymer that forms when monomers simply add together without eliminating any atoms is called _____.

2. When Si is doped with Ga, the semiconductor that results is called a _____.

3. According to Coulomb's law, the potential energy between like charges _____ with increasing distance between them.

4. Atomic solids that are held together by covalent bonds are called _____.

5. The percentage of the volume of a unit cell occupied by spheres is called the _____.

6. Solids in which the atoms or molecules that compose them are arranged in a well-ordered array are called _____.

7. Solids that are composed of ions are called _____; they have _____ melting points.

8. The energy difference between the conduction band and the valence band in a semiconductor is called the _____.

9. The two primary components of Portland cement are _____ and _____.

10. Polymers that consist of more than one type of monomer are called _____.

Problems:

1. Rank the following in order of increasing melting point:

a. NaCl, $AlCl_3$, MgS, NaBr

b. CO, CO_2, CCl_4

c. SO_2, CO_2, O_2

d. Al_2O_3, SiO_2, P_2O_3, SO_2

2. Calculate the amount of empty space in each of the closest-packing structures discussed in this chapter and relate this to the density of solids with each configuration. Using the Internet, search for metals that utilize each of the packing arrangements and confirm or reassess your conclusion.

3. Polyaramid is a term applied to polyamides that contain aromatic groups. Kevlar is an example of such a polyaramid; it is used in bulletproof vests and many other high-strength composites. The structure of Kevlar is shown below:

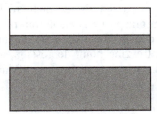

a. Draw the structure of the subunits that are used to make this polymer.

b. Is Kevlar formed in a condensation reaction or in an addition reaction? Explain your answer.

c. Kevlar is very strong because of cross-linking that occurs between the polymer strands. What intermolecular interaction do you expect to hold strands together and provide Kevlar with its unique strength? Explain your answer.

d. The strength of Kevlar is substantially reduced at high temperatures. Explain why you think that this might be.

e. In writing these questions, I wanted to provide as much information as possible, and so I searched for the molar mass of Kevlar. I couldn't find it anywhere! Why do you think that this is?

Concept Questions:

1. Consider a substance that has the band diagram shown below. The shading represents the occupation of the band with electrons. The lower-energy band is completely filled with electrons, and the higher-energy band is partially filled.

a. Based on the diagram shown, do you expect this substance to be a conductor, semiconductor, or insulator? Explain your answer.

b. What do you expect to happen to the electrical properties of this substance if all of the electrons are removed from the conduction band? Explain your answer.

c. How would you expect the conductivity from part b to change when the temperature is increased? Explain your answer.

2. Polyethylene comes in two forms: HDPE (high-density PE) and LDPE (low-density PE). LDPE is soft and waxy, while HDPE is hard and rigid. The same subunit (ethane $-\ C_2H_4$) comprises both polymers, but their structures are different. One form has branches in the chain structure, and the other does not.

a. Explain why these two forms of polyethylene have different properties and identify which polymer is more branched.

b. One of these polymers melts at 110 °C, and one of them melts at 130 °C. Which polymer melts at the higher temperature? Explain how you arrived at your answer.

c. Based on the structure of the ethene, do you think that polyethylene is produced in a condensation reaction or an addition reaction? Explain your answer.

Chapter 13: Solutions

<u>Key Learning Outcomes:</u>

- Determine Whether a Solute Is Soluble in a Solvent

- Use Henry's Law to Predict the Solubility of Gases with Increasing Pressure

- Calculate Concentrations of Solutions

- Convert between Concentration Units

- Determine the Vapor Pressure of a Solution Containing a Nonelectrolyte and Nonvolatile Solute

- Determine the Vapor Pressure of a Two-Component Solution

- Calculate Freezing Point Depression

- Calculate Boiling Point Elevation

- Determine the Osmotic Pressure

- Determine and Use the van't Hoff Factor

- Determine the Vapor Pressure of a Solution Containing an Ionic Solute

<u>Chapter Summary:</u>

  This chapter focuses on the formation and properties of solutions. You will begin by defining a solution and identifying the different components—solute and solvent—that comprise it. You will then explore the energy changes associated with solution formation, using the ideas of intermolecular forces discussed in Chapter 11. Unsaturated, saturated, and supersaturated solutions will be defined in order to explain solubility. Changes in solubility as a function of temperature and pressure will then be discussed. Next, qualitative and quantitative descriptions of solution concentration will be introduced. With a firm grasp of basic solution properties, you will be presented with the more complex ideas of colligative properties that depend upon the amount (but not identity) of solute, and include vapor pressure depression, boiling point elevation, freezing point depression, and osmotic pressure. Finally, you will learn about colloidal dispersions, which differ from true solutions because of the size of the particles.

<u>Chapter Outline:</u>

1. Antifreeze in Frogs

    a. A solution is a homogeneous mixture of two or more substances.

        i. The majority component is the solvent, and the minority component is the solute.

2. Types of Solutions and Solubility

    a. A solution is any homogeneous mixture and can be more than one gas, liquid(s) and gas(es), liquid(s) and solid(s), more than one liquid, or more than one solid.

    b. An aqueous solution is one in which water is the solvent and a solid, liquid, or gas is the solute.

    c. Solubility is the amount of a substance that will dissolve in a given amount of solvent at a specified temperature.

    d. Solutions form both because of the increased entropy that results and because of the decrease in potential energy.

        i. Entropy is a measure of energy dispersion or randomization.

        ii. The entropy, or energy dispersion, of a system tends to increase.

e. When liquids form homogeneous solutions when mixed in all proportions, they are said to be miscible.

f. Whether or not a solution forms will depend on the balance between the interactions broken and interactions formed, as well as the increased entropy.

   i. A solution always forms when the solvent–solute interactions are stronger than solute–solute and solvent–solvent interactions; both entropy and potential energy drive the formation of this type of solution.

   ii. A solution always forms when the solvent–solute interactions are about the same as the solute–solute and solvent–solvent interactions; entropy is the driving factor in this type of solution formation.

   iii. A solution will sometimes form when the solvent–solute interactions are weaker than the solute–solute and solvent–solvent interactions; whether or not the solution forms will depend on how much weaker the solvent–solute interactions are.

      1. A good rule of thumb is that "like dissolves like," meaning that solvents and solutes with similar types of intermolecular interactions will form solutions.

EXAMPLE:

Predict whether or not the following substances will form a solution based on their intermolecular interactions. It is not necessary to identify the major/minor components.

a. $CH_3CH_2CH_2CH_2CH_2CH_2CH_2CH_3$ and CH_3OH

Octane is a nonpolar covalent compound, and methanol is a polar covalent compound; these compounds will not, therefore, form a solution.

b. C_6H_6 and LiOH

Benzene is a nonpolar covalent compound, and lithium hydroxide is an ionic compound; these compounds will not, therefore, form a solution.

c. $CH_3CH_2CH_2CH_2OH$ and $CH_3CH_2CH_2CH_2CH_2CH_3$

Butanol has a polar end and a nonpolar carbon-hydrogen chain, while hexane is nonpolar, but these compounds will mix because of their similar intermolecular forces.

---

3. Energetics of Solution Formation

   a. We can think of solution formation as three energetic steps:

      i. Breaking solute–solute interactions requires energy: $\Delta H_{solute} > 0$.

      ii. Breaking solvent–solvent interactions requires energy: $\Delta H_{solvent} > 0$.

      iii. Forming new solvent–solute interactions releases energy: $\Delta H_{mix} < 0$.

$$\Delta H_{solution} = \Delta H_{solute} + \Delta H_{solvent} + \Delta H_{mix}$$

   b. The enthalpy change associated with forming a solution will be endothermic, exothermic, or thermoneutral, depending on the magnitude of the enthalpies associated with each step.

   c. For an aqueous solution, the enthalpy of hydration is $\Delta H_{hydration} = \Delta H_{solvent} + \Delta H_{mix}$, and for ionic compounds, the enthalpy of the solute is $\Delta H_{solute} = -\Delta H_{lattice}$. So for the formation of aqueous solutions:

$$\Delta H_{solution} = -\Delta H_{lattice} + \Delta H_{hydration}$$

    d.   There are three possible scenarios for aqueous solutions:

       i.   If $|-\Delta H_{\text{lattice}}| < |\Delta H_{\text{hydration}}|$, then $\Delta H_{\text{solution}} < 0$. The solution will always form.

       ii.   If $|-\Delta H_{\text{lattice}}| > |\Delta H_{\text{hydration}}|$, then $\Delta H_{\text{solution}} > 0$. The solution will only form if the enthalpy change is small and the entropy increase is dominant.

       iii.   If $|-\Delta H_{\text{lattice}}| \approx |\Delta H_{\text{hydration}}|$, then $\Delta H_{\text{solution}} \approx 0$. The solution will always form because the increase in entropy will drive it.

4.   Solution Equilibrium and Factors Affecting Solubility

    a.   Dissolution of a solid in water is an equilibrium process in which the solid and aqueous phases are in dynamic equilibrium.

       i.   The rates of dissolution and deposition are equal at equilibrium.

$$AB(s) \leftrightarrows A^+(aq) + B^-(aq)$$

       ii.   A saturated solution is one in which dynamic equilibrium is established; when additional solid is added to the solution, it will fall to the bottom as solid.

       iii.   An unsaturated solution contains less than the equilibrium amount of dissolved solute; when additional solid is added to the solution, it will dissolve until saturation is reached.

       iv.   A supersaturated solution contains more than the equilibrium amount of dissolved solute; a supersaturated solution is unstable, and solid will precipitate out of it spontaneously to form a saturated solution.

       v.   Solubility is the concentration at saturation.

    b.   The solubility of *most* solids in liquids such as water increases with increased temperature and is not affected by pressure changes.

       i.   Recrystallization is a purification technique that relies on the increased solubility of solids at higher temperatures. When the saturated solution is slowly cooled, pure crystals of the solute result.

    c.   The solubility of gases in liquids decreases with increased temperature and increases with increased pressure of the gas above the liquid.

       i.   The pressure dependence is quantified using Henry's law:

$$S_{\text{gas}} = k_H P_{\text{gas}}$$

where S_{gas} is the solubility of the gas, k_H is Henry's law constant (which is dependent on the solvent, solute, and temperature), and P_{gas} is the pressure of the gas above the solution.

EXAMPLE:

How many moles of gaseous nitrogen are in a 2.000 L container holding 0.523 L of a 0.00152 M aqueous nitrogen gas solution at 25 °C?

We need to determine the pressure of N_2 gas that must be over the solution using Henry's law. The Henry's law constant for nitrogen in water at 25 °C is 6.1×10^{-4} M/atm:

$$0.00152 \text{ M} = (6.1 \times 10^{-4} \text{ M/atm})P_{\text{gas}}$$

$$P_{\text{gas}} = 2.49 \text{ atm}$$

The pressure of the N_2 gas is related to the number of moles of N_2 gas through the ideal gas law. The volume available to the gas is the volume of the container (2.000 L) minus the volume of the solution (0.523 L): 1.477 L. Thus,

$$n = \frac{PV}{RT} = \frac{(2.49 \text{ atm})(1.477 \text{ L})}{\left(0.08206 \dfrac{\text{L} \cdot \text{atm}}{\text{K} \cdot \text{mol}}\right)(298 \text{ K})} = 0.150 \text{ mol}$$

5. Expressing Solution Concentration

    a. A qualitative description of solution concentration uses terms such as dilute (low concentration) and concentrated (high concentration).

    b. A quantitative description of solution concentration can be formulated in multiple ways.

        i. Molarity (M) is the amount of solute (in moles) in 1 L of solution:

$$M = \frac{\text{moles of solute}}{1 \text{ L of solution}}$$

        ii. Molality (m) is the amount of solute (in moles) in 1 kg of solvent:

$$m = \frac{\text{moles of solute}}{1 \text{ kg of solvent}}$$

        iii. Percent by mass (m/m%) is the mass of solute over the mass of solution:

$$\text{percent by mass} = \frac{\text{g of solute}}{1 \text{ g of solution}} \times 100\%$$

        iv. Percent by volume (v/v%) of the solute over the volume of solution and is generally used when two liquids are mixed to form a solution:

$$\text{percent by volume} = \frac{\text{L of solute}}{1 \text{ L of solution}} \times 100\%$$

        v. Parts per million (ppm) is the mass of the solute over the mass of the solution multiplied by 1,000,000:

$$\text{parts per million} = \frac{\text{g of solute}}{1 \text{ g of solution}} \times 1{,}000{,}000 \ (10^6)$$

        vi. Parts per billion (ppb) is the mass of the solute over the mass of the solution multiplied by 1,000,000,000:

$$\text{parts per billion} = \frac{\text{g of solute}}{1 \text{ g of solution}} \times 1{,}000{,}000{,}000 \ (10^9)$$

        vii. Mole fraction (χ_{solute}) is the moles of solute over total moles (solute + solvent):

$$\chi_{\text{solute}} = \frac{\text{mol of solute}}{1 \text{ mol of solution}}$$

        viii. Mole percent is the mole fraction expressed as a percentage:

$$\text{mole percent} = \chi_{\text{solute}} \times 100\%$$

EXAMPLE:

A sodium chloride solution is prepared by dissolving 2.48 g of NaCl(s) in 425 mL of H_2O(l). The aqueous solution has a density of 1.0067 g/mL. Express the solution concentration in the following units, assuming that no volume change occurs when the salt is added to the water.

a. Molarity

We need to convert the mass of NaCl given to moles of NaCl:

$$2.48 \text{ g NaCl} \times \frac{1 \text{ mol NaCl}}{58.44 \text{ g NaCl}} = 0.0424 \text{ mol NaCl}$$

This is the number of moles of solute in 0.425 L of solution, so the molarity is:

$$M = \frac{\text{moles of solute}}{1 \text{ L of solution}} = \frac{0.0424 \text{ mol NaCl}}{0.425 \text{ L}} = 0.0999 \text{ M}$$

b. Molality

Instead of dividing the moles of solute by the volume of solution, we will convert the solution volume to the mass of solution (in kg) using the density:

$$425 \text{ mL solution} \times \frac{1.0067 \text{ g solution}}{1 \text{ mL solution}} \times \frac{1 \text{ kg}}{1000 \text{ g}} = 0.428 \text{ kg solution}$$

We divide the moles of solute by the mass of the solution in order to calculate molality:

$$m = \frac{\text{moles of solute}}{1 \text{ kg of solvent}} = \frac{0.0424 \text{ mol NaCl}}{0.425 \text{ kg of solvent}} = 0.0998 \text{ m}$$

c. Percent by mass

The mass of the NaCl is given as 2.48 g, and the mass of the solution is 428 g (as calculated above). Thus, the percent by mass is:

$$\text{percent by mass} = \frac{\text{g of solute}}{\text{g of solution}} \times 100\% = \frac{2.48 \text{ g NaCl}}{428 \text{ g solution}} \times 100\% = 0.579\%$$

d. Parts per million

Again we divide the solute mass by the solution mass, but then multiply by 10^6:

$$\text{parts per million} = \frac{\text{g of solute}}{\text{g of solution}} \times 10^6 = \frac{2.48 \text{ g NaCl}}{428 \text{ g solution}} \times 10^6 = 5790 \text{ ppm}$$

6. Colligative Properties: Vapor Pressure Lowering, Freezing Point Depression, Boiling Point Elevation, and Osmotic Pressure

   a. Properties that depend on the amount of solute and not on the solute type are called colligative properties.

      i. Vapor Pressure of Solutions

         1. When a nonvolatile solute is dissolved in a solvent, the vapor pressure of the solution is lower than the vapor pressure of the pure solvent.

            a. We can explain this by considering that fewer solvent particles are at the surface of the liquid with the introduction of a solute. Since only particles at the surface are able to escape from the liquid to gas phase, a reduced number of solvent particles at the surface results in a lower vapor pressure.

     b. We can also explain this using the tendency of particles to mix. The increase in entropy that results when a solution forms causes the solution to be more stable than the pure solvent, so more energy is required to remove the solvent from the liquid phase.

2.   Raoult's law quantifies the vapor pressure of a solution using the mole fraction:

$$P_{solution} = \chi_{solvent} P^{\circ}_{solvent}$$

where P° is the vapor pressure of the pure substance. Note that if $\chi_{solvent} = 1$ (i.e., there is no solute added), then $P_{solution} = P^{\circ}_{solvent}$

3.   Vapor pressure lowering (ΔP) is the difference between the vapor pressure of the solvent and the vapor pressure of the solution:

$$\Delta P = P^{\circ}_{solvent} - P_{solution}$$

4.   Combining the expression of Raoult's law with that of vapor pressure lowering gives:

$$\Delta P = P^{\circ}_{solvent} - \chi_{solvent} P^{\circ}_{solvent} = (1 - \chi_{solvent}) P^{\circ}_{solvent}$$

And since $1 - \chi_{solvent} = \chi_{solute}$, then

$$\Delta P = \chi_{solute} P^{\circ}_{solvent}$$

This shows that the vapor pressure lowering depends only on the amount of solute dissolved in solution.

5.   Ionic solutes dissociate into ions so that the number of particles present in solution is higher than the number of solute particles dissolved.

EXAMPLE:

What is the vapor pressure of a solution prepared by dissolving 42.5 g of K_2O in enough water to prepare 1.25 L of solution at 25 °C? Assume that the density of the solution is 1.00 g/mL and that the vapor pressure of pure water is 23.8 torr at 25 °C.

We will use Raoult's Law, $P_{solution} = \chi_{solvent} P^{\circ}_{solvent}$, to determine the vapor pressure of the solution. We are given the vapor pressure of the pure solvent, so we need to first calculate the solvent mole fraction.

We can use the solution density to determine the solution mass:

$$1.25 \text{ L solution} \times \frac{1000 \text{ mL}}{1 \text{ L}} \times \frac{1.00 \text{ g solution}}{1 \text{ mL solution}} = 1250 \text{ g solution}$$

The mass of the solvent is the difference between the mass of the solution and the mass of the solute (given as 42.5 g). The mass of the solvent is then: 1250 g − 42.5 g = 1207.5 g solvent.

Converting both the solute and solvent masses to moles gives:

$$42.5 \text{ g } K_2O \times \frac{1 \text{ mol } K_2O}{94.20 \text{ g } K_2O} \times \frac{3 \text{ mol ions}}{1 \text{ mol } K_2O} = 1.3535 \text{ mol ions}$$

Notice that in order to calculate the moles of solute in the solution, we had to convert from the ionic formula to the number of ions using a conversion factor.

$$1207.5 \text{ g } H_2O \times \frac{1 \text{ mol } H_2O}{18.02 \text{ g } H_2O} = 67.009 \text{ mol } H_2O$$

Now we will calculate the mole fraction of the solute:

$$\chi_{solute} = \frac{\text{moles ions}}{\text{moles ions} + \text{moles of } H_2O} = \frac{1.3535 \text{ mol ions}}{1.3535 \text{ mol ions} + 67.009 \text{ mol } H_2O} = 0.0198$$

The mole fraction of water (the solvent) is simply one minus the mole fraction of ions (solute):

$$\chi_{H_2O} = 1 - \chi_{solute} = 1 - 0.0198 = 0.9802$$

We can now calculate the vapor pressure of the solution using Raoult's law:

$$P_{solution} = \chi_{solvent} P^o_{solvent} = (0.9802)(23.8 \text{ torr}) = 23.3 \text{ torr}$$

6. Ideal solutions are those that obey Raoult's law exactly. In these solutions, the energy associated with the solvent–solute interactions are the same as solvent–solvent and solute–solute interactions. For a two-component system where both components are volatile:

$$P_A = \chi_A P^o_A$$

$$P_B = \chi_B P^o_B$$

$$P_{total} = P_A + P_B$$

7. Deviations from ideal behavior occur because the interactions between the solvent and solute are not equal to solute–solute or solvent–solvent interactions.

   a. A positive deviation (i.e., higher vapor pressure) from ideal behavior occurs when the solute–solvent interactions are weaker than solute–solute and solvent–solvent interactions.

   b. A negative deviation (i.e., lower vapor pressure) from ideal behavior occurs when the solute–solvent interactions are stronger than solute–solute and solvent–solvent interactions.

EXAMPLE:

Calculate the total pressure above 1.0 L of a 1.42 M $CH_3OH(aq)$ solution at 25 °C. The vapor pressure of pure methanol at 25 °C is 127.1 torr, and the vapor pressure of pure water is 23.8 torr at 25 °C. You may assume that the density of the solution is 1.00 g/mL and that the solution behaves ideally.

We note that both water and methanol are volatile (and nonelectrolytes), so we will apply Raoult's law to both solution components.

We first need to calculate the mass of the solute in the solution so that we can determine the mass of the solvent. We will assume that we have 1.00 L of the solution.

$$1.00 \text{ L solution} \times \frac{1.42 \text{ mol } CH_3OH}{1 \text{ L solution}} \times \frac{32.04 \text{ g } CH_3OH}{1 \text{ mol } CH_3OH} = 45.50 \text{ g } CH_3OH$$

We find the mass of the solution using the density:

$$1.00 \text{ L solution} \times \frac{1000 \text{ mL}}{1 \text{ L}} \times \frac{1 \text{ g solution}}{1 \text{ mL solution}} = 1000 \text{ g solution}$$

The mass of the solvent is the difference between these values, so there are 954.50 g of solvent (H_2O).

We will now calculate the moles of the solute and moles of solvent:

$$1.00 \text{ L solution} \times \frac{1.42 \text{ mol CH}_3\text{OH}}{1 \text{ L solution}} = 1.42 \text{ mol CH}_3\text{OH}$$

$$954.50 \text{ g H}_2\text{O} \times \frac{1 \text{ mol H}_2\text{O}}{18.02 \text{ g H}_2\text{O}} = 52.97 \text{ mol H}_2\text{O}$$

Now we will calculate the mole fraction of each:

$$\chi_{\text{CH}_3\text{OH}} = \frac{\text{moles of CH}_3\text{OH}}{\text{moles of CH}_3\text{OH} + \text{moles of H}_2\text{O}} = \frac{1.42 \text{ mol CH}_3\text{OH}}{1.42 \text{ mol CH}_3\text{OH} + 52.97 \text{ mol H}_2\text{O}} = 0.0261$$

The mole fraction of water is simply one minus the mole fraction of methanol:

$$\chi_{\text{H}_2\text{O}} = 1 - \chi_{\text{CH}_3\text{OH}} = 1 - 0.0261 = 0.9739$$

We can now calculate the vapor pressure of the solvent and solution using Raoult's law:

$$P_{\text{H}_2\text{O}} = \chi_{\text{H}_2\text{O}} P^{\text{o}}_{\text{H}_2\text{O}} = (0.9739)(23.8 \text{ torr}) = 23.2 \text{ torr}$$

$$P_{\text{CH}_3\text{OH}} = \chi_{\text{CH}_3\text{OH}} P^{\text{o}}_{\text{CH}_3\text{OH}} = (0.0261)(127.1 \text{ torr}) = 3.32 \text{ torr}$$

The total pressure is the sum of the partial pressures, so:

$$P_{\text{total}} = P_{\text{CH}_3\text{OH}} + P_{\text{H}_2\text{O}} = (3.32 + 23.2) \text{ torr} = 26.5 \text{ torr}$$

---

    ii.   Freezing Point Depression, Boiling Point Elevation, and Osmosis

        1.   Because the vapor pressure of a solution is lower than that of the solvent, the freezing point will be lower and the boiling point will be higher for a solution.

                a.   The freezing point depression depends on the solvent properties:

$$\Delta T_{\text{f}} = m \times K_{\text{f}}$$

where m is the molality of the solute and K_{f} is the freezing point depression constant for the solvent (in units of °C/m). Note that K_{f} does not depend on the solute.

EXAMPLE:

Calculate the freezing point of a solution that results when 2.5 g of glucose ($C_6H_{12}O_6$) is dissolved in 500 mL of water.

We will first calculate the molality of the solution:

$$\frac{2.5 \text{ g C}_6\text{H}_{12}\text{O}_6}{500 \text{ mL H}_2\text{O}} \times \frac{1 \text{ mol C}_6\text{H}_{12}\text{O}_6}{180.16 \text{ g C}_6\text{H}_{12}\text{O}_6} \times \frac{1 \text{ mL H}_2\text{O}}{1 \text{ g H}_2\text{O}} \times \frac{1000 \text{ g}}{1 \text{ kg}} = 0.02775 \ m \ \text{C}_6\text{H}_{12}\text{O}_6 \text{ solution}$$

We next calculate the change in the freezing point using the freezing point depression constant for water, which is 1.86 °C/m:

$$\Delta T_{\text{f}} = m \times K_{\text{f}} = (0.02775 \ m) \times (1.86 \ {}^{\text{o}}\text{C}/m) = 0.0516 \ {}^{\text{o}}\text{C}$$

The normal freezing point of water is 0.00 °C, so the freezing point of the solution is –0.05 °C.

---

2. The boiling point elevation also depends on a constant that is determined by the solvent identity:

$$\Delta T_b = m \times K_b$$

where m is the molality of the solution and K_b is the boiling point elevation constant for the solvent (in units of °C/m). Note that K_b does not depend on the solute.

EXAMPLE:

Calculate the boiling point of a solution that results when 2.5 g of glucose ($C_6H_{12}O_6$) is dissolved in 500 mL of water. Note that this is the same solution as in the previous example, so we don't need to calculate the molality again.

The boiling point elevation constant for water is 0.512 °C/m, so the change in the boiling point is:

$$\Delta T_b = m \times K_b = (0.02775\ m) \times (0.512\ °C/m) = 0.0142\ °C$$

The normal boiling point is 100.00 °C, so the boiling point of the solution is 100.01 °C.

3. Osmosis is the flow of solvent from a solution of lower solute concentration (higher solvent concentration) to higher solute concentration (lower solvent concentration).

▶ When considering the direction of solvent flow, remember that the region of *high solvent* concentration is the region of *low solute* concentration. In osmosis, the solutions are trying to equalize the concentrations.

    a. Osmosis occurs because of nature's tendency to mix.

    b. A semipermeable membrane allows some substances to pass through but not others (usually based on size differences).

    c. Osmosis is quantified as the external pressure that needs to be applied in order to stop osmotic flow through a semipermeable membrane. This pressure is the osmotic pressure (Π):

$$\Pi = MRT$$

where M is the molarity, R is the gas constant, and T is the absolute temperature.

EXAMPLE:

Calculate the osmotic pressure required to keep the sugar solution from the previous two examples from flowing through a semipermeable membrane at 25 °C. Assume that the volume of the solution is equal to the volume of the water used.

We first need to calculate the molarity of the solution.

$$\frac{2.5\ g\ C_6H_{12}O_6}{500\ mL\ solution} \times \frac{1\ mol\ C_6H_{12}O_6}{180.16\ g\ C_6H_{12}O_6} \times \frac{1000\ mL}{1\ L} = 0.02775\ M\ C_6H_{12}O_6\ solution$$

We can now calculate the osmotic pressure of this solution:

$$\Pi = MRT = (0.02775\ M) \times \left(0.08206 \frac{L \cdot atm}{mol \cdot K}\right) \times (298.15\ K) = 0.68\ atm$$

7. Colligative Properties of Strong Electrolyte Solutions

    a. For ionic solutions, the number of particles depends on the number of ions produced when the ionic solute is dissolved.

        i. The number of particles is represented by the van't Hoff factor (i):

$$i = \frac{\text{moles of particles in solution}}{\text{moles of formula units dissolved}}$$

        ii. Boiling point elevation, melting point depression, and osmotic pressure expressions must be adjusted for ionic solutions to include the van't Hoff factor.

$$\Delta T_f = im \times K_f$$

$$\Delta T_b = im \times K_b$$

$$\Pi = iMRT$$

EXAMPLE:

Calculate the boiling point of a solution that results when 2.5 g of NaCl is dissolved in 500 mL of water. We will first calculate the molality of the solution:

$$\frac{2.5 \text{ g NaCl}}{500 \text{ mL H}_2\text{O}} \times \frac{1 \text{ mol NaCl}}{58.44 \text{ g NaCl}} \times \frac{1 \text{ mL H}_2\text{O}}{1 \text{ g H}_2\text{O}} \times \frac{1000 \text{ g}}{1 \text{ kg}} = 0.08556 \ m \text{ NaCl solution}$$

The boiling point elevation constant for water is 0.512 °C/m, so the change in the boiling point is:

$$\Delta T_b = i \times m \times K_b = 2 \times (0.08556 \ m) \times (0.512 \text{ °C}/m) = 0.0876 \text{ °C}$$

The normal boiling point is 100.00 °C, so the boiling point of the solution is 100.09 °C.

    b. Bodily fluids are categorized based on the osmotic pressure of cells.

        i. Hyperosmotic solutions are those with an osmotic pressure greater than body fluids. Cells shrink when placed in hyperosmotic solutions.

        ii. Hyposmotic solutions are those with an osmotic pressure less than body fluids. Cells swell in hyposmotic solutions.

        iii. Isosmotic (or isotonic) solutions have an osmotic pressure equal to that of body fluid.

Fill in the Blank:

1. When two substances are soluble in each other in all proportions, they are _____.

2. _____ relates the solubility of a gas to the pressure of the gas above a solution of it.

3. A(n) _____ solution is one in which water is the solvent.

4. _____ is the pressure that must be applied in order to stop the flow of solvent through a semipermeable membrane.

5. The boiling point of a solution is _____ than the boiling point of the pure solvent.

6. _____ is a measure of energy randomization or energy dispersal.

7. The vapor pressure of a solution with a nonvolatile solute is _____ than the vapor pressure of the pure solvent.

8. A(n) _____ solution is one where the solid phase and the aqueous phase are in equilibrium with one another.

9. A solution is a(n) _____ mixture of two or more substances.

10. A(n) _____ follows Raoult's law exactly.

11. A real solution that has solvent–solute interactions that are stronger than solute–solute and solvent–solvent interactions will deviate in a _____ from Raoult's law.

12. The _____ is the sum of the heat of hydration and the negative lattice energy.

13. The _____ must be included in colligative property calculations for solutions of ionic substances.

14. The solubility of gases _____ with an increase in temperature and _____ with an increase in the pressure of the gas over the solution.

15. The _____ of a solution is the amount of solute (in moles) over the mass of the solvent (in kilograms).

Problems:

1. Which of the following substances do you expect to dissolve to a significant extent in water? Explain your choices, and suggest a different solvent for those that you expect not to dissolve in water.

    a. $NaCl(s)$

    b. $CO_2(g)$

    c. $CH_3CH_2CH_2CH_2CH_2CH_3(l)$

    d. $LiOH(g)$

    e. NO_3^-

2. Ammonium chloride is a common cold pack ingredient.

    a. Solid ammonium chloride is added to an ammonium chloride solution. Some of the solid dissolves while the rest falls to the container floor. Was the original solution saturated, unsaturated, or supersaturated? What about the solution after adding ammonium chloride? Explain your answer.

    b. If ammonium chloride is used for a cold pack, what are the relative magnitudes of the heat of hydration and lattice energy?

    c. The lattice energy of ammonium chloride is smaller than that of sodium chloride. Why do you think that this is?

3. What is the molar solubility of O_2 in water when 5.0 mol of O_2 is pumped into a 1.0 L container holding 525 mL of water at 25 °C? Henry's law constant for O_2 is 0.0013 M/atm. *Hint:* You will need to use the ideal gas law and express the moles of gas in terms of the solubility.

4. The density of a 20% by volume solution of ethanol, $CH_3CH_2OH(l)$, in water is 0.9687 g/mL.

    a. Suggest why the density of the ethanol solution is less than the density of water.

    b. Given that the density of pure ethanol is 0.7893 g/mL, do you expect this solution to behave ideally? If not, will it deviate in a positive or negative fashion?

    c. Calculate the molarity of the solution.

    d. Calculate the molality of the solution.

    e. Calculate the percent by mass of ethanol in the solution.

    f. Calculate the mole fraction of ethanol in the solution.

    g. Calculate the mole fraction of water in the solution.

5. The vapor pressure of liquid ammonia is 10.2 atm at 25 °C.

    a. Calculate the partial pressure of ammonia over of a 1.2 M solution of ammonia in water at 25 °C. Assume that the density of the ammonia solution is the same as the density of water.

    b. Calculate the partial pressure of water over this same solution.

    c. Do you expect that the actual vapor pressure of ammonia over the solution will be higher or lower than the value that you calculated in part a? Explain your answer.

6. Calculate the vapor pressure, freezing point, boiling point, and osmotic pressure of a solution prepared by dissolving the following substances in 1.0 L of ethanol at 25 °C. The vapor pressure of pure ethanol is 59.0 torr. Assume that no volume change occurs upon dissolution and that liquid volumes are additive.

    a. 5.45 g $CaCl_2$

    b. 32.7 mL H_2O

    c. 89.25 mg of LiCl

    d. 63.8 g of SiO_2

    e. 398 mL of a 0.10 M methanol (CH_3OH) solution

7. Calculate the amount of salt that you would need to dissolve in water in order to change its temperature by 1 °C. Based on your answer, do you think that salt is added to cooking water in order to change its temperature or for flavor? Justify your answer.

Concept Questions:

1. When defining solution types, every phase could be combined with every other phase except gases and solids. Why do you think that these phases are not included?

2. There is a rule of thumb for the solubility of alcohols in water. It states that when less than six carbon atoms are present in the alcohol, it will be soluble in water, and if there are more than six carbon atoms in the alcohol, it will be soluble in hexane. Explain the reasoning behind this rule of thumb.

3. Explain why the solubility of gases decreases with increasing temperature using the Boltzmann distribution of molecular speeds.

4. The K_f values listed in your book vary a great deal. For example, carbon tetrachloride has a freezing point depression constant of 29.9 °C/m, while diethyl ether has a freezing point depression constant of 1.79 °C/m.

    a. Which of these solutions will have a larger freezing point change when 1.0 mol of NaCl is dissolved in equal volumes of each of them? (Assume that the solutions form.)

    b. Suggest one reason for the large difference in K_f values.

    c. The boiling point elevation constants do not vary by such a large amount. Suggest one reason for this observation.

5. We generally assume that the volume of a solution is equal to the volume of the solvent.

    a. This is based on the idea that the solute–solvent interactions are approximately equal to the solvent–solvent interactions. Why does this assumption lead to a negligible volume change?

    b. Is this always a valid assumption? If so, why? If not, suggest a system that would deviate from this assumption and state the direction of the deviation.

    c. What do you expect to be true about the vapor pressures of solutions that have a volume that is smaller than the solvent volume?

6. Rank the following in order of increasing solubility in water: $MgCl_2$, MgO, $RbBr_2$. Explain.

7. NaCl(s) is added to two beakers. One beaker contains liquid pentane, $CH_3CH_2CH_2CH_2CH_3$, and one beaker contains liquid pentanol, $CH_3CH_2CH_2CH_2CH_2OH$.

    a. In which beaker do you expect more NaCl to dissolve? Explain.

    b. Both substances will have a new boiling point upon the addition of NaCl. Explain why the boiling point changes.

    c. The boiling point changes by a different amount for the two beakers. Explain why this is and which solution will have a higher boiling point elevation.

8. There are pumps in your cells that operate to remove sodium ions from the cell. These pumps operate against the concentration gradient, meaning that they pump sodium ions from a region of low concentration to a region of high concentration. Is this process energetically favorable? Explain your answer using the concepts of entropy discussed in this chapter.

9. Consider two nonideal solutions: one made by mixing water with ethanol (CH_3CH_2OH) and one made by mixing water with nitric acid (HNO_3).

    a. Does the ethanol/water solution deviate from Raoult's law in a positive or negative fashion?

    b. Based on your answer in part a, do you think that the actual change in the boiling point for the solution compared to the solvent will be greater or less than in an ideal solution?

    c. Does the nitric acid/water solution deviate from Raoult's law in a positive or negative fashion?

    d. Based on your answer in part c, do you think that the actual change in the boiling point for the solution compared to the solvent will be greater or less than for an ideal solution?

    e. Do you expect these solutions to have a similar or different change in their boiling point elevations? Explain your answer.

10. Soap solutions are great because they help to dissolve dirt and oil in water; this is how your laundry detergent gets your clothes clean. Explain how soap works to dissolve nonpolar substances such as dirt in a polar solvent such as water.

Chapter 14: Chemical Kinetics

Key Learning Outcomes:

- Express Reaction Rates

- Determine the Order, Rate Law, and Rate Constant of a Reaction

- Use Graphical Analysis of Reaction Data to Determine Reaction Order and Rate Constants

- Determine the Concentration of a Reactant at a Given Time

- Work with the Half-Life of a Reaction

- Use the Arrhenius Equation to Determine Kinetic Parameters

- Determine Whether a Reaction Mechanism Is Valid

Chapter Summary:

Chemical kinetics is the study of the rates of chemical reactions and the underlying physical processes that define them. In this chapter, you will learn to appreciate the broad spectrum of reaction speeds, types, and molecular events. First, you will carry out the most basic reaction rate calculation, the determination of an average rate. Then you will learn a more sophisticated calculation: the determination of the instantaneous rate. Experimentally, the method of initial rates is used to determine the rate law of a chemical reaction, which you will learn how to derive and interpret. You will then be able to use the rate law (rate as a function of reactant concentration) and the integrated rate law (concentration as a function of time) in conjunction with experimental data to determine the order of a reaction. The pathway of a chemical reaction is described using a series of elementary steps, which taken together comprise the reaction mechanism. You will learn to identify possible reaction mechanisms using the experimental rate law and your understanding of chemical phenomena. The temperature dependence of the reaction rate will be examined in the context of collision theory. You will learn how to use an Arrhenius plot to determine the activation energy and the collision requirements of a particular chemical reaction. Finally, you will learn how a catalyst alters a reaction pathway to one with a lower barrier and therefore causes the reaction to occur faster.

Chapter Outline:

1. Catching Lizards

    a. The rate of a chemical reaction is a measure of how fast the reaction occurs.

2. Rates of Reaction and the Particulate Nature of Matter

    a. Chemical reactions can be thought of as resulting from particle collisions.

    b. The rate of a reaction depends on such factors as concentration, temperature, and the structure and relative orientation of reacting particles.

3. Defining and Measuring the Rate of a Chemical Reaction

    a. The rate is equal to the ratio of the change in reactant (or product) concentration to the change in time:

$$\text{rate} = -\frac{\Delta[\text{reactant}]}{\text{time}} = \frac{\Delta[\text{product}]}{\text{time}}$$

Since $\Delta[\text{reactant}]$ is always negative and Δ time is positive, the negative sign in the definition involving reactants results in a rate that is always a positive number.

b. The rate reflects the reaction stoichiometry. For the reaction:

$$aA + bB \rightarrow cC + dD$$

the rate is:

$$\text{rate} = -\frac{\Delta[A]}{\Delta t} = -\frac{\Delta[B]}{\Delta t} = \frac{\Delta[C]}{\Delta t} = \frac{\Delta[D]}{\Delta t}$$

▶ The inclusion of the stoichiometric coefficients ensures that each reaction has a single defined rate associated with it.

c. Reactant concentrations decrease over time, while product concentrations increase over time. Thus, a negative sign is included for reactant changes only.

d. The average rate of a reaction can be calculated using any two data sets.

e. The average rate of a reaction decreases over time as the reactants convert to products.

EXAMPLE:

Consider the generic chemical reaction: $\quad A + B \rightarrow 2C$

a. Using the data in the table below, calculate the average reaction rate over three different time intervals:

Experiment	[A] (M)	Time (s)
1	0.052	0.00
2	0.024	0.10
3	0.014	0.20

The rate is given by the expression:

$$\text{rate} = -\frac{\Delta[A]}{\Delta t} = -\frac{\Delta[B]}{\Delta t} = \frac{1}{2}\frac{\Delta[C]}{\Delta t}$$

So the rate of the reaction for the first time interval is:

$$\text{rate} = -\frac{[A]_2 - [A]_1}{t_2 - t_1} = -\frac{0.024 - 0.052}{0.10 - 0.00} = 0.28 \text{ M/s}$$

The rate of the reaction for the second time interval is:

$$\text{rate} = -\frac{[A]_3 - [A]_2}{t_3 - t_2} = -\frac{0.014 - 0.024}{0.20 - 0.10} = 0.10 \text{ M/s}$$

The rate of the reaction for the overall time interval is:

$$\text{rate} = -\frac{[A]_3 - [A]_1}{t_3 - t_1} = -\frac{0.014 - 0.052}{0.20 - 0.00} = 0.19 \text{ M/s}$$

b. Comment on the differences between the three reaction rates calculated.

The reaction rate for the first time interval is fastest and for the second time interval is slowest. This is consistent with the idea that the reaction rate decreases over time as a consequence of a decrease in the amount of reactant(s) available for reaction. The reaction rate calculated for the third time interval is the least meaningful since it is an average over a longer period of time—the reaction rate is not actually 0.19 M/s over this whole range.

c. Predict the concentration of B after 0.10 s if the initial concentration is 0.052 M (the same as the initial concentration of A).

Since the rate of the reaction is the same for all reactants and products and the reaction stoichiometry indicates a 1:1 ratio of A to B, the concentration of B will change at the same rate as the concentration of A changes. Since we begin with the same amount of A and B, at some time later the amounts of A and B will remain the same. Therefore, the concentration of B after 0.10 s will be 0.024 M.

This can also be solved mathematically:

$$-\frac{[A]_2 - [A]_1}{t_2 - t_1} = -\frac{[B]_2 - [B]_1}{t_2 - t_1} \quad \rightarrow \quad -\frac{0.024 - 0.052}{0.10 - 0.00} = -\frac{[B]_2 - 0.052}{0.10 - 0.00}$$

---

f. The instantaneous rate of a reaction is the rate at a given point in time and can be calculated using the slope of the line tangent to the plot of concentration versus time at the specified time.

EXAMPLE:

Again, consider the generic chemical reaction: $A + B \rightarrow 2C$

Data is collected over a period of time and plotted on the graph as shown below.

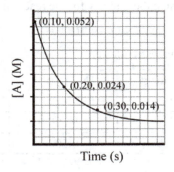

Calculate the instantaneous rate at 0.20 s.

We can draw the line tangent to the plotted data at 0.20 s and select two points on that line:

Using the two points selected, we can calculate the slope of the line: $\frac{0.013 - 0.033}{0.15 - 0.05} = -0.20$ M/s, which gives an instantaneous rate of 0.20 M/s at 0.20 s.

---

g. Reaction rates are measured experimentally using a number of techniques such as polarimetry, spectroscopy, and pressure changes.

h. The reaction rate *cannot* be predicted from the balanced chemical reaction alone.

4. The Rate Law: The Effect of Concentration on Reaction Rate

a. For a generic reaction with a single reactant, $A \rightarrow B$, the rate law relates the rate of the reaction to the concentration of the reactant raised to some power.

$$\text{Rate} = k[A]^n$$

The (temperature-dependent) proportionality constant k is called the rate constant, and the exponent n is the order of the reaction.

i. When $n = 0$, the reaction is zero order and the rate of the reaction is independent of the concentration of A:

$$\text{rate} = k \text{ (units of } k\text{: M/time)}$$

ii. When $n = 1$, the reaction is first order and the rate of the reaction is directly proportional to the concentration of A:

$$\text{rate} = k[A] \ (\text{units of } k: \text{time}^{-1})$$

iii. When $n = 2$, the reaction is second order and the rate of the reaction is proportional to the concentration of A squared:

$$\text{rate} = k[A]^2 \ (\text{units of } k: M^{-1} \text{ time}^{-1})$$

b. Determining the Order of a Reaction

i. Rate laws cannot be determined directly from the reaction stoichiometry; they must be determined by experiment.

ii. The method of initial rates measures the instantaneous rate at the beginning of the reaction for several different starting conditions. The data is used to determine the exponent (i.e., reactant order) in the rate law.

EXAMPLE:

The method of initial rates is used to determine the rate law for the reaction:

$$2N_2O_5 \rightarrow 4NO_2 + O_2$$

At a constant temperature, the following data was collected:

Experiment	$[N_2O_5]$ (M)	rate (M/s)
1	0.100	5.1×10^{-5}
2	0.200	1.0×10^{-4}
3	0.300	1.5×10^{-4}

a. Determine the rate law for the reaction.

The rate law is of the form: rate $= k[N_2O_5]^n$. We can determine the value of n by using the ratio of the rate laws for any two data points:

$$\frac{\text{rate}_1}{\text{rate}_2} = \frac{k[N_2O_5]_1^n}{k[N_2O_5]_2^n} \quad \rightarrow \quad \frac{0.100 \ M \cdot s^{-1}}{0.200 \ M \cdot s^{-1}} = \frac{k(5.1 \times 10^{-5} \ M)^n}{k(1.0 \times 10^{-4} \ M)^n}$$

Since the rate constant does not change (constant temperature), we can cancel it out. We can also collect the concentration terms and simplify:

$$\frac{0.100 \ M \cdot s^{-1}}{0.200 \ M \cdot s^{-1}} = \left(\frac{5.1 \times 10^{-5} \ M}{1.0 \times 10^{-4} \ M} \right)^n \quad \rightarrow \quad 0.500 = (0.51)^n \quad \rightarrow \quad n = 1$$

In the example above, it is simple to determine the value of n by inspection. For more complicated problems, it is sometimes easier to use logarithms to solve:

$$\log(0.50) = \log(0.51)^n$$
$$\log(0.50) = n\log(0.51)$$
$$n = \frac{\log(0.50)}{\log(0.51)}$$
$$n = \frac{-0.30103}{-0.29243}$$
$$n = 1.0294 \approx 1$$

So the rate law is: rate $= k \ [N_2O_5]$.

b. What is the order of the reaction with respect to N_2O_5?

The reaction is first order with respect to N_2O_5.

c. Calculate the rate constant.

The rate constant can be calculated by using one of the data sets in the rate law:

$$5.1 \times 10^{-5} \ M \cdot s^{-1} = k \times 0.100 \ M \quad \rightarrow \quad k = \frac{5.1 \times 10^{-5} \ M \cdot s^{-1}}{0.100 \ M}$$

So $k = 5.1 \times 10^{-4} \ s^{-1}$ (be sure to include correct units).

---

c. Reaction Order for Multiple Reactants

i. For more complicated reactions, such as:

$$a A + b B \rightarrow c C + d D$$

the rate is given by the general rate law:

$$\text{rate} = k[A]^m[B]^n$$

where m is the order of the reaction with respect to A, n is the order of the reaction with respect to B, and $(m + n)$ is the overall reaction order.

▶ The determination of n and m is the same as in the above example with an additional step.

EXAMPLE:

The following data was collected for the reaction of NO with Cl_2:

$$2NO + Cl_2 \rightarrow 2NOCl$$

Experiment	[NO] (M)	[Cl$_2$] (M)	Rate (M/min)
1	0.100	0.100	0.18
2	0.100	0.200	0.35
3	0.200	0.200	1.45

a. Determine the rate law for the reaction.

The rate law is of the form: rate = $k[NO]^n[Cl_2]^m$. We can determine m using the first and second experiments (notice that since [NO] remains constant, it cancels out of the expression):

$$\frac{\text{rate}_1}{\text{rate}_2} = \frac{k[NO]_1^n[Cl_2]_1^m}{k[NO]_2^n[Cl_2]_2^m} \quad \rightarrow \quad \frac{0.18 \ M/min}{0.35 \ M/min} = \frac{k(0.100 \ M)^n(0.100 \ M)^m}{k(0.100 \ M)^n(0.200 \ M)^m}$$

We can now collect like terms and simplify:

$$0.51 = (0.500)^m \quad \rightarrow \quad m = 1$$

And now we can use the second and third experiments to determine n (this time [Cl$_2$] remains unchanged and therefore cancels out of the expression):

$$\frac{\text{rate}_2}{\text{rate}_3} = \frac{k[NO]_2^n[Cl_2]_2^m}{k[NO]_3^n[Cl_2]_3^m} \quad \rightarrow \quad \frac{0.35 \ M/min}{1.45 \ M/min} = \frac{k(0.100 \ M)^n(0.200 \ M)}{k(0.200 \ M)^n(0.200 \ M)}$$

We now collect like terms and simplify:

$$0.24 = (0.500)^n \quad \rightarrow \quad n = 2$$

So the rate law is: rate $= k[NO]^2[Cl_2]$.

b. Calculate the rate constant for the reaction.

We can calculate the rate constant using one set of data points and the rate law that we have just identified:

$$1.45 \text{ M/min} = k(0.200 \text{ M})^2(0.200 \text{ M}) \quad \rightarrow \quad k = 181 \ (M^2 \cdot min)^{-1}$$

5. The Integrated Rate Law: The Dependence of Concentration on Time

    a. The integrated rate law for a chemical reaction is the relationship between concentrations of reactants and time. In the following expressions, $[A]_0$ is the initial concentration of A and $[A]_t$ is the concentration of A after some time, t.

        i. Zero-order integrated rate law:

$$[A]_t = -kt + [A]_0$$

        ii. First-order integrated rate law:

$$\ln[A]_t = -kt + \ln[A]_0$$

        iii. Second-order integrated rate law:

$$\frac{1}{[A]_t} = kt + \frac{1}{[A]_0}$$

▶ All of these expressions can be compared to the equation of a line with $x = t$ and $y = [A]_t$ (zero order), $y = \ln[A]_t$ (first order), or $y = 1/[A]_t$ (second order). Thus, a best-fit analysis can be performed to determine the reaction order.

EXAMPLE:

Use the plotted data shown below to determine the reaction order (n), the rate constant (k), the initial concentration ($[A]_0$), and the concentration of A after 60 s ($[A]_{60}$):

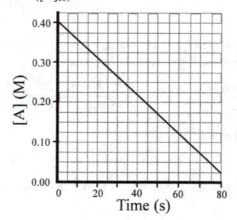

Since the data are linear when the concentration is plotted versus time, it is a zeroth-order reaction. We can compare the integrated rate law to the equation of a line ($y = mx + b$) to determine the rate constant and the initial concentration:

$$[A]_t = -kt + [A]_0$$

$$y = [A]_t, \ m \ (\text{slope}) = -k, \ x = t, \text{ and } b \ (y\text{-intercept}) = [A]_0$$

Inspection of the graph shows that the slope is $\frac{\Delta y}{\Delta x} = \frac{(0.40 - 0.125)}{(0.00 - 60)} = -0.00458$ M/s and the y-intercept is 0.40 M.

So, the rate constant, k, is 0.0046 M/s and the initial concentration is 0.40 M. We can find the concentration of A after 60 seconds by solving the integrated rate law:

$$[A]_{60} = -(0.00458 \text{ M/s})(60 \text{ s}) + 0.40 \text{ M} \quad \rightarrow \quad [A]_{60} = 0.125 \text{ M}$$

We could also obtain the concentration after 60 s from the graph and would arrive at the same answer.

---

    b.   The half-life of a reaction is the time required for the concentration of a reactant to be reduced to one half of its original amount ($[A]_t = \frac{1}{2}[A]_0$).

        i.   Zero-order reaction half-life:

$$t_{1/2} = \frac{[A]_0}{k}$$

        The half-life decreases as the reactant concentration decreases.

        ii.   First-order reaction half-life:

$$t_{1/2} = \frac{0.693}{k}$$

        The half-life is independent of the reactant concentration.

        iii.   Second-order reaction half-life:

$$t_{1/2} = \frac{1}{k[A]_0}$$

        The half-life increases as the reactant concentration decreases.

EXAMPLE:

Carbon-14 decays via first-order kinetics and has a half life of 5730 years. If 1.0 g of ^{14}C is produced in a campfire, how many years will it take for only 7.8×10^{-3} g to remain?

We can solve this problem by calculating the number of half-lives required. To do this, we take the starting amount, 1.0 g, and divide by two until we have 0.0078 g remaining:

$$1.0/2 \rightarrow 0.50/2 \rightarrow 0.25/2 \rightarrow 0.125/2 \rightarrow 0.0625/2 \rightarrow 0.03125/2 \rightarrow 0.015625/2 \rightarrow 0.0078125$$

Seven half-lives are required to reduce 1.0 g of ^{14}C to 0.0078 g of ^{14}C. Seven half-lives is a total time of 7×5730 years = 40,110 years.

EXAMPLE:

Dinitrogen pentoxide, N_2O_5, undergoes a decomposition reaction according to first-order kinetics with a half-life of 4.03×10^3 s.

a.   What is the rate constant for the decomposition of N_2O_5?

    We can solve for the rate constant by using the half-life equation and the given value:

$$t_{1/2} = \frac{0.693}{k} \quad \rightarrow \quad 4.03 \times 10^3 \text{ s} = \frac{0.693}{k} \quad \rightarrow \quad k = 1.72 \times 10^{-4} \text{ s}^{-1}$$

b. How much N_2O_5 will remain after two days, if you start with 10.0 g?

In order to solve this problem, we need to determine how many half-lives two days are:

$$2 \text{ days} \times \frac{24 \text{ hr}}{1 \text{ day}} \times \frac{60 \text{ min}}{1 \text{ hr}} \times \frac{60 \text{ s}}{1 \text{ min}} \times \frac{1 \text{ half-life}}{4030 \text{ s}} = 43 \text{ half-lives}$$

So if we begin with 10 g, we will end up with $10/2^{43}$ g or 1.14×10^{-12} g.

6. The Effect of Temperature on Reaction Rate

    a. A good rule of thumb is that the rate of a reaction approximately doubles for each increase in temperature of 10 °C.

    b. Rates depend on temperature because the rate constant, k, is temperature-dependent.

    c. The Arrhenius equation relates k to the temperature through an exponential factor:

    $$k = Ae^{-E_a/RT}$$

    where k is the rate constant at a temperature T (in Kelvin), E_a is the activation energy (typically reported in J/mol or kJ/mol), A is the frequency factor, and R is the gas constant (with SI units).

    i. The activation energy (E_a) is the minimum amount of kinetic energy that reactants must possess in order that their collisions result in reaction. It is the difference in energy between the reactants and the transition state. The higher the activation energy is, the slower the reaction rate will be.

        1. The activated complex is a high-energy, transitory species with only partial bonds, which exists at the transition state. It either goes on to form products or falls apart into the original reactants.

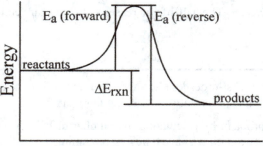

Reaction Progress

    ii. The frequency factor (A) can be thought of as the number of times the reactant(s) approach(es) the barrier in a given time period. The higher the value of A, the greater the reaction rate.

    iii. The exponential factor ($\exp^{-Ea/RT}$) is the fraction of molecules with enough energy to overcome the barrier (the activation energy). The exponential factor is highly dependent on the temperature.

    d. Increasing the temperature increases the energy of reacting species, which increases the fraction of reactant species with enough energy to react upon collision. This is easily seen by inspecting the Boltzmann distribution for the fraction of molecules with energy greater than a certain value (for $T_2 > T_1$).

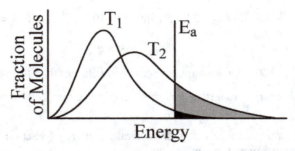

e. When the rate constant is measured at a variety of temperatures, an Arrhenius plot can be used to find the activation energy and frequency factor for a given reaction.

$$\ln k = -\frac{E_a}{R}\left(\frac{1}{T}\right) + \ln A$$

▶ A plot of $\ln k$ (y-axis) versus $1/T$ (x-axis) yields a straight line with slope $= -E_a/R$ and y-intercept $= \ln A$.

The two-point form of the equation can be derived:

$$\frac{\ln k_1}{\ln k_2} = -\frac{E_a}{R}\left(\frac{1}{T_1} - \frac{1}{T_2}\right)$$

EXAMPLE:

The conversion of cyclopropane to propene occurs at 300 °C with a rate constant of 2.41×10^{-10} s^{-1} and occurs at 400 °C with a rate constant of 1.16×10^{-6} s^{-1}.

a. Calculate the activation energy of the reaction.

We need to convert the temperature to the Kelvin scale, and then we can use the Arrhenius equation, along with the given information:

$$\text{For the first set of conditions: } \ln(2.41 \times 10^{-10}) = \frac{-E_a}{R}\left(\frac{1}{573.15}\right) + \ln A$$

$$\text{For the second set of conditions: } \ln(1.16 \times 10^{-6}) = \frac{-E_a}{R}\left(\frac{1}{673.15}\right) + \ln A$$

Now we can subtract the second equation from the first equation and simplify:

$$\ln(2.41 \times 10^{-10}) - \ln(1.16 \times 10^{-6}) = -\frac{E_a}{8.3145}\left(\frac{1}{573.15}\right) - \left(-\frac{E_a}{8.3145}\left(\frac{1}{673.15}\right)\right)$$

$$\ln\frac{2.41 \times 10^{-10}}{1.16 \times 10^{-6}} = -\frac{E_a}{8.3145}\left(\frac{1}{573.15} - \frac{1}{673.15}\right)$$

$$-8.4791 = \frac{E_a}{8.3145}(-0.00025919)$$

$$E_a = 272000 \text{ J/mol or } 272 \text{ kJ/mol.}$$

b. Calculate the rate constant at 500 °C.

We have solved for the activation energy, but we have not solved for the frequency factor. We will therefore need to use the two-point form of the Arrhenius equation as we did above:

$$\ln\frac{k}{2.41 \times 10^{-10}} = -\frac{272000}{8.3145}\left(\frac{1}{773.15} - \frac{1}{673.15}\right) \quad \rightarrow \quad \ln\frac{k}{2.41 \times 10^{-10}} = 6.28575$$

$$\ln k = 6.28575 + \ln(2.41 \times 10^{-10}) \quad \rightarrow \quad \ln k = -15.8604$$

$$k = e^{-15.8604}$$

So $k = 1.29 \times 10^{-7}\,\text{s}^{-1}$ at 500 °C, which makes sense—it is larger than the value of k at 300 °C, a lower temperature.

c. Sketch the potential energy diagram for this reaction given that the overall reaction is exothermic with a $\Delta H° = -33$ kJ/mol.

We can sketch the reaction profile using the activation energy (reaction barrier) calculated and the enthalpy of reaction given, assuming a one-step conversion:

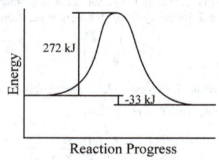

f. The frequency factor, A, can be interpreted using the collision model.

   i. The collision model assumes that a reaction occurs when reactant molecules collide in the proper orientation with enough energy to overcome the barrier. The orientation criterion is embedded in the frequency factor:

$$A = pz$$

where p is the orientation factor and z is the collision frequency.

   1. Using the collision model, the Arrhenius equation becomes:

$$k = pze^{-E_a/RT}$$

   2. The orientation factor, which assumes values of 1 or less, takes into account the fact that some reactions involving complex species need a specific collision geometry in order to occur, thereby reducing the probability of a reaction.

   3. The collision frequency is the number of collisions per unit time.

EXAMPLE:

If all of the reactions are carried out at the same temperature, which reaction do you expect to have the largest value of A?

a. $H_2 + Cl_2 \rightarrow 2HCl$

b. $Na^+ + Cl^- \rightarrow NaCl$

c. $CH_2{=}CH_2 + H_2 \rightarrow CH_3{-}CH_3$

Since all of the reactions are carried out at the same temperature, we can assume that the total number of collisions are similar (temperature is proportional to kinetic energy). This means that the main factor influencing the value of A is the orientation factor, p. We can conclude that reaction b, $Na^+ + Cl^- \rightarrow NaCl$, will have the largest A value since it is a reaction of two spherically symmetric species that can collide with any orientation and have a successful reaction outcome. Note that we have ignored attractive/repulsive forces and differences in mass in this problem for simplicity.

7. Reaction Mechanisms

    a. Reactions are usually written as an overall reaction equation and do not provide information about how the reaction progresses.

    b. A reaction mechanism is a sequence of simple, elementary reactions that add up to give the overall reaction.

        i. Elementary reactions show the microscopic (molecular) events as they are thought to happen.

    c. Reaction intermediates are species that are formed in one elementary step and consumed in a subsequent elementary step. Intermediates do not appear in the overall chemical reaction.

    d. The molecularity of an elementary step is the number of species reacting in that step—it is the overall order of the elementary reaction.

        i. A unimolecular reaction involves a single reactant species.

        ii. A bimolecular reaction involves two reactant species.

        iii. A termolecular reaction involves three reactant species.

        ▶ Note that while the reaction order has no obvious physical meaning, the molecularity does.

    e. Very few termolecular (and no higher-order molecularity) reactions exist; this is presumably because the probability of three or more reactant species colliding simultaneously with the correct orientation is very small.

    f. The rate law for an elementary reaction can be deduced directly from the balanced reaction.

    g. The rate-determining step is the slowest step in a reaction mechanism and determines the rate of the overall reaction.

    ▶ The slowest step may be first, or it may follow one or more fast steps (equilibria).

EXAMPLE:

The reaction between nitrogen dioxide and fluorine gas is believed to occur via the following mechanism:

$$NO_2 + F_2 \rightarrow NO_2F + F \qquad \text{slow}$$

$$NO_2 + F \rightarrow NO_2F \qquad \text{fast}$$

a. What is the overall reaction that is occurring?

The overall reaction is the sum of the steps with any species that appear on both sides eliminated:

$$2NO_2 + F_2 \rightarrow 2NO_2F$$

b. Identify any intermediates in the reaction mechanism.

A reaction intermediate is any species that is produced in one step of the reaction mechanism and is subsequently consumed. F is produced in the first step, consumed in the second step, and does not appear in the overall reaction—F is therefore a reaction intermediate.

c. What is the rate law predicted by the reaction mechanism?

The rate law can be written directly from the slow elementary step:

$$\text{rate} = k[NO_2][F_2]$$

d. What is the molecularity of the first step?

The first step is bimolecular since there are two species that must collide in that step.

EXAMPLE:

The reaction between hydrochloric acid and oxygen gas is believed to occur in three steps:

$$HCl + O_2 \rightarrow HOOCl$$

$$HOOCl + HCl \rightarrow 2HOCl$$

$$HOCl + HCl \rightarrow H_2O + Cl_2$$

If the experimentally observed rate law is rate = $k[HCl][O_2]$, what is the rate-determining step of the reaction mechanism?

Since the rate law for an elementary reaction can be deduced directly from the balanced reaction, we see that the first step is the slow, or rate-determining, step:

$$HCl + O_2 \rightarrow HOOCl \qquad \text{rate} = k_1[HCl][O_2]$$

$$HOOCl + HCl \rightarrow 2HOCl \qquad \text{rate} = k_2[HOOCl][HCl]$$

$$HOCl + HCl \rightarrow H_2O + Cl_2 \qquad \text{rate} = k_3[HOCl][HCl]$$

EXAMPLE:

Consider the reaction mechanism shown below.

$$NO + NO \rightleftharpoons N_2O_2 \qquad \text{fast equilibrium}$$

$$N_2O_2 + O_2 \rightarrow 2NO_2 \qquad \text{slow}$$

a. What is the overall reaction?

The overall reaction is the sum of the steps with any species that appear on both sides eliminated:

$$2NO + O_2 \rightarrow 2NO_2$$

b. What is the rate law for the reaction?

The rate law is determined from the slow step:

$$\text{rate} = k_2[N_2O_2][O_2]$$

but the rate law should not include intermediates like N_2O_2. So we need to consider the equilibrium expression:

$$\text{rate}_{\text{forward}} = k_1[NO][NO]$$

$$\text{rate}_{\text{reverse}} = k_{-1}[N_2O_2]$$

At equilibrium the reverse and forward rates are equal, so rate$_{\text{forward}}$ = rate$_{\text{reverse}}$ and we solve for $[N_2O_2]$:

$$k_1[NO][NO] = k_{-1}[N_2O_2]$$

$$\frac{k_1}{k_{-1}} = \frac{[N_2O_5]}{[NO][NO]} \quad \rightarrow \quad [N_2O_5] = \frac{k_1}{k_{-1}}[NO]^2$$

Substituting this into the rate expression gives:

$$\text{rate} = \frac{k_2 k_1}{k_{-1}}[O_2][NO]^2$$

Since k_2, k_1, and k_{-1} are all constants, we can replace them with a single constant, k:

$$\text{rate} = k[O_2][NO]^2$$

a. Reaction mechanisms are never proven to be true but are only shown to agree with experimental evidence. In order for a proposed mechanism to be reasonable, it must:

    i. Have elementary steps that add up to give the correct overall reaction.

    ii. Give a rate law that is consistent with the experimentally determined rate law.

        1. The rate law should be written in terms of reactant species that appear in the overall reaction. Intermediate species do not appear in the rate law.

    iii. Be physically plausible.

EXAMPLE:

The reaction between NO_2 and F_2 was studied. The experimentally determined rate law is:

$$2NO_2 + F_2 \rightarrow 2NO_2F \qquad \text{rate} = k[NO_2][F_2]$$

Which of the following mechanisms is most likely?

a. $2NO_2 + F_2 \rightarrow 2NO_2F$ single step

b. $NO_2 + F_2 \rightarrow NO_2F + F$ fast

    $NO_2 + F \rightarrow NO_2F$ slow

c. $F_2 \rightarrow 2F$ slow

    $F + NO_2 \rightarrow NO_2F$ fast

d. $NO_2 + F_2 \rightarrow NO_2F + F$ slow

    $NO_2 + F \rightarrow NO_2F$ fast

We can first look at the rate laws for each of the reaction mechanisms:

a. rate $= k[NO_2]^2[F_2]$

b. rate $= k[NO_2][F]$ (*note that F is an intermediate)

c. rate $= k[F_2]$

d. rate $= k[NO_2][F_2]$

We see that the rate law for mechanism d is consistent with the experimentally determined rate law. Since it also adds up to the overall reaction and appears to be physically reasonable (both elementary steps are bimolecular), mechanism d is most likely.

---

8. Catalysis

    a. A catalyst is a substance, added in nonstoichiometric (small) quantities, that increases the rate of a chemical reaction without being consumed by it.

        i. A homogeneous catalyst exists in the same phase as the reacting species.

        ii. A heterogeneous catalyst exists in a different phase as the reacting species.

    b. Catalysts often increase reaction rates by providing an alternate reaction mechanism (pathway) with an activation energy that is lower than the uncatalyzed reaction.

EXAMPLE:

The decomposition of hydrogen peroxide to produce water and oxygen is an exothermic reaction:

$$2\ H_2O_2 \rightarrow O_2 + 2\ H_2O$$

The reaction can be catalyzed using iodide ions; the catalyzed reaction is believed to occur via the following two-step mechanism:

$$H_2O_2 + I^- \rightarrow OI^- + H_2O \qquad \text{slow}$$

$$H_2O_2 + OI^- \rightarrow O_2 + H_2O + I^- \qquad \text{fast}$$

Sketch the potential energy diagram of the catalyzed and uncatalyzed reaction using the information given.

The uncatalyzed reaction can be drawn as our starting point. We are given that the reaction is exothermic, so the products are lower in energy than the reactants. Assuming a one-step dissociation process:

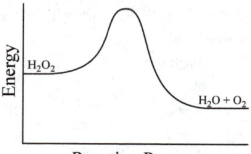

We can then sketch in the catalyzed reaction. It occurs in two steps, and we know that the activation energy of the two steps is lower than that of the uncatalyzed reaction. Further, we know that the activation energy of the first step is larger than the activation energy of the second step because the first step is slower.

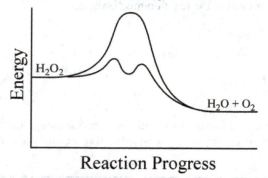

c. Enzymes are biological catalysts that increase the rate of biochemical reactions that occur in organisms.

   i. Enzymes are very selective in the reactions that they catalyze.

   ii. The active site of an enzyme is the region of the enzyme (E) that binds to the reactant molecule, called the substrate (S).

   iii. Enzyme kinetics are generally described by the following mechanism:

   $$E + S \rightleftharpoons ES \qquad \text{fast equilibrium}$$

   $$ES \rightarrow E + P \qquad \text{slow}$$

Fill in the Blank:

These problems are intended to ensure that you are familiar with the common terms and their definitions from this chapter.

1. The relationship between the reaction rate and the concentration of the reactant species is given by the _____.

2. In a(n) _____-order reaction, the half-life of the reacting species is independent of concentration.

3. The _____ limits the overall reaction rate and therefore determines the rate law for a given chemical reaction.

4. A good rule of thumb to remember is the rate of a reaction _____ with each temperature increase of 10 °C.

5. The instantaneous rate can be calculated by finding the slope of the line _____ to the concentration versus time data at a given point in time.

6. Reaction mechanisms can never be _____ but can only be shown to be consistent with experimental evidence.

7. A(n) _____ speeds up a chemical reaction by providing a lower-energy pathway, but it is not consumed in the reaction.

8. The integrated rate law is an expression of the concentration as a function of _____.

9. A(n) _____ is a species that is produced in an elementary step, consumed in a subsequent step, and does not appear in the overall chemical reaction.

10. _____ is the study of the rates of chemical reactions and the underlying microscopic steps that comprise them.

11. The reactant concentration _____ over time in a chemical reaction.

12. In a second-order reaction, the half-life of the reacting species _____ as the concentration of the reactant decreases.

13. The _____ is given by the sum of the exponents of the reacting species in the rate law.

14. The rate law for a(n) _____ can be written directly from its balanced reaction.

15. The _____ is the minimum amount of energy required in a collision in order for a reaction to occur.

16. _____ are biological catalysts that catalyze specific biochemical reactions.

17. The frequency factor contains two terms: _____, which determines the fraction of collisions with the correct geometry, and _____, which indicates the total number of collisions.

18. When a plot of _____ versus time is linear, the reaction can be identified as second order.

19. The rate of a chemical reaction has units of _____ over time.

20. The _____ indicates the number of species that must come together in a single elementary step.

21. A first-order reaction is one in which the rate is _____ the concentration of the reacting species.

22. A(n) _____ provides a molecular view of the individual steps that comprise a reaction.

23. The primary reason that temperature affects reaction rates is that the _____ increases with an increase in temperature.

24. The slope of an Arrhenius plot can be used to determine the _____ of a reaction.

25. Reaction rates _____ as the concentration of reactants decrease over time.

Problems:

1. Use the reaction shown below and the experimental data gathered in the table to determine the rate law for the reaction, the rate for the fourth experiment, and the rate constant.

$$2H_2(g) + 2NO(g) \rightarrow N_2(g) + 2H_2O(g)$$

Experiment	[H_2] (M)	[NO] (M)	Rate (M/s)
1	0.010	0.020	1.64×10^{-6}
2	0.020	0.020	3.28×10^{-6}
3	0.020	0.040	6.56×10^{-6}
4	0.020	0.080	

2. Which of the following mechanisms do you think is most plausible for the reaction in question 1? Provide a 1–2 sentence justification for your answer.

Mechanism 1:	$H_2 + NO \rightarrow H_2O + N$	slow
	$N + NO \rightarrow N_2 + O$	fast
	$O + H_2 \rightarrow H_2O$	fast
Mechanism 2:	$H_2 + 2NO \rightarrow N_2O + H_2O$	slow
	$N_2O + H_2 \rightarrow N_2 + H_2O$	fast
Mechanism 3:	$2NO \leftrightarrows N_2O_2$	fast equilibrium
	$N_2O_2 + H_2 \rightarrow N_2O + H_2O$	slow
	$N_2O + H_2 \rightarrow N_2 + H_2O$	fast

3. Consider the reaction of hydrogen and iodine gases to form hydrogen iodide:

$$H_2(g) + I_2(g) \rightarrow 2HI(g)$$

This reaction is thought to occur via the following reaction mechanism:

$I_2(g) \leftrightarrows 2I(g)$	fast equilibrium
$H_2(g) + 2I(g) \rightarrow 2HI(g)$	slow

Determine the rate law and sketch the potential energy diagram for the reaction based on the mechanism given.

4. 2-Butene has two isomers: cis-2-butene and trans-2-butene. The cis conformation is 4 kJ/mol higher in energy than the trans form, and therefore, the trans form is dominant in a mixture of the compound. Cis-2-butene is converted to trans-2-butene using iodine gas as a catalyst according to the following five-step mechanism:

First step: I_2 dissociates into 2I atoms.

Second step: I bonds to a carbon atom of cis-2-butene, breaking the carbon-carbon double bond.

Third step: The carbon-carbon bond, now a single bond, rotates freely.

Fourth step: The I atom dissociates from its carbon, re-creating the carbon-carbon double bond.

Fifth step: I_2 is regenerated from 2I atoms.

Given that the uncatalyzed reaction occurs in a single step with an activation energy of 262 kJ/mol and that the third step of the catalyzed reaction has the highest activation energy (115 kJ) of the steps in the catalyzed mechanism, sketch the potential energy diagram for the catalyzed and uncatalyzed reactions.

5. Phenol acetate reacts with water to form acetic acid and phenol. This reaction is carried out, and the concentration of phenol acetate is monitored over time:

[phenol acetate] (M)	Time (s)
0.55	0
0.42	15
0.31	30
0.23	45
0.17	60
0.12	75
0.082	90

Graph the data in the following three ways:

1. [phenol acetate] versus time
2. 1/[phenol acetate] versus time
3. ln[phenol acetate] versus time

Use these plots to determine the order of the reaction with respect to the concentration of phenol acetate.

6. Consider an exothermic reaction with an activation energy of 125 kJ/mol and a frequency factor of $4.0 \times 10^{13} \, s^{-1}$.

a. What is the rate constant of the reaction at 25 °C?

b. What will be the rate constant at 25 °C if a catalyst is used that reduces the activation energy to 75 kJ/mol?

c. At what temperature would the uncatalyzed reaction have to be carried out in order for the rate constant to be equal to that of the catalyzed reaction at 25 °C?

7. The rate of bacterial hydrolysis of fish muscle is twice as great at 2.2 °C as it is at −1.1 °C.

a. Estimate the value of the activation barrier assuming that the frequency factor is temperature-independent.

b. What is the relationship between the sensitivity of fish to storage conditions and the answer that you calculated in part a?

c. What would the activation barrier at −1.1 °C have to be in order to affect the same change in the rate as the temperature increase to 2.2 °C does?

8. Radon is a radioactive noble gas that is sometimes found in basements of homes. Radon-222 decays according to first-order kinetics and has a half-life of 3.82 days. If 1.0 L of air contains 2.5×10^{13} atoms of radon, how many atoms of radon-222 will remain in 1.0 L of air in your basement after 30 days?

9. The thermal decomposition of phosphine into hydrogen and phosphorus is a first-order reaction with a half-life of 35 s at 680 °C.

$$4PH_3(g) \rightarrow P_4(g) + 6H_2(g)$$

a. How much time, in minutes, is required for 90% of the sample to decompose?

b. Do you expect that this reaction occurs in a single elementary step?

10. The isomerization of cyclopropane, C_3H_6, to propene, $CH_2=CHCH_3$, is believed to occur via formation of an excited cyclcopropane molecule as shown in the equation:

$$C_3H_6 + C_3H_6 \rightarrow C_3H_6 + C_3H_6{}^* \qquad \text{slow}$$

$$C_3H_6{}^* \rightarrow CH_2=CHCH_3 \qquad \text{fast}$$

a. What is the rate law for the reaction using this mechanism?

b. Identify any intermediates in the reaction.

c. This reaction has a rate constant of 5.85×10^{-4} $M^{-1}s^{-1}$ at 485 °C. If you start with 1.05 M of cyclopropane, what fraction will have reacted after 2.0 hrs?

d. How will increasing the concentration of cyclopropane affect the value of the rate constant?

11. The decomposition of N_2O to N_2 and O_2 is a first-order reaction with a half-life of 3580 min at 730 °C. If the initial pressure of N_2O is 2.1 atm at 730 °C, calculate the total gas pressure inside the vessel after one half-life. Assume that the total volume and temperature remain constant.

Concept Questions:

1. A successful chemical reaction requires that reacting species collide, that they collide with the correct geometry, and that the reacting species have sufficient energy to overcome the reaction barrier. Discuss how temperature affects each of these criteria.

2. A series of three experiments was carried out on a single reaction at two different temperatures. The results of the experiments are plotted below.

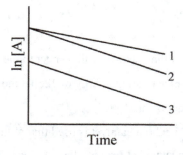

Time

Explain how the experimental conditions of 1, 2, and 3 are different and how you can tell.

3. A glow stick can be "saved" by putting it in the freezer. How does this "save" it?

4. The reaction order can be an integer or fraction, a positive or negative number. The molecularity, however, must be a positive integer. Explain why this difference exists.

5. Consider the combustion of ethane (C_2H_6):

$$2C_2H_6(g) + 7O_2(g) \rightarrow 4CO_2(g) + 6H_2O(g)$$

Do you think that this reaction occurs in a single elementary step or multiple steps? Explain the reasoning behind your answer in 2–3 sentences.

6. In this chapter, we learned that increasing the temperature increases the rate of a chemical reaction. Some reactions are reversible; that is, they go in the forward and in the reverse direction. If heating up the reaction always increases the rate, then we should expect that both the forward and reverse reactions are sped up if the reaction is reversible. Using an Arrhenius plot, predict which is more sensitive to temperature, the forward or reverse reaction, for a reaction that is exothermic in the forward direction.

7. Termolecular reactions usually involve the association or combination of two particles with the help of a third, as shown in the reaction below.

$$2A + M \rightarrow B + M$$

In this reaction, M is a particle whose role is to remove the excess energy that is produced upon the bond formation that results in B. M is not a spectator since its presence is necessary in order for the reaction to occur. Using the information you have learned in this chapter, explain why it is reasonable that most termolecular reactions are of this type.

8. Three mechanisms are proposed for the reaction between carbon monoxide and nitrogen dioxide:

   Mechanism 1:　　　　$NO_2 + CO \rightarrow CO_2 + NO$

   Mechanism 2:　　　　$NO_2 + NO_2 \rightarrow NO + NO_3$　　　　slow
   　　　　　　　　　　$NO_3 + CO \rightarrow NO_2 + CO_2$　　　　fast

   Mechanism 3:　　　　$NO_2 + NO_2 \leftrightarrows NO + NO_3$　　　　fast equilibrium
   　　　　　　　　　　$NO_3 + CO \rightarrow NO_2 + CO_2$　　　　slow

   Suggest an experiment that would allow you to differentiate between the three mechanisms.

9. Rate laws can be written directly from the rate-determining elementary step. We can understand this by considering the concentration as being directly related to the number of collisions between reacting species. For example, the rate law: rate = $k[A][B]$ suggests that the rate increases as the number of A and B molecules increase due to the increase in the number of collisions. Use the same type of logic to explain the exponent in the rate law: rate = $k[A]^2$.

10. An important aspect of catalysts is that they are not consumed in the reaction that they catalyze. For example, a single enzyme can be used to catalyze millions of reactions. Suggest why this might be a biologically important aspect to enzyme kinetics.

11. Substance X is found to be carcinogenic and have a half-life of 1.4 days. Considering the relationship between half-life and reaction order, after 1 week will substance X have a higher concentration if it decays via zero-, first-, or second-order kinetics?

Chapter 15: Chemical Equilibrium

Learning Goals:

- Express Equilibrium Constants for Chemical Equations

- Manipulate the Equilibrium Constant to Reflect Changes in the Chemical Equation

- Relate K_p and K_c

- Write Equilibrium Expressions for Reactions Involving a Solid or a Liquid

- Find Equilibrium Constants from Experimental Concentration Measurements

- Predict the Direction of a Reaction by Comparing Q and K

- Calculate Equilibrium Concentrations from the Equilibrium Constant and One or More Equilibrium Concentrations

- Find Equilibrium Concentrations from Initial Concentrations and the Equilibrium Constant

- Calculate Equilibrium Partial Pressures from the Equilibrium Constant and Initial Partial Pressures

- Find Equilibrium Concentrations from Initial Concentrations in Cases with a Small Equilibrium Constant

- Determine the Effect of a Concentration Change on Equilibrium

- Determine the Effect of a Volume Change on Equilibrium

- Determine the Effect of a Temperature Change on Equilibrium

Chapter Summary:

Up until this point in the textbook, we have mostly written chemical reactions as proceeding in a single direction, from reactants to products. In this chapter, you will learn that chemistry is not generally that simple. Reactions often proceed in both the forward and reverse directions; when the rates of the forward and reverse reactions are equal, a dynamic equilibrium is established where the concentrations of all chemical species no longer appear to change. The position of the equilibrium is quantified by the equilibrium constant, which is based on the product/reactant ratio of equilibrium concentrations and pressures. You will learn how to understand the meaning of the equilibrium constant qualitatively and how to calculate its value quantitatively. By comparing nonequilibrium conditions to equilibrium conditions, you will learn to predict the direction in which a chemical reaction will proceed in order to establish equilibrium. Given the value of the equilibrium constant, you will be able to calculate equilibrium concentrations for all species using exact and approximate methods. Finally, you will learn to utilize Le Châtelier's principle to understand and predict the ways that equilibrium positions change when systems are pushed away from equilibrium.

Chapter Outline:

1. Fetal Hemoglobin and Equilibrium

    a. The concentrations of reactants and products are described by an equilibrium constant.

2. The Concept of Dynamic Equilibrium

    a. Reversible reactions are reactions that proceed in both the forward and reverse directions. Two oppositely facing arrows are used to denote reversible reactions:

    $$\text{Reactants} \leftrightarrows \text{Products}$$

    b. A dynamic equilibrium for a chemical reaction is the condition at which the rates of the forward and reverse reactions are equal.

i.   The concentrations of reactants and products do not change once equilibrium is established, even though both forward and reverse reactions continue to occur.

3.   The Equilibrium Constant (K)

   a.   At equilibrium, it is not the concentrations of species but rather the rates of the forward and reverse reactions that are equal.

   b.   The equilibrium constant (K) quantifies equilibrium product and reactant concentrations.

   c.   The equilibrium constant (K) is the ratio of product concentrations (in units of molarity) raised to their stoichiometric coefficients divided by the reactant concentrations (in units of molarity) raised to their stoichiometric coefficients. K is unitless.

   d.   The law of mass action gives the relationship between the balanced chemical reaction and the equilibrium constant:

$$a\text{A} + b\text{B} \rightleftharpoons c\text{C} + d\text{D}$$

$$K = \frac{[C]^c[D]^d}{[A]^a[B]^b}$$

▶   Each reaction has its own characteristic equilibrium constant that is determined experimentally and varies only with temperature. Further, the value of K does not depend on the initial composition of the reaction mixture.

EXAMPLE:

Write equilibrium constants for the following reactions:

a. $N_2O_4(g) \rightleftharpoons 2NO_2(g)$

$$K = \frac{[NO_2]^2}{[N_2O_4]}$$

b. $PCl_5(g) \rightleftharpoons PCl_3(g) + Cl_2(g)$

$$K = \frac{[PCl_3][Cl_2]}{[PCl_5]}$$

c. $CH_4(g) + 2O_2(g) \rightleftharpoons 2H_2O(g) + CO_2(g)$

$$K = \frac{[H_2O]^2[CO_2]}{[CH_4][O_2]^2}$$

d. $CH_3CO_2H(aq) + NH_3(aq) \rightleftharpoons NH_4^+(aq) + CH_3CO_2^-(aq)$

$$K = \frac{[CH_3CO_2^-][NH_4^+]}{[CH_3CO_2H][NH_3]}$$

   e.   There are three regions in which K can be qualitatively interpreted:

      i.   When $K \gg 1$, the reaction favors product formation at equilibrium.

      ii.   When $K \ll 1$, the reaction favors reactant formation at equilibrium.

iii. When $K \approx 1$, neither the product nor the reactant formation is favored at equilibrium and the reaction proceeds about half way.

▶ Note that an equilibrium constant of 1 does not mean that the concentrations of reactants and products are equal; it only means that the ratio of concentrations raised to their proper coefficients is equal to one.

EXAMPLE:

Determine if each of the following reactions favor the formation of products, reactants, or neither.

a. $2H_2O(l) \leftrightarrows H_3O^+(aq) + OH^-(aq)$ $K = 1 \times 10^{-14}$ at 25 °C

We see here that K is a number that is much smaller than 1, so this is a reactant-favored reaction.

b. $N_2(g) + O_2(g) \leftrightarrows 2NO(g)$ $K = 5.0 \times 10^{-4}$ at 1627 °C

We see again that K is a number that is much less than 1, so this is a reactant-favored reaction.

c. $ClNO_2(g) + NO(g) \leftrightarrows NO_2(g) + ClNO(g)$ $K = 1.3 \times 10^4$ at 25 °C

In this case, K is a number much larger than 1, so this is a product-favored reaction.

---

f. When manipulating a chemical equation, the value of the equilibrium constant changes in predictable ways:

i. When the reaction is reversed, the new equilibrium constant is the inverse of the original equilibrium constant.

ii. When the reaction is multiplied by a factor, the new equilibrium constant is the original equilibrium constant raised to the power of the factor.

iii. When reactions are added together to give an overall reaction, the equilibrium constant of the overall reaction is the product of the equilibrium constants for each step in the reaction.

EXAMPLE:

Determine the equilibrium constant for the reaction $HCl(g) \leftrightarrows \frac{1}{2}Cl_2(g) + \frac{1}{2}H_2(g)$ given that the equilibrium constant for the reaction $Cl_2(g) + H_2(g) \leftrightarrows 2HCl(g)$ is 4×10^{31} at 300 K.

We first must reverse the reaction and take the inverse of the equilibrium constant:

$$2HCl(g) \leftrightarrows Cl_2(g) + H_2(g) \qquad K = \frac{1}{4 \times 10^{31}} = 2.5 \times 10^{-32}$$

Now we need to multiply the stoichiometric coefficients by ½ and raise K to the ½ power:

$$HCl(g) \leftrightarrows \frac{1}{2}Cl_2(g) + \frac{1}{2}H_2(g) \qquad K = (2.5 \times 10^{-32})^{\frac{1}{2}} = 2 \times 10^{-16}$$

---

4. Expressing the Equilibrium Constant in Terms of Pressure

a. Up until now, we have expressed the equilibrium constant in terms of concentration. When the equilibrium constant is expressed in these terms, it is called K_c.

b. For a gas-phase reaction, we can express the equilibrium constant in terms of partial pressures and will give the equilibrium constant the symbol K_p.

$$a\text{A}(g) + b\text{B}(g) \rightleftharpoons c\text{C}(g) + d\text{D}(g)$$

$$K_p = \frac{P_C^c P_D^d}{P_A^a P_B^b}$$

EXAMPLE:

Write equilibrium constants in terms of pressure for the following reactions:

a. $N_2O_4(g) \rightleftharpoons 2NO_2(g)$

$$K_p = \frac{P_{NO_2}^2}{P_{N_2O_4}}$$

b. $PCl_5(g) \rightleftharpoons PCl_3(g) + Cl_2(g)$

$$K_p = \frac{P_{PCl_3} P_{Cl_2}}{P_{PCl_5}}$$

c. $CH_4(g) + 2O_2(g) \rightleftharpoons 2H_2O(g) + CO_2(g)$

$$K_p = \frac{P_{H_2O}^2 P_{CO_2}}{P_{CH_4} P_{O_2}^2}$$

---

c. K_c and K_p are equal only when there is no change in the number of gas-phase particles during the reaction. When the number of gas-phase particles changes by Δn, K_p, and K_c are related through the expression:

$$K_p = K_c (RT)^{\Delta n}$$

▶ This relationship can be derived using the ideal gas law ($P = (n/V)RT$).

EXAMPLE:

Determine K_p for the following reactions:

a. $PCl_3(g) + Cl_2(g) \rightleftharpoons PCl_5(g)$ $\qquad\qquad$ $K_c = 98.3$ at 400 K

There is 1 mole of gas on the product side of the reaction and 2 moles of gas on the reactant side, so the change in moles of gas, Δn, is -1. We can use this in the expression for K_p:

$$K_p = 98.30(0.0821 \times 400)^{-1}$$

$$K_p = 2.99 \text{ (at 400 K)}$$

b. $C_2H_6(g) + Cl_2(g) \rightleftharpoons C_2H_5Cl(g) + HCl(g)$ $\qquad$ $K_c = 0.10$ at 283 K

There are 2 moles of gas on the product side of the reaction and 2 moles of gas on the reactant side, so the change in moles of gas, Δn, is 0. We can use this in the expression for K_p:

$$K_p = 0.10(0.0821 \times 283)^0$$

$$K_p = 0.10 \text{ (at 283 K)}$$

We see that in this case $K_p = K_c$ because the number of moles of gas didn't change in the course of the reaction.

c. $N_2O_4(g) \leftrightarrows 2NO_2(g)$ $\qquad\qquad$ $K_c = 11$ at 373 K

There are 2 moles of gas on the product side of the reaction and 1 mole of gas on the reactant side, so the change in moles of gas, Δn, is 1. We can use this in the expression for K_p:

$$K_p = 11(0.0821 \times 373)^1$$

$K_p = 340$ (at 373 K)

d. $SO_2(g) + \frac{1}{2}O_2(g) \leftrightarrows SO_3(g)$ $\qquad\qquad$ $K_c = 2.2$ at 700 K

There is 1 mole of gas on the product side of the reaction and 1.5 moles of gas on the reactant side, so the change in moles of gas, Δn, is $-\frac{1}{2}$. We can use this in the expression for K_p:

$$K_p = 2.2(0.0821 \times 700)^{-1/2}$$

$K_p = 0.29$ (at 700 K)

---

d. The units of K are generally dropped because they are variable; units of molarity are assumed for the terms in K_c, and units of atm are assumed for the terms in K_p.

5. Heterogeneous Equilibria: Reactions Involving Solids and Liquids

   a. Concentrations of pure solids and liquids do not change in the course of a reaction as long as there is always some of the pure liquid and/or solid present.

   b. Pure liquids and solids are not included in the equilibrium constant expression.

      i. Since the concentrations of pure solids and liquids do not change in the course of a reaction, they are not incorporated into the value of K_c or K_p.

EXAMPLE:

Write equilibrium constants for the following reactions:

a. $CO_2(g) + H_2(g) \leftrightarrows CO(g) + H_2O(l)$

$$K_c = \frac{[CO]}{[CO_2][H_2]}$$

b. $AgCl(s) \leftrightarrows Ag^+(aq) + Cl^-(aq)$

$$K_c = [Ag^+][Cl^-]$$

c. $CuO(s) + H_2(g) \leftrightarrows Cu(s) + H_2O(g)$

$$K_c = \frac{[H_2O]}{[H_2]}$$

---

6. Calculating the Equilibrium Constant from Measured Equilibrium Concentrations

   a. The equilibrium constant can be calculated by substituting equilibrium concentrations into the equilibrium constant expression for a reaction and solving for K.

   b. The value of the equilibrium constant is always the same for a given temperature, but the equilibrium concentration of reactant and product species can vary (as a result of initial concentrations).

EXAMPLE:

Calculate the equilibrium constant, K_c, at 100 °C for the reaction $N_2O_4(g) \leftrightarrows 2NO_2(g)$ given the following equilibrium concentrations:

a. $[N_2O_4] = 1.45$ M, $[NO_2] = 3.99$ M

First, we will write down the equilibrium constant expression for the reaction:

$$K = \frac{[NO_2]^2}{[N_2O_4]}$$

We can now compute K by using the equilibrium concentrations given in the problem:

$$K = \frac{3.99^2}{1.45} = 11.0$$

b. $[N_2O_4] = 1.5 \times 10^{-4}$ M, $[NO_2] = 4.1 \times 10^{-2}$ M

Since we already have the equilibrium constant expression, we can simply use the new equilibrium values to solve for K:

$$K = \frac{(4.1 \times 10^{-2})^2}{(1.5 \times 10^{-4})} = 11.0$$

c. $[N_2O_4] = 8.95 \times 10^8$ M, $[NO_2] = 9.92 \times 10^4$ M

$$K = \frac{(9.92 \times 10^{-4})^2}{8.95 \times 10^{-8}} = 11.0$$

d. $[N_2O_4] = 3.86$ M, $[NO_2] = 0.652$ M

$$K = \frac{(0.652)^2}{3.86} = 0.110$$

We see from these four examples that the equilibrium constant is always the same value when the equilibrium concentrations are measured at 100 °C for this reaction.

EXAMPLE:

The following reaction is carried out with initial concentrations of $[PCl_5] = 4.580$ M, $[Cl_2] = 5.870$ M, and $[PCl_3] = 1.280$ M.

$$PCl_3(g) + Cl_2(g) \leftrightarrows PCl_5(g)$$

Given that the equilibrium concentration of PCl_5 is 5.847 M, calculate K_c.

First, we see that the concentration of product, $PCl_5(g)$, has increased from 4.580 M to 5.847 M. So the reaction has shifted from left to right in order to reach equilibrium.

We can use a table of the concentrations to solve this problem:

Reaction Condition	$[PCl_3]$ (M)	$[Cl_2]$ (M)	$[PCl_5]$ (M)
Initial	1.280	5.870	4.580
Change	$-x$	$-x$	$+x$
Equilibrium	$1.280 - x$	$5.870 - x$	$4.580 + x$

The number of moles that react is x (which can be applied here because we assume that the volume is constant). x comes from the reaction stoichiometry: for every molecule of PCl_3 that reacts, one molecule of Cl_2 will react and one molecule of PCl_5 will be produced.

We are given that the equilibrium concentration of PCl_5 is 5.847 M, so we can solve for x:

$$4.580 + x = 5.847 \quad \rightarrow \quad x = 1.267$$

From this we can calculate the equilibrium concentrations of PCl_5 and Cl_2:

$$[PCl_3] = 1.280 - x = 1.280 - 1.267 = 0.013 \text{ M}$$

$$[Cl_2] = 5.870 - x = 5.870 - 1.267 = 4.603 \text{ M}$$

Now that we have the equilibrium concentrations, we can substitute them into the equilibrium constant expression to solve for K_c:

$$K_c = \frac{[PCl_5]}{[PCl_3][Cl_2]}$$

$$K_c = \frac{5.87}{0.013 \times 4.603}$$

$$K_c = 98$$

---

7. The Reaction Quotient: Predicting the Direction of Change

    a. The expression for the reaction quotient, Q, is of the same form as the equilibrium constant, K, but incorporates nonequilibrium concentrations and can be used to predict the direction in which a chemical reaction proceeds:

        i. If $Q < K$, the reaction will proceed toward products (from left to right) until equilibrium is established.

        ii. If $Q > K$, the reaction will proceed toward reactants (from right to left) until equilibrium is established.

        iii. If $Q = K$, the reaction is at equilibrium.

EXAMPLE:

Consider the reaction between sulfur dioxide and nitrogen dioxide:

$$SO_2(g) + NO_2(g) \rightleftharpoons SO_3(g) + NO(g) \qquad K_c = 8.8 \text{ at } 1000 \text{ K}$$

Given the following concentrations, predict the direction in which the reaction will proceed:

a. $[SO_2] = 0.640$ M, $[NO_2] = 0.124$, $[SO_3] = 0.185$ M, $[NO] = 0.875$ M

We will start by writing the expression for Q and then inserting the concentration values into it:

$$Q = \frac{[SO_3][NO]}{[SO_2][NO_2]}$$

$$Q = \frac{(0.185)(0.875)}{(0.640)(0.124)}$$

$$Q = 2.04$$

We see that $Q_c < K_c$, so the reaction will proceed toward products until equilibrium is reached.

b. $[SO_2] = 0.0173$ M, $[NO_2] = 0.581$, $[SO_3] = 4.89$ M, $[NO] = 1.582$ M

We will use the same expression for Q and calculate using the revised concentrations:

$$Q = \frac{4.89 \times 1.582}{0.0173 \times 0.581} = 770$$

In this case, $Q_c > K_c$ so the reaction will proceed in the reverse direction toward reactants until equilibrium is reached.

c. $P_{SO_2} = 4.87$ atm, $P_{NO_2} = 4.87$ atm, $P_{SO_3} = 10.4$ atm, $P_{NO} = 10.4$ atm

We need to rewrite Q in terms of partial pressures and calculate K_p:

$$Q = \frac{P_{SO_3} P_{NO_3}}{P_{SO_2} P_{NO_3}}$$

$$K_p = K_c (RT)^{\Delta n} \quad \rightarrow \quad K_p = K_c (RT)^0 \quad \rightarrow \quad K_p = K_c$$

Now we can use our expression for Q as before:

$$Q = \frac{10.4 \times 10.4}{4.87 \times 4.87} = 4.56$$

Since $Q_p < K_p$, the reaction will proceed toward products until equilibrium is established.

---

8.   Finding Equilibrium Concentrations
   a.   Given the equilibrium constant and all but one of the equilibrium concentrations, the final equilibrium concentration can be easily identified.

EXAMPLE:

The equilibrium constant for the reaction $N_2O_4(g) \leftrightarrows 2NO_2(g)$ is 11.0 at 373 K. Calculate the equilibrium concentration of NO_2 given that the equilibrium concentration of N_2O_4 is 4.87 M.

We use the equilibrium constant expression and rearrange it to solve for the unknown variable:

$$K = \frac{[NO_2]^2}{[N_2O_4]} \quad \rightarrow \quad K[N_2O_4] = [NO_2]^2 \quad \rightarrow \quad \sqrt{K[N_2O_4]} = [NO_2]$$

And now we use the values given to solve for $[NO_2]$ at equilibrium:

$$[NO_2] = \sqrt{11.0 \times 4.87} = 7.32 \text{ M}$$

---

   b.   Given the equilibrium constant and the initial concentrations or pressures, an ICE table can be used to determine equilibrium concentrations, using the variable x to represent the change in concentration required to reach equilibrium.

EXAMPLE:

The production of nitrogen monoxide gas from nitrogen and oxygen gases is reactant-favored at 2500 K:

$$N_2(g) + O_2(g) \leftrightarrows 2NO(g) \qquad\qquad K_c = 5.0 \times 10^{-4} \text{ at 2500 K}$$

If you set up this reaction with a 2.5 M sample of $N_2(g)$ and a 2.5 M sample of $O_2(g)$, what will be the concentration of $NO(g)$ when the reaction reaches equilibrium?

We will set up an ICE table for this reaction using the information given and the reaction stoichiometry:

Reaction Condition	$[N_2]$ (M)	$[O_2]$ (M)	$[NO]$ (M)
Initial	2.5	2.5	0
Change	$-x$	$-x$	$+2x$
Equilibrium	$2.5 - x$	$2.5 - x$	$2x$

We see in the table that for each mole of nitrogen gas that reacts, 1 mole of oxygen gas will react and 2 moles of nitrogen monoxide gas will be produced. Further, since only $N_2(g)$ and $O_2(g)$ are present initially, the reaction must shift to the right to reach equilibrium.

Now we will write the equilibrium constant expression for the reaction and then insert the equilibrium concentrations from the ICE table into it:

$$K = \frac{[NO_2]^2}{[N_2O_4]}$$

$$5.0 \times 10^{-4} = \frac{(2x)^2}{(2.5-x)(2.5-x)}$$

$$5.0 \times 10^{-4} = \frac{(2x)^2}{(2.5-x)^2}$$

We can take the square root of both sides of this equation in order to simplify the expression:

$$\sqrt{5.0 \times 10^{-4}} = \frac{(2x)}{(2.5-x)}$$

We will now rearrange the expression and solve for x:

$$0.055902 - 0.022361x = 2x$$
$$0.055902 - 2.022361 = x$$
$$x = 0.02764$$

Since the concentration of NO at equilibrium is equal to $2x$, $[NO] = 0.055$ M.

EXAMPLE:

CO_2 and H_2 are placed in a closed container and react at a constant temperature according to the equation below. Calculate the equilibrium concentrations of all species when the initial pressures are $P_{CO_2} = 2.00$ atm and $P_{H_2} = 1.00$ atm.

$$CO_2(g) + H_2(g) \leftrightarrows CO(g) + H_2O(g) \qquad K_p = 0.64 \text{ at } 900 \text{ K}$$

We will set up our ICE table using partial pressures and express changes in terms of pressure. This is reasonable since the pressure is directly proportional to the number of moles when the volume and temperature are held constant.

Reaction Condition	$[CO_2]$ (atm)	$[H_2]$ (atm)	$[CO]$ (atm)	$[H_2O]$ (atm)
Initial	2.00	1.00	0	0
Change	$-x$	$-x$	$+x$	$+x$
Equilibrium	$2.00 - x$	$1.00 - x$	x	x

We can now write the equilibrium constant expression and then use the partial pressures of all species at equilibrium in order to solve for x:

$$K_p = \frac{P_{CO} P_{H_2O}}{P_{CO_2} P_{H_2}}$$

$$0.64 = \frac{(x)(x)}{(2.00 - x)(1.00 - x)}$$

We will now multiply out the terms in the denominator to get all values in the numerator of the equation:

$$0.64 = \frac{x^2}{(2.00 - 3.00x + x^2)}$$

$$1.28 - 1.92x + 0.64x^2 = x^2$$

Next, we will put our expression in the form needed to use the quadratic formula:

$$1.28 - 1.92x - 0.36x^2 = 0$$

$$0.36x^2 + 1.92x - 1.28 = 0$$

We have $a = 0.36$, $b = 1.92$, and $c = -1.28$ for the quadratic formula:

$$x = \frac{-1.92 \pm \sqrt{(1.92)^2 - 4(0.36)(-1.28)}}{2(0.36)}$$

$$x = 0.60 \text{ or } x = -5.9$$

The solution $x = -5.9$ atm is physically unreasonable since it is negative, and the product gases cannot have negative pressures. The answer is, therefore, $x = 0.60$ atm. Now we need to put this back into our expressions for the equilibrium partial pressures to answer the question asked:

$$P_{CO_2} = 2.00 - 0.60 = 1.40 \text{ atm}$$

$$P_{H_2} = 1.00 - 0.60 = 0.40 \text{ atm}$$

$$P_{CO} = 0.60 \text{ atm}$$

$$P_{H_2O} = 0.60 \text{ atm}$$

To check the answer, plug these pressures into the equilibrium ratio: $(0.60 \times 0.60)/(1.40 \times 0.40) = 0.64$.

c. When the equilibrium constant is very small, the reaction will not proceed very far toward products, and an approximation can be used to determine the equilibrium concentrations.

   i. The approximation is valid only when <5% of the original reactant concentration reacts.

   ii. Throughout the remainder of this book, this approximation will be referred to as the "x is small" approximation.

   ▶ In cases where the equilibrium constant is 100 times smaller than the initial concentration of the reactant, this approximation is generally valid.

EXAMPLE:

Acetic acid partially dissociates in water according to the equation:

$$HC_2H_3O_2(aq) + H_2O(l) \rightleftharpoons C_2H_3O_2^-(aq) + H_3O^+(aq) \quad K = 1.8 \times 10^{-5} \text{ at } 25\ °C$$

If you begin with a 0.476 M solution of acetic acid, what will be the concentrations of all species when the solution reaches equilibrium?

We will set up an ICE table for the reaction. In the ICE table, we will ignore H_2O since pure liquids do not appear in the equilibrium constant expression:

Reaction Condition	$[HC_2H_3O_2]$ (M)	$[H_2O]$ (M)	$[C_2H_3O_2^-]$ (M)	$[H_3O^+]$ (M)
Initial	0.476	–	0	0
Change	$-x$	–	$+x$	$+x$
Equilibrium	$0.476 - x$	–	x	x

We can now write the equilibrium constant expression and then use the concentrations of all species at equilibrium in order to solve for x:

$$K = \frac{[C_2H_3O_2^-][H_3O^+]}{[HC_2H_3O_2]}$$

$$1.8 \times 10^{-5} = \frac{(x)(x)}{(0.476 - x)}$$

Here we could rearrange and solve the quadratic formula. However, since K is so small in this case, we know that the reaction will not go very far to form products; we make the approximation that $0.476 - x$ will be about 0.476 and solve the much simpler expression:

$$1.8 \times 10^{-5} = \frac{x^2}{0.476}$$

$$8.568 \times 10^{-6} = x^2$$

$$x = 0.00293 \text{ M}$$

In order to determine if our approximation is valid, we will calculate the percent of the original concentration that has reacted:

$$\frac{0.00293}{0.476} \times 100\% = 0.615\%$$

Since this is much less than 5%, our approximation is valid and we can calculate the equilibrium concentrations of all species:

$$[HC_2H_3O_2] = 0.473 \text{ M} - x = 0.473 \text{ M} - 0.00293 \text{ M} = 0.47 \text{ M}$$

$$[C_2H_3O_2^-] = x = 0.0029 \text{ M}$$

$$[H_3O^+] = x = 0.0029 \text{ M}$$

---

    iii.   When >5% of the original reactant concentration reacts, the method of successive approximation can be used.

EXAMPLE:

Calculate the percentage of phosphorous pentachloride that dissociates when 0.05 mol of PCl_5 is placed in a closed container at 250 °C and 2.00 atm pressure. Note that the volume of the container does not change during the course of the reaction.

$$PCl_5(g) \rightleftharpoons PCl_3(g) + Cl_2(g) \qquad K_p = 1.78 \text{ at } 250 \text{ °C}$$

Since the volume of the container is not permitted to change, the pressure in the container will change over time. As the given quantity is in moles, it will be simpler to use the K_c expression, so we need to calculate the initial concentration of PCl_5 and the value of K_c at 250 °C:

$$\frac{n}{V} = \frac{P}{RT} \quad \rightarrow \quad \frac{n}{V} = \frac{2.00 \text{ atm}}{0.0821 \dfrac{\text{L} \cdot \text{atm}}{\text{mol} \cdot \text{K}} \cdot 523 \text{ K}} \quad \rightarrow \quad \frac{n}{V} = 0.046578 \text{ M}$$

$$K_p = K_c(RT)^{\Delta n} \quad \rightarrow \quad \frac{K_p}{(RT)^{\Delta n}} = K_c \quad \rightarrow \quad K_c = \frac{1.78}{(0.0821 \cdot 523)^1} \quad \rightarrow \quad K_c = 0.041455$$

Now we can set up an ICE table for the reaction:

Reaction Condition	[PCl$_5$] (M)	[PCl$_3$] (M)	[Cl$_2$] (M)
Initial	0.04658	0	0
Change	$-x$	$+x$	$+x$
Equilibrium	$0.04658 - x$	x	x

Now we will insert these values into the equilibrium constant expression:

$$K = \frac{[PCl_3][Cl_2]}{[PCl_5]}$$

$$0.04155 = \frac{(x)(x)}{(0.04658 - x)}$$

We will approximate the value of $0.04658 - x$ to be 0.04658 and solve for x:

$$0.04155 = \frac{x^2}{(0.04658)}$$

$$0.0019354 = x^2$$

$$x = 0.04399$$

This is much larger than 5% of the initial PCl_5 concentration—it is actually 94.4%. So we cannot stop our calculation here. Instead, we will use this value to get a new approximation. Instead of letting $0.04658 - x = 0.04658$, we will insert our calculated value of x:

$$0.04155 = \frac{x^2}{(0.04658 - 0.04399)} \qquad \rightarrow \qquad x = 0.01037$$

We will again substitute this value into the denominator of our equilibrium constant expression because it is very different from our previous result:

$$0.04155 = \frac{x^2}{(0.04658 - 0.01037)} \qquad \rightarrow \qquad x = 0.03879$$

Substituting in with our new value of x gives:

$$0.04155 = \frac{x^2}{(0.04658 - 0.03873)} \qquad \rightarrow \qquad x = 0.01806$$

Substituting in with our new value of x gives:

$$0.04155 = \frac{x^2}{(0.04658 - 0.01806)} \qquad \rightarrow \qquad x = 0.03442$$

We will continue to substitute in the new value of x until our answers converge to a single value. The numerous substitutions are not shown here – only the beginning and end of the process are shown.

Substituting in with our new value of x gives:

$$0.04155 = \frac{x^2}{(0.04658 - 0.02789)} \qquad \rightarrow \qquad x = 0.02787$$

Substituting in with our new value of x gives:

$$0.04155 = \frac{x^2}{(0.04658 - 0.02787)} \qquad \rightarrow \qquad x = 0.02788$$

So we can now calculate the equilibrium concentration values:

$$[PCl_5] = 0.0187 \text{ M}$$

$$[PCl_3] = 0.0279 \text{ M}$$

$$[Cl_2] = 0.0279 \text{ M}$$

The % dissociation of $PCl_5 = x/[PCl_5]_{initial} \times 100 = 0.0279/0.04658 \times 100 = 59.9\%$

Note that in this problem, the quadratic formula would have been a much simpler (and faster) way to solve for x. A good rule of thumb is that the method of successive approximations is more efficient only when the percent dissociation calculated with the first approximation is around 10% or less.

---

9. Le Châtelier's Principle: How a System at Equilibrium Responds to Disturbances

   a. Le Châtelier's principle states that when a chemical system at equilibrium is disturbed, the system shifts in a direction that minimizes the disturbance.

      i. The system will find a new equilibrium position after a disturbance.

   b. Concentration changes affect the equilibrium position in a manner predicted by the reaction quotient, Q.

      i. An increase in the reactant concentrations results in a shift toward products.

      ii. An increase in the product concentrations results in a shift toward reactants.

      iii. A decrease in the reactant concentrations results in a shift toward reactants.

      iv. A decrease in the product concentrations results in a shift toward products.

EXAMPLE:

Aqueous iron(III) ions are yellow and form a red, complex ion ($Fe(SCN)^{2+}$) when combined with the colorless thiocynate ion (SCN^-):

$$Fe^{3+}(aq) + SCN^-(aq) \leftrightarrows Fe(SCN)^{2+}(aq)$$

How will the color of the solution change when the following disturbances are imposed on the equilibrium system?

a. A solution of $Fe(NO_3)_3$ is added to the reaction.

When $Fe(NO_3)_3$ is added to the reaction, the concentration of Fe^{3+} increases. Since Fe^{3+} is a reactant, the reaction will shift to the right (toward the formation of products) until equilibrium is reestablished. The formation of more products will result in a solution that is more red.

b. Solid KSCN is added to the solution.

When KSCN is added to the reaction, the concentration of SCN^- ions increases. Since SCN^- is a reactant, the reaction will shift toward the formation of products until equilibrium is reestablished. The formation of more products will result in a solution that is more red.

c. Fe^{3+} ions are removed from the flask.

Since $Fe^{3+}(aq)$ is a reactant, its removal will result in the formation of more reactants in order to reestablish equilibrium. The reduction in the product concentration will result in a solution that is less red.

---

c. The effect of a change in volume or pressure change on equilibrium depends on the number of gas-phase species on each side of the equation.

   i. A decrease in volume (or increase in pressure) will cause the equilibrium to shift toward the side of the reaction with fewer gas-phase particles.

   ii. An increase in volume (or decrease in pressure) will cause the equilibrium to shift toward the side of the reaction with more gas-phase particles.

   iii. If there is the same number of gas-phase particles on either side of the reaction, the equilibrium position will not change when the volume (or pressure) is changed.

   iv. When an inert gas is added to a reaction vessel, the partial pressures of all species remain unchanged and there is, therefore, no change in the equilibrium position.

   ▶ You can think of a change in volume as another example of a change in concentration since an increase in volume is actually a decrease in concentration. (From the [rearranged] ideal gas law, $P/RT = n/V$)

EXAMPLE:

When solid carbon is placed in a closed container with water vapor, hydrogen gas and carbon monoxide gas are produced according to the following equation:

$$C(s) + H_2O(g) \leftrightharpoons CO(g) + H_2(g)$$

The reaction is allowed to reach equilibrium. Predict how the concentration of water vapor will change when each of the following adjustments is made:

a. Carbon monoxide gas is added to the reaction vessel.

Since $CO(g)$ is a product in the reaction, its addition will cause the reaction to shift toward reactants, increasing the concentration of water vapor.

b. The volume of the container is doubled.

When the volume of the container is doubled, the pressure is cut in half. In order for the equilibrium to be reestablished, the pressure will increase through the formation of more gas molecules. Since the product side of the reaction has more gas particles, the reaction will shift to form more products until equilibrium is established and thereby reduce the concentration of water vapor.

c. 5.0 mol of argon gas is added to the container.

When an inert gas such as $Ar(g)$ is added to the container, the partial pressures of the gases remain unchanged. Since there is no change in gas partial pressures, the system is not moved from the equilibrium position and the concentration of water vapor will thus remain unchanged.

d. Solid carbon is added to the reaction vessel.

When solid carbon is added to the container, the equilibrium does not shift since the concentration of a solid is never changed (so long as there is some available). Consequently, the concentration of water vapor will remain unchanged.

e. Liquid water is added to the reaction vessel.

The addition of liquid water to the reaction vessel will not affect the equilibrium position since the vapor pressure of water is temperature dependent and will not change.

---

d. When the temperature of a reaction is changed, the equilibrium shifts *and K* changes.

   i. In order to predict the direction of change (and the impact on *K*), we can think of heat as either a reactant or product in the reaction.

      1. In an exothermic reaction, heat can be treated as a product. Thus, increasing the temperature will cause the reaction to shift toward the reactants, and decreasing the temperature will cause the reaction to shift toward the products.

$$a\text{A} + b\text{B} \leftrightharpoons c\text{C} + d\text{D} + \text{"heat"}$$

      2. In an endothermic reaction, heat can be thought of as a reactant. Thus, increasing the temperature will cause the reaction to shift toward the products, and decreasing the temperature will cause the reaction to shift toward the reactants.

$$\text{"heat"} + a\text{A} + b\text{B} \leftrightharpoons c\text{C} + d\text{D}$$

EXAMPLE:

The reaction of hydrogen gas with iodine gas to form hydrogen iodide is endothermic:

$$H_2(g) + I_2(g) \leftrightharpoons 2HI(g) \qquad \Delta H = +52 \text{ kJ/mol}$$

How will placing the reaction vessel in an ice bath change the concentration of HI at equilibrium?

Since the reaction is endothermic, we can consider heat to be a reactant in the reaction. If we place the reaction vessel in ice water, heat will flow from the reaction vessel into the surroundings, meaning that heat will be removed from the system. Removing heat (a reactant) will cause the reaction to shift toward the reactants until equilibrium is reestablished; the concentration of HI will therefore decrease.

Fill in the Blank:

1. Pure solids and liquids do not appear in the equilibrium constant expression because their concentrations _____.

2. K_p and K_c are equal to one another one when the _____ does not change in the course of a chemical reaction.

3. In an exothermic reaction, raising the temperature will cause the reaction to shift toward the _____.

4. The _____ is an expression similar to the equilibrium constant expression but may utilize nonequilibrium concentrations.

5. Dynamic equilibrium is the condition for a chemical reaction under which the rate of the forward reaction is _____ to the rate of the reverse reaction.

6. When *K* is much larger than 1, the reaction favors the formation of _____ at equilibrium.

7. When a chemical reaction is reversed, the value of the equilibrium constant is _____.

8. The _____ is a way to quantify the relationships between concentrations of reactants and products at equilibrium.

9. Le Châtelier's principle states that a system at equilibrium, when disturbed, will shift in the direction that _____ the disturbance.

10. A reaction that can proceed in both the forward and reverse directions is said to be _____.

Problems:

1. Use the reactions provided below to determine the value of the equilibrium constant, K_c, for the reaction of hydrosulfuric acid and water:

$$2H_2S(aq) + 3H_2O(l) \leftrightarrows 3H_3O^+(aq) + S^{2-}(aq) + HS^-(aq)$$

Reaction 1: $H_2S(aq) + H_2O(l) \leftrightarrows H_3O^+(aq) + HS^-(aq)$ $K_c = 1.1 \times 10^{-3}$

Reaction 2: $H_3O^+(aq) + S^{2-}(aq) \leftrightarrows HS^-(aq) + H_2O(l)$ $K_c = 1.0 \times 10^{19}$

2. At 500 °C the equilibrium constant for the formation of ammonia is measured:

$$N_2(g) + 3H_2(g) \leftrightarrows 2NH_3(g) \qquad K_p = 0.0610$$

If analysis of the concentrations of all species finds that 3.0 mol of N_2, 2.0 mol of H_2, and 0.50 mol of NH_3 are present in a 1.0-L reaction flask, is the reaction at equilibrium? If not, in which direction will the reaction proceed in order to reach equilibrium?

3. When PCl_5 is put into a closed 2.0-L container at 556 K, it decomposes according to the reaction:

$$PCl_5(g) \leftrightarrows Cl_2(g) + PCl_3(g) \qquad K_p = 4.96$$

   a. What is the value of K_c for this reaction?

   b. If the total pressure in the reaction vessel is 0.50 atm, what fraction of PCl_5 has decomposed at equilibrium?

   c. If the total pressure is 1.00 atm, what fraction of PCl_5 has decomposed at equilibrium?

   d. How do your answers in part b and c relate to Le Châtelier's principle?

4. A 200.0 g sample of NH_4HS is placed into a 5.0-L reaction vessel at 25 °C. Assume that the only reaction that takes place is:

$$NH_4HS(s) \leftrightarrows NH_3(g) + H_2S(g) \qquad K_p = 0.110$$

   a. What are the equilibrium partial pressures of NH_3 and H_2S?

   b. What percentage of NH_4HS has decomposed at this temperature?

   c. What would the equilibrium partial pressures of NH_3 and H_2S be if 0.500 mol of NH_3 is added to the flask after it has reached equilibrium?

5. Solid carbon will reaction with carbon dioxide gas at 1123 K:

$$C(s) + CO_2(g) \leftrightarrows 2CO(g) \qquad K_p = 0.065$$

   a. Does the reaction favor products, reactions, or neither?

   b. When the reaction comes to equilibrium in a 5.0 L flask at 1123 K, the total pressure in the flask is measured to be 1.00 atm. How much carbon dioxide (in mol) must have been placed in the reaction flask to start the reaction?

   c. 1.0 g C, 0.240 atm CO_2, and 0.053 atm CO are placed into the flask and it is sealed. In which direction will the reaction proceed in order to reach equilibrium?

   d. What is the value of K_c for the reaction at 1123 K?

e.   If the equilibrium constant for the decomposition of $COCl_2$ is:

$$COCl_2(g) \leftrightharpoons CO(g) + Cl_2(g) \quad K_p = 170$$

What is the equilibrium constant for the reaction below?

$$C(s) + CO_2(g) + Cl_2(g) \leftrightharpoons 2COCl_2$$

6.   The decomposition of ammonia into hydrogen and nitrogen gases is reactant favored at 900 K:

$$2NH_3(g) \leftrightharpoons N_2(g) + 3H_2(g) \quad K_p = 0.0076$$

   a.   Calculate the partial pressure of nitrogen gas produced when 0.025 atm of ammonia is allowed to decompose.

   b.   Calculate the partial pressure of nitrogen gas produced when 7.85 atm of ammonia is added to the equilibrium mixture.

   c.   Why is it appropriate to use the "x is small" approximation in part b but not in part a?

7.   The photosynthesis reaction is endothermic (the energy from the sun is used) and results in the creation of sugar:

$$6CO_2(g) + 6H_2O(l) \leftrightharpoons C_6H_{12}O_6(aq) + 6O_2(g) \quad \Delta H = +2802 \text{ kJ}$$

Suppose that this reaction is at equilibrium in a closed container. How will the equilibrium shift if the following changes are made?

   a.   The partial pressure of O_2 is increased.

   b.   The system is compressed.

   c.   The temperature is increased.

   d.   Helium gas is added to the container at constant pressure.

   e.   Helium gas is added to the container at constant volume.

   f.   Liquid H_2O is added.

   g.   The partial pressure of CO_2 is decreased.

8.   The Haber process is the reverse of the reaction in Problem 6:

$$N_2(g) + 3H_2(g) \leftrightharpoons 2NH_3(g) \quad \Delta H < 0$$

If the reaction is allowed to come to equilibrium, how will the equilibrium shift in response to the following changes?

   a.   The volume of the container is increased.

   b.   The temperature of the reaction vessel is increased.

   c.   The amount of gaseous molecular hydrogen is doubled.

   d.   A catalyst is added to the reaction flask.

9.   Silver bicarbonate dissolves in water according to the equilibrium equation:

$$AgHCO_3(s) + H_2O(l) \leftrightharpoons Ag^+(aq) + HCO_3^-(aq)$$

The bicarbonate ion can further dissociate:

$$HCO_3^-(aq) + H_2O(l) \leftrightharpoons H_3O^+(aq) + CO_3^{2-}(aq)$$

   a.   List all of the species that are present at equilibrium when $AgHCO_3$ is dissolved in water.

   b.   How would the addition of hydrochloric acid affect the amount of silver bicarbonate present in the reaction mixture?

   c.   Adding which of the following compounds would result in a decrease in the concentration of H_3O^+ in the equilibrium solution: NaOH, $NaHCO_3$, or NaCl?

Concept Questions:

1. In this chapter, we have learned that a chemical reaction that has reached completion may actually still be occurring—but in both directions. Since concentrations and reaction rates do not change after equilibrium is established, this can be hard to believe. Suggest an experiment that would allow us to confirm that reactions at equilibrium are, in fact, dynamic processes.

2. A reaction, A + B $\leftrightarrows$ C, is studied by a group of students in the laboratory. The first experiment shows that $K = 1$. It is suggested that no more experiments need to be done because the concentrations will always be equal. What is wrong with this suggestion?

3. The definition of equilibrium states that the rates of the forward and reverse reactions are equal at equilibrium. For the general reaction:

$$aA(aq) + bB(aq) \leftrightarrows cC(aq)$$

Show that $K = k_1/k_{-1}$ using the definition of equilibrium in terms of reaction rates. *Hint:* Start by assuming that this is a one-step reaction and that the rate constant for the forward reaction is k_1 and the rate of the reverse reaction is k_{-1}.

4. Le Châtelier's principle predicts how changes in reactant and product concentrations will cause the equilibrium position to shift. Explain how concentration changes affect reaction rates and use the expression $K = k_1/k_{-1}$ to verify your predictions using Le Châtelier's principle.

5. The discussion of Le Châtelier's principle in this chapter considers changes in temperature as if heat were a product or reactant in the reaction. In fact, the effects of temperature changes are more complicated than the effects of concentration changes because the value of the equilibrium constant is temperature dependent.

   Sketch a potential energy diagram for an exothermic reaction, and then use an Arrhenius plot to examine how temperature affects the rates of the forward and reverse reactions. Using this analysis and your expression for K from the previous problem, explain the way that temperature affects the position of the equilibrium for an exothermic reaction.

6. Analyze the effects of temperature changes on equilibrium positions using the Boltzmann distribution of molecular speeds. First, look at the fraction of reactant species that will have sufficient energy to overcome a reaction barrier at two different temperatures. Estimate the ratio of reactive species at the two temperatures:

$$\frac{\text{fraction of molecules with energy above } E_a \text{ at } T_1}{\text{fraction of molecules with energy above } E_a \text{ at } T_2}$$

Finally, determine how this ratio will change when the activation energy is changed.

Chapter 16: Acids and Bases

Key Learning Outcomes:

-   Identify Brønsted–Lowry Acids and Bases and Their Conjugates

-   Determine Relative Acid Strength from Molecular Structure

-   Use K_w in Calculations

-   Calculate pH from $[H_3O^+]$ or $[OH^-]$

-   Find the pH of a Weak Acid Solution

-   Find the Acid Ionization Constant from pH

-   Find the Percent Ionization of a Weak Acid

-   Find the pH in Mixtures of Weak Acids

-   Find the $[OH^-]$ and pH of a Strong Base Solution

-   Find the $[OH^-]$ and pH of a Weak Base Solution

-   Determine Whether an Anion Is Basic or Neutral

-   Determine the pH of a Solution Containing an Anion Acting as a Base

-   Determine Whether a Cation Is Acidic or Neutral

-   Determine the Overall Acidity or Basicity of Salt Solutions

-   Find the pH of a Polyprotic Acid Solution

-   Find the $[H_3O^+]$ in Dilute H_2SO_4 Solutions

-   Find the Concentration of the Anions for a Weak Diprotic Acid Solution

Chapter Summary:

In this chapter, a specific type of equilibrium, acid and base ionization, will be explored in detail. First you will learn to identify acids and bases using the Arrhenius and Brønsted–Lowry definitions. Using the reactions that describe the behavior of acids and bases in water, you will learn how to write acid and base dissociation constants; these will then be used to calculate equilibrium concentrations of all species in acid and base solutions. You will learn about the pH scale and how the autoionization of water relates to all acid and base solutions. Once you are comfortable with equilibrium calculations for acids and bases, you will learn how to carry out similar calculations on the conjugate pairs of acids and bases; both a quantitative and qualitative treatment of salt solutions will then be considered. Acids that are able to donate more than one proton will be explored next. After a thorough quantitative treatment, you will learn how to compare acid strengths based on molecular structure. Finally, Lewis acids and bases will be explained.

Chapter Outline:

1. Batman's Basic Blunder

2. The Nature of Acids and Bases

    a. Acids have a sour taste, dissolve many metals, turn litmus red, and neutralize bases.

        i. Common acids include HCl, H_2SO_4, HNO_3, and $HC_2H_3O_2$.

        ii. Many naturally occurring acids are carboxylic acids, which contain the $-CO_2H$ group.

    b. Bases have a bitter taste, feel slippery, turn litmus blue, and neutralize acids.

   i.   Common bases include NaOH, KOH, NaHCO$_3$, and NH$_3$.

   ii. Alkaloids are organic bases found in plants.

3. Definitions of Acids and Bases

   a. According to the Arrhenius definition, acids produce H$^+$ when dissolved in water and bases produce OH$^-$ when dissolved in water.

      i.   When acids dissolve in water, we don't write H$^+$; instead, we write H$_3$O$^+$ (hydronium ion) to indicate that a proton is too attracted to polar water molecules to exist freely (unbound) in water.

      ii. Acids and bases react together to produce water and a neutral salt. This is what we are referring to when we say that acids neutralize bases and vice versa.

   b. According to the Brønsted–Lowry definition, acids donate protons while bases accept protons.

$$HA + B \leftrightarrows BH^+ + A^-$$

      i.   Acids and bases always appear together in a Brønsted–Lowry reaction.

      ii.   When an acid (HA) donates a proton, it becomes a conjugate base (A$^-$); when a base (B) accepts a proton, it becomes a conjugate acid (BH$^+$).

      iii. The acid and its conjugate base constitute a conjugate acid–base pair.

      iv.   An amphoteric species is one that can serve as an acid or a base in a chemical reaction. Example: H$_2$O

      v.   The acid–base reaction need not occur in water.

   ▶ The Brønsted–Lowry definition encompasses all Arrhenius acids and bases but expands the class of bases to include such species as NH$_3$, F$^-$, and CO$_3^{2-}$. In all cases, Brønsted–Lowry bases must contain a lone electron pair to bind to the H$^+$ ion.

EXAMPLE:

Write the acid–base reaction that will occur between each pair of species. Identify the conjugate acid–base pairs in each equation.

a. Ammonia with water

$$NH_3(aq) + H_2O(l) \leftrightarrows NH_4^+(aq) + OH^-(aq)$$

Ammonia is the base, and ammonium ion is its conjugate acid. Water is an acid here, and the hydroxide ion is its conjugate base.

b. Ammonia with hydrochloric acid

$$NH_3(aq) + HCl(aq) \leftrightarrows NH_4^+(aq) + Cl^-(aq)$$

In this reaction, ammonia is the base, and the ammonium ion is its conjugate acid. Hydrochloric acid is the acid, and the chloride ion is its conjugate base.

We can write this reaction in an alternate way that emphasizes the strength of hydrochloric acid. Since hydrochloric acid is a strong acid, it will dissociate completely in water to produce chloride ions and hydronium ions:

$$NH_3(aq) + H_3O^+(aq) + Cl^-(aq) \leftrightarrows NH_4^+(aq) + H_2O(l) + Cl^-(aq)$$

We see that chloride ions are spectator ions in this reaction. Ammonia is again the base, while the ammonium ion is its conjugate acid. H$_3$O$^+$ is the acid, while H$_2$O is its conjugate base.

c. Carbonic acid with dimethylamine, $(CH_3)_2NH$

$$H_2CO_3(aq) + (CH_3)_2NH(aq) \leftrightarrows HCO_3^-(aq) + (CH_3)_2NH_2^+(aq)$$

Carbonic acid is the acid, while the bicarbonate ion is its conjugate base. Dimethylamine is the base, and the dimethyl ammonium ion is its conjugate acid.

d. Sodium acetate with ammonium chloride

$$NaC_2H_3O_2(aq) + NH_4Cl(aq) \leftrightarrows HC_2H_3O_2(aq) + NH_3(aq) + NaCl(aq)$$

It is difficult to see what the conjugate acid base pairs are here, so we will rewrite the equation without the spectator ions:

$$C_2H_3O_2^-(aq) + NH_4^+(aq) \leftrightarrows HC_2H_3O_2(aq) + NH_3(aq)$$

We can see that the acetate ion is the base, while acetic acid is its conjugate acid. The ammonium ion is the acid, with ammonia as its conjugate base.

---

4. Acid Strength and Molecular Structure

    a. Binary acids consist of hydrogen and one other element and have the general form of H_nX. Two competing factors affect the strength of binary acids.

        i. The strength of a binary acid depends on bond polarity; the greater the electronegativity of X, the more polar the H—X bond, and the stronger the acid.

        ii. The strength of a binary acid depends on the H—X bond strength; the weaker the H—X bond (the larger the size of X), the stronger the acid.

    b. Oxyacids contain a hydrogen atom bonded to oxygen and have the general form of H—O—Y, where Y is a generic group.

        i. The more electronegative the Y group is, the more polarized the H—O bond will be and the stronger the acid.

        ii. The more oxygen atoms attached to Y, the more electron density will be pulled from the hydrogen atom and the stronger the acid will be.

5. Acid Strength and the Acid Ionization Constant (K_a)

    a. The strength of an acid depends on the equilibrium position of its ionization in water:

$$HA(aq) + H_2O(l) \leftrightarrows H_3O^+(aq) + A^-(aq)$$

    b. The equilibrium constant for this reaction is called the acid dissociation or acid ionization constant, K_a, and has the form:

$$K_a = \frac{[H_3O^+][A^-]}{[HA]}$$

EXAMPLE:

Write equations for dissociation and the acid ionization constants for the following weak acids:

The acid dissociation reaction will be the reaction of the weak acid, with water serving as a base. The acid ionization constant will be the concentrations of the products over the concentrations of reactants; water is not included in the acid ionization constants because it is a pure liquid.

a. Hydrofluoric acid, HF

$$HF(aq) + H_2O(l) \leftrightarrows H_3O^+(aq) + F^-(aq) \qquad K_a(HF) = \frac{[H_3O^+][F^-]}{[HF]}$$

b. Oxalic acid, $H_2C_2O_4$

$$H_2C_2O_4(aq) + H_2O(l) \rightleftharpoons H_3O^+(aq) + HC_2O_4^-(aq) \qquad K_a(H_2C_2O_4) = \frac{[H_3O^+][HC_2O_4^-]}{[H_2C_2O_4]}$$

c. Benzoic acid, $C_6H_5CO_2H$

$$C_6H_5CO_2H(aq) + H_2O(l) \rightleftharpoons H_3O^+(aq) + C_6H_5CO_2^-(aq) \qquad K_a(C_6H_5CO_2H) = \frac{[H_3O^+][C_6H_5CO_2^-]}{[C_6H_5CO_2H]}$$

---

   c. Thus, the strength of an acid depends on the value of K_a, which is an indication of the position of the equilibrium.

   ▶ Describing an acid as strong or weak is NOT the same as describing it as concentrated or dilute.

      i. Strong acids are species that completely ionize in water; the equilibrium position of a strong acid lies far to the right.

        1. Strong acids have such large equilibrium constants ($K_a \gg 1$) that we write their dissociation using a single forward arrow:

$$HA(aq) + H_2O(l) \rightarrow H_3O^+(aq) + A^-(aq)$$

        2. A strong acid has a very weak conjugate base.

      ii. Weak acids only partially ionize in water ($K_a < 10^{-2}$ or lower) so that both the ionized and nonionized forms of the acid exist in water.

        The extent to which an acid ionizes depends on the attraction of the anion (conjugate base) to the H^+. The more stable the conjugate base, the weaker its attraction for H^+, and the stronger the acid from which it is derived; that is, the forward (and not reverse) reaction is preferred.

   d. Monoprotic acids have only one ionizable proton, diprotic acids have two ionizable protons, and triproptic acids have three ionizable protons.

6. Autoionization of Water and pH

   a. Water is amphoteric, which means that it can serve as an acid or as a base.

   b. The autoionization of water is the reaction of water with itself in an acid/base manner:

$$H_2O(l) + H_2O(l) \rightleftharpoons H_3O^+(aq) + OH^-(aq)$$

      i. The equilibrium constant for this reaction is called the ion product constant of water, K_w:

$$K_w = [H_3O^+][OH^-] = 1 \times 10^{-14} \text{ at } 25\,°C$$

   c. In pure water $[H_3O^+] = [OH^-] = 10^{-7}$ M at $25\,°C$.

   d. In an aqueous solution, the equilibrium constant for the autoionization of water remains constant (equilibrium constants change only when the temperature changes); this is true even when an acid or a base is dissolved in water.

      i. In an acidic solution $[H_3O^+] > [OH^-]$

      ii. In a basic solution $[H_3O^+] < [OH^-]$

      iii. In a neutral solution $[H_3O^+] = [OH^-]$

e. The pH scale is used to compare the acidity of solutions:

$$pH = -\log[H_3O^+]$$

    i. In an acidic solution pH < 7

    ii. In a basic solution pH > 7

    iii. In a neutral solution pH = 7

    iv. The scale extends from < 0 to > 14.

f. The pOH scale is used to compare the basicity of solutions:

$$pOH = -\log[OH^-]$$

g. Using the equilibrium constant of water, we can derive a relationship between pH and pOH:

$$-\log K_w = -\log[H_3O^+] - \log[OH^-]$$

or

$$pH + pOH = 14 \text{ at } 25\ °C$$

h. pK_a is used to quantify acid strength; the smaller the value of the pK_a the stronger the acid.

EXAMPLE:

Calculate the hydronium ion concentration and the pH of the following solutions, and determine whether the solutions are acidic, basic, or neutral.

a. $[OH^-] = 4.63 \times 10^{-3}$ M

The hydroxide ion concentration and the hydronium ion concentration are related by the ion product constant of water, $[H_3O^+][OH^-] = 1 \times 10^{-14}$:

$$1 \times 10^{-14}/4.63 \times 10^{-3}\ M = [H_3O^+] = 2.16 \times 10^{-12}\ M$$

The pH can then be calculated by taking the negative log of the hydronium ion concentration:

$$pH = -\log(2.16 \times 10^{-12}) = 11.666$$

which is a pH that is greater than 7, meaning that the solution is basic.

b. $[OH^-] = 1.89 \times 10^{-8}$ M

Again, we will calculate the hydronium ion concentration using the ion product constant of water:

$$1 \times 10^{-14} = (1.89 \times 10^{-8}) \times [H_3O^+]$$

$$[H_3O^+] = 5.291 \times 10^{-7}\ M$$

The pH can then be calculated by taking the negative log of the hydronium ion concentration:

$$pH = -\log(5.291 \times 10^{-7}) = 6.276$$

A pH that is less than 7 means that the solution is acidic.

---

7. Finding the $[H_3O^+]$ and pH of Strong and Weak Acid Solutions

    a. In most acid solutions, the concentration of hydronium ions from the autoionization of water is negligible. Note that this is not true when the acid is sufficiently dilute or sufficiently weak.

    b. The concentration of hydronium ions in a strong acid solution is equal to the concentration of the strong acid.

EXAMPLE:

What is the pH of the following solutions?

a. 1.0×10^{-3} M HCl

HCl is a strong acid, so it will completely dissociate in water; the concentration of hydronium ions in water will therefore be 1.0×10^{-3} M. We can easily see that the pH of this solution is 3.00 because the log function is the inverse of raising ten to the power of a number (i.e., $\log_{10} 10^x = x$).

b. 1.25×10^{-2} M HClO$_4$

HClO$_4$ is a strong acid and will dissociate completely in water to give a hydronium ion concentration of 1.25×10^{-2} M. This is a number between 10^{-1} M and 10^{-2} M, so the pH should be between 1 and 2. Using a calculator, we find:

$$pH = -\log(1.25 \times 10^{-2}) = 1.90$$

c. 6×10^{-9} M HNO$_3$

HNO$_3$ is another strong acid, so it will dissociate completely in water to give a hydronium ion concentration of 6×10^{-9} M. We can calculate the pH of the solution from this value (pH = 8.22), but this cannot be the pH of the solution since it is basic! Because the nitric acid is so dilute, the hydronium ion contribution from the autoionization of water is much larger than that from the acid. This solution will actually have a hydronium ion concentration of:

$$1.0 \times 10^{-7} \text{ M (from water)} + 6 \times 10^{-9} \text{ M (from HNO}_3) = 1.06 \times 10^{-7} \text{ M}$$

which gives a pH of:

$$pH = -\log(1.06 \times 10^{-7}) = 6.97$$

d. 4.51×10^{-2} M NaOH

Since NaOH is a strong base, it will completely dissociate in water to give a hydroxide ion concentration of 4.51×10^{-2} M. We can use this concentration to calculate the pOH:

$$pOH = -\log(4.51 \times 10^{-2}) = 1.3458$$

This is reasonable since 4.51×10^{-2} is a number between 10^{-2} and 10^{-1}.

The pH can then be calculated by subtracting the pOH from 14.000 (we have three decimal places corresponding to three significant figures):

$$pH = 14.000 - 1.346 = 12.6542$$

Our answer should have three significant figures after the decimal point, so the pH is 12.654.

---

   c.   In a weak acid solution, the concentration of hydronium ions must be solved according to the equilibrium equation using an ICE table as in the previous chapter.

▶   This type of problem is just a simple example of the type of problem from Chapter 16.

EXAMPLE:

Calculate the pH of a 0.425 M solution of boric acid, H$_3$BO$_3$. The K_a of boric acid is 5.4×10^{-10}.

We first need to write out the ionization equation for boric acid.

$$H_3BO_3(aq) + H_2O(l) \leftrightharpoons H_3O^+(aq) + H_2BO_3^-(aq) \qquad K_a = \frac{[H_3O^+][H_2BO_3^-]}{[H_3BO_3]}$$

We can use an ICE table to solve for the equilibrium concentrations:

Reaction Condition	$[H_3BO_3]$ (M)	H_2O	$[H_2BO_3^-]$ (M)	$[H_3O^+]$ (M)
Initial	0.425	—	0	0
Change	$-x$	—	$+x$	$+x$
Equilibrium	$0.425 - x$	—	x	x

We can now substitute values into the acid ionization constant expression:

$$5.4 \times 10^{-10} = \frac{x \cdot x}{(0.425 - x)}$$

We can compare the acid ionization constant of boric acid with the initial concentration of boric acid to determine if we can approximate $0.425 - x \approx 0.425$. Since the K_a value is 9 orders of magnitude smaller than the initial acid concentration, we are safe in making the approximation.

$$5.4 \times 10^{-10} \approx \frac{x \cdot x}{0.425}$$

$$x = \sqrt{2.295 \times 10^{-10}}$$

$$x = 1.515 \times 10^{-5} = [H_3O^+]$$

Since x is equal to the hydronium ion concentration at equilibrium, we simply need to calculate the pH:

$$pH = -\log(1.515 \times 10^{-5}) = 4.82$$

This value makes sense because we have a very weak acid.

---

d. The percent ionization of an acid is equal to the concentration of the conjugate base at equilibrium or the hydronium ion concentration at equilibrium over the initial acid concentration expressed as a percentage:

$$\frac{[A^-]_{eq}}{[HA]_{initial}} \times 100\% = \frac{[H_3O^+]_{eq}}{[HA]_{initial}} \times 100\% = \text{Percent Ionization}$$

e. The greater the concentration of a weak acid solution, the smaller the percent ionization; that is, the ionization is suppressed.

▶ Note that strong acids are 100% ionized, regardless of concentration.

EXAMPLE:

Calculate the percent ionization for the boric acid solution in the previous example.

We calculated the hydronium ion concentration and the conjugate base concentration in the boric acid solution as 1.515×10^{-5} M. The percent ionization is then:

$$\frac{1.515 \times 10^{-5} \text{ M } H_2BO_3^-}{0.425 \text{ M } H_3BO_3} \times 100\% = 0.0036\% \text{ Ionized}$$

---

f. When a strong acid and a weak acid of similar concentrations are mixed, the weak acid can usually be neglected in the equilibrium expression.

> ► This is similar to ignoring the autoionization of water in an acid ionization problem. The idea is that the weak acid contributes an amount of hydronium ion that is negligible compared with the amount contributed by the strong acid.

   g. When two weak acids are mixed, the weaker acid can be neglected if the K_a differs by at least two orders of magnitude.

> ► This is similar to the criteria for using the "x is small" approximation.

EXAMPLE:

Calculate the hydronium ion concentration in a solution containing 5.45 mL of a 0.145 M HCl(aq) and 5.32 mL of a 0.029 M HIO(aq). (K_a(HIO) $= 2.3 \times 10^{-11}$).

We can see from the acid ionization constant of hypoiodous acid that it is very weak. Since the concentration of hypoiodous acid is very dilute, the contribution of hydronium ions from the weak acid is negligible compared to the strong acid contribution. Since HCl is a strong acid, it will completely ionize and we can calculate the number of moles of hydronium ions that will be produced from the HCl:

$$5.45 \text{ mL} \times \frac{1 \text{ L}}{1000 \text{ mL}} \times \frac{0.145 \text{ mol HCl}}{1 \text{ L}} \times \frac{1 \text{ mol H}_3\text{O}^+}{1 \text{ mol HCl}} = 7.903 \times 10^{-4} \text{ mol H}_3\text{O}^+$$

The concentration of hydronium ions in the solution will be the number of moles of hydronium ions (calculated above) divided by the total volume:

$$\frac{7.903 \times 10^{-4} \text{ mol H}_3\text{O}^+}{(5.45 + 5.32) \text{ mL}} \times \frac{1000 \text{ mL}}{1 \text{ L}} = 0.07338 \text{ M H}_3\text{O}^+$$

Our answer should have three significant figures, so the final hydronium concentration is 0.0734 M.

---

   8. Finding the [OH⁻] and pH of Strong and Weak Base Solutions

      a. When bases are dissolved in solution, they react with water according to the equilibrium expression:

$$\text{B}(aq) + \text{H}_2\text{O}(l) \rightleftharpoons \text{HB}^+(aq) + \text{OH}^-(aq)$$

      b. The equilibrium constant for this reaction is called the base dissociation or base ionization constant, K_b, and has the form:

$$K_b = \frac{[\text{HB}^+][\text{OH}^-]}{[\text{B}]}$$

      c. The strength of a base depends on the value of K_b, which is an indication of the position of the equilibrium.

         i. Strong bases have such large K_b values that we write their dissociation using a single forward arrow:

$$\text{B}(aq) + \text{H}_2\text{O}(l) \rightarrow \text{HB}^+(aq) + \text{OH}^-(aq)$$

            1. Most strong bases are hydroxides of Group 1A and Group 2A metals.

$$\text{M(OH)}_x(aq) \rightarrow \text{M}^{x+}(aq) + x\text{OH}^-(aq)$$

         ii. Weak bases only partially ionize in water, so that both the neutral and protonated form of the base exist in water.

            1. The most common weak bases are ammonia and derivatives of ammonia, that is, amines.

    d. The process of calculating the concentration of hydroxide ions and the pH of basic solutions is exactly the same as for acid solutions, with the added step of converting the pOH to pH.

▶ Again, this is just a simple equilibrium problem of the type examined in Chapter 16.

EXAMPLE:

Calculate the pH of a solution prepared by dissolving 9.7 g of sodium hydroxide in enough water to yield 250 mL of solution.

This is a simple conversion factor problem because sodium hydroxide is a strong base that will completely dissociate into sodium ions and hydroxide ions. First, we find the moles of hydroxide ions in the solution:

$$9.7 \text{ g} \times \frac{1 \text{ mol NaOH}}{39.998 \text{ g}} \times \frac{1 \text{ mol OH}^-}{1 \text{ mol NaOH}} = 0.2425 \text{ mol OH}^-$$

This is the number of moles in 250 mL, so the concentration of hydroxide ions is:

$$\frac{0.2425 \text{ mol OH}^-}{250 \text{ mL}} \times \frac{1000 \text{ mL}}{1 \text{ L}} = 0.97004 \text{ M OH}^-$$

The hydronium ion concentration can be found by dividing the ion product constant of water by the hydroxide ion concentration and then taking the negative log of that value:

$$[H_3O^+] = \frac{1.0 \times 10^{-14}}{0.97004 \text{ M OH}^-} = 1.0309 \times 10^{-14}$$

$$pH = -\log[H_3O^+] = 13.987$$

Our answer should have two significant figures, so the pH of this solution is 13.99.

This value makes sense because we have a reasonably concentrated solution of a strong base.

EXAMPLE:

Calculate the pH of a 4.5 M nicotine, $C_{10}H_{14}N_2$, solution. The K_b for nicotine is 1.0×10^{-6}.

First, we need to write the base ionization equation:

$$C_{10}H_{14}N_2(aq) + H_2O(l) \rightleftharpoons C_{10}H_{14}N_2H^+(aq) + OH^-(aq) \qquad K_b = \frac{[C_{10}H_{14}N_2H^+][OH^-]}{[C_{10}H_{14}N_2]}$$

Now we can use an ICE table to determine the equilibrium concentrations of all species:

Reaction Condition	$[C_{10}H_{14}N_2]$ (M)	$[H_2O]$ (M)	$[C_{10}H_{14}N_2H^+]$ (M)	$[OH^-]$ (M)
Initial	4.5	—	0	0
Change	$-x$	—	$+x$	$+x$
Equilibrium	$4.5 - x$	—	x	x

We can now substitute values into the base ionization constant expression:

$$1.0 \times 10^{-6} = \frac{(x)(x)}{(4.5 - x)}$$

As the concentration of nicotine and the base ionization constant differ by 4 orders of magnitude, we can approximate $4.5 - x \approx 4.5$:

$$1.0 \times 10^{-6} = \frac{(x)(x)}{4.5}$$

$$x = \sqrt{4.5 \times 10^{-6}} = 0.002121 \, M = [OH^-]$$

This is the concentration of hydroxide ions in the solution. In order to calculate the pH, we need to use the ion product constant for water to find the hydroxide ions first.

$$[H_3O^+] = \frac{1.0 \times 10^{-14}}{0.0.002121 \, M \, OH^-} = 4.714 \times 10^{-12} \, M$$

The pH is then:

$$pH = -\log(4.714 \times 10^{-12}) = 11.3266$$

Our answer should have two significant figures, so the pH of the solution is 11.33.

This value makes sense because we have a concentrated solution of a weak base.

(Alternatively, after finding $x = [OH^-]$, we could have solved for pOH and then calculated the pH).

---

9. The Acid–Base Properties of Ions and Salts

    a. Salts are ionic substances that dissociate in water to produce anions and cations.

    b. If the anion of the salt is a weak base (conjugate base of a weak acid), the pH of the solution will be greater than 7.

        i. The anion can interact with water as a base:

$$A^-(aq) + H_2O(l) \rightarrow HA(aq) + OH^-(aq)$$

        ii. The equilibrium constant for this reaction will be the K_b value of the conjugate base:

$$K_b = \frac{[HA][OH^-]}{[A^-]}$$

            1. The value of K_b for a base is related to K_a of its conjugate acid through the water ionization constant:

$$K_b \times K_a = K_w$$

EXAMPLE:

Calculate the pH of a solution of a 1.05 M NaCN given that the K_a of HCN is 4.9×10^{-10}.

In water, NaCN dissociates to $Na^+(aq)$ and $CN^-(aq)$. The latter ion, a weak base, can react with water.

We first need to calculate the K_b value of CN^-:

$$K_b(CN^-) = \frac{K_w}{K_a(HCN)}$$

$$K_b = \frac{1.0 \times 10^{-14}}{4.9 \times 10^{-10}} = 2.0408 \times 10^{-5}$$

Now we can write the base ionization equation and use an ICE table to find equilibrium concentrations:

$$CN^-(aq) + H_2O(l) \leftrightarrows HCN(aq) + OH^-(aq)$$

Reaction Condition	$[CN^-]$ (M)	$[H_2O]$ (M)	$[HCN]$ (M)	$[OH^-]$ (M)
Initial	1.05	—	0	0
Change	$-x$	—	$+x$	$+x$
Equilbrium	$1.05 - x$	—	x	x

We will now write the base ionization expression for the reaction:

$$K_b = \frac{[CN^-][OH^-]}{[HCN]}$$

$$2.0408 \times 10^{-5} = \frac{(x)(x)}{(1.05 - x)}$$

As the concentration of cyanide ion and the base ionization constant differ by 5 orders of magnitude, we can approximate $1.05 - x \approx 1.05$:

$$2.0408 \times 10^{-5} = \frac{(x)(x)}{1.05}$$

$$x = \sqrt{2.143 \times 10^{-5}} = 0.0046291 = [OH^-]$$

This is the concentration of hydroxide ions in the solution. In order to calculate the pH, we need to use the ion product constant for water to find the concentration of hydronium ions first.

$$[H_3O^+] = \frac{10^{-14}}{0.0046291 \text{ M OH}^-} = 2.1602 \times 10^{-12}$$

The pH is then:

$$pH = -\log(2.1602 \times 10^{-12}) = 11.6655$$

Our answer should have two significant figures, so the pH of the solution is 11.67.

This value makes sense because we have a relatively concentrated solution of a weak base.

---

    c.   If the cation is a weak acid (conjugate acid of a weak base), the pH of the solution will be less than 7.

        i.   Weakly acidic solutions also form when small, highly charged metals are dissolved in water.

EXAMPLE:

Will a 1.0 M solution of CH_3NH_3Cl be acidic, basic, or neutral?

We need to consider each component of the salt. Cl^- is a very weak base (the conjugate of a strong acid, HCl), so its effect on the pH of the solution is negligible. $CH_3NH_3^+$ is a moderately weak acid, meaning that it will interact with water to produce CH_3NH_2 and H_3O^+. The presence of H_3O^+ will result in an acidic solution.

---

    d.   When the cation of a salt serves as a weak acid and the anion of the salt behaves as a weak base, the pH of the solution will depend on the strength of the weak acid and base; that is, we compare the K_a and K_b values.

    e.   If neither the cation nor the anion acts as an acid or base, the solution will have a neutral pH.

EXAMPLE:

Will a solution containing 10.0 g of ammonium lactate, $NH_4C_3H_5O_3$, be acidic, basic, or neutral?

The K_b of ammonia, NH_3, is 1.76×10^{-5} and the K_a of lactic acid, $HC_3H_5O_3$, is 1.4×10^{-5}.

Both the cation and the anion of the salt are weak:

$$NH_4^+(aq) + H_2O(l) \leftrightharpoons NH_3(aq) + H_3O^+(aq) \qquad K_a(NH_4^+)$$

$$C_3H_5O_3^-(aq) + H_2O(l) \leftrightharpoons HC_3H_5O_3(aq) + OH^-(aq) \quad K_b(C_3H_5O_3^-)$$

We need to calculate the K_a of the ammonium ion and the K_b of the lactate.

The K_a of ammonium, NH_4^+, is:

$$K_a(NH_4^+) = \frac{K_w}{K_b(NH_3)}$$

$$K_a = \frac{1.0 \times 10^{-14}}{1.76 \times 10^{-5}} = 5.7 \times 10^{-10}$$

and the K_b of the lactate ion, $C_3H_5O_3^-$, is:

$$K_b(C_3H_5O_3^-) = \frac{K_w}{K_a(HC_3H_5O_3)}$$

$$K_b = \frac{1.0 \times 10^{-14}}{1.4 \times 10^{-5}} = 7.1 \times 10^{-10}$$

We can see from these values that because the K_b value of the lactate ion is slightly larger than the K_a value of ammonium, this equilibrium will be favored and the solution will be slightly basic. Note that this solution will be only slightly basic because the values of the equilibrium constants are so similar.

---

10. Polyprotic Acids

    a. Acids with more than one ionizable proton will lose protons in successive steps; each step (or reaction) will have its own equilibrium constant, K_a.

        i. The K_a values get smaller with each successive deprotonation; that is, $K_{a1} > K_{a2} > K_{a3}$. This is because it becomes increasingly difficult for a proton to be stripped from an anion.

    b. The pH of a polyprotic acid solution is calculated like any other mixture of acid solutions.

EXAMPLE:

What is the pH of a 0.052 M H_2SO_4 aqueous solution?

Since sulfuric acid is a strong acid, the dissociation of the first ionizable proton will give a hydronium ion concentration that is equal to the sulfuric acid concentration, or 0.052 M H_3O^+.

$$H_2SO_4(aq) + H_2O(l) \rightarrow HSO_4^-(aq) + H_3O^+(aq)$$

The acid ionization constant for the second proton, K_{a2}, is 1.2×10^{-2}, which is large enough compared to the concentration of sulfuric acid that we should not neglect its contribution.

We will write the acid ionization equation for the bisulfate ion and set up an ICE table:

$$HSO_4^-(aq) + H_2O(l) \leftrightharpoons SO_4^{2-}(aq) + H_3O^+(aq)$$

Reaction Condition	$[HSO_4^-]$ (M)	$[H_2O]$ (M)	$[SO_4^{2-}]$ (M)	$[H_3O^+]$ (M)
Initial	0.052	—	0	0.052
Change	$-x$	—	$+x$	$+x$
Equilibrium	$0.052 - x$	—	x	$0.052 + x$

Note that the initial bisulfate and hydronium ion concentrations are due to the dissociation of the original sulfuric acid solution.

The reaction will proceed in the forward direction because there is no sulfate ion present initially.

Now we will write the acid ionization constant for the reaction:

$$K_a = \frac{[SO_4^{2-}][H_3O^+]}{[HSO_4^+]}$$

$$1.2 \times 10^{-2} = \frac{(x)(0.052 + x)}{(0.052 - x)}$$

We need to use the quadratic formula, so we will first expand this expression to get the proper form:

$$0.000624 - 0.012x = 0.052x + x^2$$

$$x^2 + 0.064x - 0.000624 = 0$$

We see that $a = 1$, $b = 0.064$, and $c = -0.000624$ for use in the quadratic formula:

$$x = \frac{-0.064 \pm \sqrt{(0.064)^2 - 4(1)(-0.000624)}}{2(1)}$$

$$x = 0.0086 \text{ or } x = -0.0726$$

We cannot have a negative concentration, so we see that the concentration of hydronium ions at equilibrium is $0.052 + 0.0086 = 0.0606$ M.

The pH is then:

$$pH = -\log(0.0606) = 1.2175$$

Our answer should have two significant figures, so the pH of the solution is 1.22.

EXAMPLE:

Calculate the pH of a 0.0451 M solution of oxalic acid, $H_2C_2O_4$, given that $K_{a1} = 5.9 \times 10^{-2}$ and $K_{a2} = 6.4 \times 10^{-5}$.

We will start by writing the acid ionization equation for oxalic acid:

$$H_2C_2O_4(aq) + H_2O(l) \leftrightharpoons HC_2O_4^-(aq) + H_3O^+(aq)$$

Using an ICE table, we find the equilibrium concentrations:

Reaction Condition	$[H_2C_2O_4]$ (M)	$[H_2O]$ (M)	$[HC_2O_4^-]$ (M)	$[H_3O^+]$ (M)
Initial	0.0451	—	0	0
Change	$-x$	—	$+x$	$+x$
Equilibrium	$0.0451 - x$	—	x	x

We will now write the base ionization expression for the reaction:

$$K_a = \frac{[HC_2O_4^-][H_3O^+]}{[H_2C_2O_4]}$$

$$5.9 \times 10^{-2} = \frac{(x)(x)}{(0.0451 - x)}$$

We will use the quadratic formula to solve this problem since the original concentration and first acid ionization constant are so similar (i.e., we cannot approximate by ignoring x in the denominator):

$$0.0026609 - 0.059x = x^2$$

$$x^2 + 0.059x - 0.0026609 = 0$$

We see that $a = 1$, $b = 0.059$, and $c = -0.0026609$ for use in the quadratic formula:

$$x = \frac{-0.059 \pm \sqrt{(0.059)^2 - 4(1)(-0.0026609)}}{2(1)}$$

$$x = 0.0299 \text{ or } x = -0.0889$$

We cannot have a negative concentration, so we see that the concentration of hydronium ions at equilibrium is 0.0299 M.

We do not need to consider the ionization of the second proton because we already have a large hydronium ion concentration (limiting the extent to which the second ionization will occur) and the acid ionization constant for $HC_2O_4^-$ is very small.

The pH is therefore:

$$pH = -\log(0.0299) = 1.524$$

Our answer should have two significant figures, so the pH of the solution is 1.52.

---

11. Lewis Acids and Bases

   a.   A Lewis acid is an electron pair acceptor, and a Lewis base is an electron pair donor.

      i.   Lewis acids must have an empty orbital (or rearrange to provide an empty orbital) in order to accept the Lewis base electron pair.

      ii.   Small, highly charged metals form acidic solutions in water because they act as Lewis acids.

      iii.   Lewis acid–base reactions form an adduct, a single species with a coordinate covalent bond.

   ▶   This definition greatly expands the classes of acids.

Fill in the Blank:

1.   A(n) _____ is an acid that ionizes in successive steps.

2.   The _____ is the ratio of the ionized acid at equilibrium over the initial acid concentration expressed as a percentage.

3.   Acids are electron pair acceptors according to the _____ definition.

4.   A Brønsted-Lowry base is a(n) _____.

5.   In pure water, the hydronium ion concentration is _____ and the pH is _____.

6.   A strong acid is a species that completely _____ when dissolved in water.

7. In general, the stronger the acid, the _____ the conjugate base.

8. The equilibrium constant for the ionization reaction of a weak acid is called the _____.

9. The more electronegative the atom attached to hydrogen in a binary acid, the _____ the acid (other factors being equal).

10. When a salt containing an anion that is the conjugate base of a weak acid is dissolved in water, the solution will be _____.

11. According to the Arrhenius definition, an acid is a(n) _____.

12. NH_4^+ is the conjugate _____ of NH_3.

13. Small, highly charged metal cations form _____ solutions.

14. The strength of an oxyacid increases with a(n) _____ number of oxygen atoms.

15. A(n) _____ substance is one that can act as an acid or a base.

Problems:

1. Formic acid, $HCHO_2$, has a K_a value of 1.8×10^{-4}. 100.4 g of formic acid are used to prepare 10. L of a stock acid solution.

   a. What is the acid ionization equation for formic acid?

   b. Identify the conjugate acid/base pairs in the equation from part a.

   c. What is the pH of the stock solution?

   d. What is the pOH of the stock solution?

   e. What is the percent ionization of formic acid in the stock solution?

   f. A solution for use in a laboratory is made by taking a 150.0-mL aliquot of the stock solution and diluting it to 1.0 L.

      i. What is the pH of the new solution?

      ii. What is the percent ionization of the new solution?

   g. Considering your answers to parts c–f, is the preparation of a stock solution a wise use of time when the acid is weak?

2. A 0.529 M solution of an unknown acid is prepared, and the pH is measured to be 4.82. What is the acid ionization constant value of the unknown acid?

3. An aqueous solution of ethylamine, $C_2H_5NH_2$, of an unknown concentration is found on a laboratory shelf. The pH of the solution is found to be 8.53. Given that the K_b of ethylamine is 5.6×10^{-4}, calculate the concentration of the solution.

4. What mass of nitrous acid must be dissolved in 145 mL of water in order to obtain a solution with a pH of 5.2? The K_a value for nitrous acid is 4.6×10^{-4}.

5. Hypochlorous acid has an acid ionization constant of 3.5×10^{-8}. What are the concentrations of all species at equilibrium in a 0.0015 M solution of hypochlorous acid, and what is the pH of the solution?

6. Calculate the pH of a 1.45 M solution of hydrosulfuric acid given that the K_{a1} value is 8.9×10^{-8} and the K_{a2} value is 1×10^{-19}.

7. Salicylic acid is used in the manufacture of aspirin. A saturated solution of salicylic acid contains 2.2 g of the acid per liter of solution and has a pH of 2.43. The molar mass of salicylic acid is 138 g/mol.

   a. What is the K_a value of salicylic acid?

   b. What is the percent ionization of the acid?

8. Acetic acid and sodium bicarbonate are combined in equal amounts in 1.0 L of water.

   a. Write the balanced chemical reaction for the combination of these in water.

   b. Label the conjugate acid–base pairs in the reaction from part a.

   c. Given that the acid ionization constant for acetic acid is 1.8×10^{-5} and the acid ionization constant for carbonic acid is 4.2×10^{-7}, predict whether or not a reaction will occur when the acetic acid and sodium bicarbonate are mixed.

9. The bicarbonate ion, HCO_3^-, is amphoteric. The K_a value is 5.6×10^{-11} and the K_b value is 1.7×10^{-9}.

   a. Will a solution of sodium bicarbonate be acidic or basic? (You should be able to answer this without doing any calculations.)

   b. Calculate the pH and percent ionization of a 1.0 M bicarbonate ion solution assuming that it behaves as an acid.

   c. Calculate the pH and percent ionization of a 1.0 M bicarbonate ion solution assuming that it behaves as a base.

   d. Use your results in parts b and c to justify or refute the answer that you gave in part a.

   e. Sodium bicarbonate will "fizz" when combined with vinegar. Write the chemical reaction for this process, and explain the observation in the context of your answers to the previous parts of this question.

10. A 0.025 M solution of an unknown base has a pH of 10.09.

    a. What are the hydronium ion and hydroxide ion concentrations of the solution?

    b. Calculate the base ionization constant for the unknown base.

    c. Is the base strong, moderately weak, or very weak? Justify your answer based on the base ionization constant.

    d. What is the percent ionization of the base?

    e. Is your answer to part c consistent with your answer to part d? Explain.

11. Ascorbic acid is a diprotic acid with a $K_{a1} = 6.8 \times 10^{-5}$ and a $K_{a2} = 2.7 \times 10^{-12}$.

    a. List all species that are present in a solution of ascorbic acid at equilibrium.

    b. What is the pH of a solution that contains 5.0 mg of ascorbic acid per milliliter of solution?

12. How many milliliters of a strong monoprotic acid solution at a pH of 4.12 must be added to 528 mL of a solution of the same acid at a pH of 5.76 in order to prepare a solution with a pH of 5.34? Assume that the volumes are additive.

13. When Cr^{3+} is dissolved in water, a slightly acidic solution results.

    a. Write the balanced acid ionization equation for Cr^{3+} in water.

    b. What is the pH of a solution that has 16.52 g of $Cr(NO_3)_3$ dissolved to give 175 mL of solution? The K_a value of Cr^{3+} is 1.6×10^{-4}.

14. Explain why H—OCN is a stronger acid than H—CN.

15. Ethene, CH_2CH_2, contains two carbons with a double bond between them. Explain how and why ethene is a Lewis acid.

Concept Questions:

1. Explain the autoionization of water using the Arrhenius, Brønsted–Lowry, and Lewis definitions of acids and bases.

2. Use Le Châtelier's principle to explain why the percent ionization of a weak acid is larger for a dilute solution of the acid while the hydronium ion concentration is larger for a more concentrated solution.

3. In which of the following do you expect the percent ionization of 1.0 mol of NH_4Cl to be greatest: 500 mL water, 1000 mL water, or 1000 mL of a 0.10 M NH_3 solution? Explain.

4. Explain, using resonance structures, why you think that so many naturally occurring acids contain the carboxylic acid ($-CO_2H$) group.

5. Which of the following do you expect to have the largest K_b value? Explain your answer in one or two sentences.

    a. H_3PO_4 b. $H_2PO_4^-$ c. HPO_4^{2-} d. PO_4^{2-}

6. Which of the following will result in the largest hydronium ion concentration when dissolved in water?

    a. CO_2 b. Al_2O_3 c. Na_2O d. NaOH

7. Consider 1.0 M solutions of the following oxyacids: H_3PO_4, H_2SO_4, and $HClO_4$. Which will have the largest pH? Explain.

8. Explain why the hydronium ion is the strongest acid that can exist in water.

9. The strong acids sulfuric, nitric, perchloric, hydrochloric, hydrobromic, and hydroiodic acids actually have varying strengths. For example, hydroiodic acid is stronger than hydrochloric acid.

    a. The strength of these acids cannot be compared when they are dissolved in water. Explain why.

    b. Suggest an experiment that would allow you to test the strengths of these acids.

    c. Using what you have learned about acid strengths, predict the relative strengths of these acids.

10. Explain why the concentration of hydronium ions from the autoionization of water can be neglected for a 0.0010 M HCl solution using the rules for significant figures.

11. Explain why the pH scale is used rather than the hydronium ion concentration to report acidity. *Hint:* Many acids and bases are relatively weak.

12. Most weak bases are derivatives of ammonia called amines. Explain why you think that this observation is true based on the structure of an amine.

13. Pure, elemental boron is very rare; it is more commonly found as borax. Look up the structure and chemistry of borax, and explain why you think that this is so, using concepts of Lewis acid–base chemistry.

Chapter 17: Aqueous Ionic Equilibrium

Key Learning Outcomes:

- Calculate the pH of a Buffer Solution

- Use the Henderson–Hasselbalch Equation to Calculate the pH of a Buffer Solution

- Calculate the pH Change in a Buffer Solution after the Addition of a Small Amount of Strong Acid or Base

- Use the Henderson–Hasselbalch Equation to Calculate the pH of a Buffer Solution Composed of a Weak Base and Its Conjugate Acid

- Determine Buffer Range

- Determine pH's for a Strong Acid–Strong Base Titration

- Determine pH's for a Weak Acid–Strong Base Titration

- Calculate Molar Solubility from K_{sp}

- Calculate K_{sp} from Molar Solubility

- Calculate Molar Solubility in the Presence of a Common Ion

- Determine the Effect of pH on Solubility

- Predict Precipitation Reactions by Comparing Q and K_{sp}

- Find the Minimum Required Reagent Concentration for Selective Precipitation

- Find the Concentrations of Ions Left in Solution after Selective Precipitation

- Work with Complex Ion Equilibria

Chapter Summary:

   Now that you are familiar with general equilibrium reactions and acid/base equilibrium reactions in particular, we will explore more complex equilibrium reactions. First, you will learn about buffer solutions, which are weak acid/conjugate base or weak base/conjugate acid solutions that resist pH changes. You will learn how and why buffer systems resist pH changes, how to calculate the pH of a buffer, and how to determine the buffer capacity and range. Your understanding of solutions containing acid/base conjugate pairs will allow you to carry out a titration experiment, to sketch the resulting titration curve and to identify important regions of that curve. The dissolution of an "insoluble" salt will then be explored; you will learn how to calculate the solubility of a salt and how that solubility changes upon the addition of acids and/or bases. With your understanding of solubility, you will learn how to qualitatively identify the presence of certain ions in solution. Finally, you will look at complex ion equilibria, with an emphasis on how complex ion formation affects solubility and how pH affects complex ion formation.

Chapter Outline:

1. The Danger of Antifreeze

2. Buffers: Solutions That Resist pH Change

   a. Buffers are solutions that resist pH changes upon the addition of small amounts of acid or base.

      i. Buffers contain significant amounts of a weak acid and a salt of its conjugate base (or a weak base and a salt of its conjugate acid).

    ii.   Buffers resist pH changes by converting strong acids into weak acids and strong bases into weak bases:

$$H_3O^+(aq) + A^-(aq) \leftrightarrows H_2O(l) + HA(aq)$$

$$OH^-(aq) + HA(aq) \leftrightarrows H_2O(l) + A^-(aq)$$

    iii.   Weak acids and bases do not ionize to a significant extent in water so that the hydronium ion concentration is not changed dramatically even after a strong acid or base has been added.

    iv.   The pH of a buffer solution can be calculated in the same way as in the acid/base ionization problems from the last chapter.

        1.   This is an equilibrium problem in which a common ion is present.

        2.   The common ion effect refers to the suppression of an equilibrium due to the presence of one of the product species.

EXAMPLE:

A buffer solution is prepared by adding 10.4 g of sodium hypochlorite to 1.23 L of a 0.105 M hypochlorous acid solution. What is the pH of the buffer? The K_a of HClO is 2.9×10^{-8}.

First, we will determine the concentration of sodium hypochlorite that will initially be present:

$$10.4 \text{ g NaClO} \times \frac{1 \text{ mol NaClO}}{74.44 \text{ g NaClO}} = 0.1397 \text{ mol NaClO}$$

This is dissolved in 1.23 L of solution, so the concentration of NaClO is 0.1136 M.

Now we will write the acid ionization equation and use an ICE table to determine the equilibrium concentrations of all species:

$$HClO(aq) + H_2O(l) \leftrightarrows ClO^-(aq) + H_3O^+(aq)$$

Reaction Condition	[HClO] (M)	[H_2O] (M)	[ClO^-] (M)	[H_3O^+] (M)
Initial	0.105	—	0.1136	0
Change	$-x$	—	$+x$	$+x$
Equilibrium	$0.105 - x$	—	$0.1136 + x$	x

We will now write the acid ionization expression for the reaction and substitute the appropriate values:

$$K_a = \frac{[ClO^-][H_3O^+]}{[HClO]}$$

$$2.9 \times 10^{-8} = \frac{(0.1136 + x)(x)}{(0.105 - x)}$$

Because the acid ionization constant is very small (10^{-8}) and the presence of hypochlorite ions is initially high, the "x is small" approximation should be valid:

$$2.9 \times 10^{-8} = \frac{(0.1136)(x)}{(0.105)}$$

We find that $x = 2.680 \times 10^{-8}$, which is equal to the hydronium ion concentration at equilibrium. The pH is then equal to 7.572.

---

v. A buffer calculation can be simplified using the K_a expression:

$$K_a = \frac{[H_3O^+][A^-]}{[HA]}$$

Taking the negative log of both sides:

$$-\log K_a = -\log[H_3O^+] + -\log\frac{[A^-]}{[HA]}$$

We recognize that $-\log K_a$ is the pK_a and the $-\log[H_3O^+]$ is the pH:

$$pK_a = pH + -\log\frac{[A^-]}{[HA]}$$

Rearranging to solve for pH:

$$pH = pK_a + \log\frac{[A^-]}{[HA]}$$

This is called the *Henderson–Hasselbalch* equation and can be used to simplify buffer calculations. This expression is always valid (no approximation has been made) but is most useful when the percent ionization of a weak acid is less than 5% (i.e., when the "x is small" approximation is valid). An example is when $[HA]_{equilibrium} = [HA]_{initial} - x \approx [HA]_{initial}$, and $[A^-]_{equilibrium} = [A^-]_{initial} + x \approx [A^-]_{initial}$:

$$pH = pK_a + \log\frac{[A^-]_{initial}}{[HA]_{initial}}$$

EXAMPLE:

Calculate the pH of the buffer system from the previous example using the Henderson–Hasselbalch equation.

In order to use the Henderson–Hasselbalch equation, we first need to calculate the pK_a of the acid:

$$pK_a = -\log K_a$$

$$pK_a = -\log(2.9 \times 10^{-8}) = 7.5376.$$

Using the initial concentrations of the acid and conjugate base, we have:

$$pH = 7.5376 + \log\frac{(0.1136)}{(0.105)} = 7.57$$

We are assuming that the final concentration of acid/base is equal to the initial concentration of acid/base since the amount that ionized will be so small ("x is small").

The pH of the solution is 7.57, which is exactly the same as the pH calculated in the previous problem.

EXAMPLE:

Calculate the pH of a buffer system prepared by dissolving 2.24 g of dimethylamine, $(CH_3)_2NH$, and 5.08 g of $(CH_3)_2NH_2Cl$ to prepare 545 mL of solution. The K_b of dimethylamine is 5.4×10^{-4}.

The following equilibrium will be established:

$$(CH_3)_2NH(aq) + H_2O(l) \rightleftharpoons (CH_3)_2NH_2^+(aq) + OH^-(aq)$$

We will first calculate the concentration of the weak base and its conjugate acid:

$$2.24 \text{ g } (CH_3)_2 NH \times \frac{1 \text{ mol } (CH_3)_2 NH}{45.086 \text{ g } (CH_3)_2 NH} = 0.04968 \text{ mol } (CH_3)_2 NH$$

$$5.08 \text{ g } (CH_3)_2 NH_2Cl \times \frac{1 \text{ mol } (CH_3)_2 NH_2Cl}{81.544 \text{ g } (CH_3)_2 NH_2Cl} = 0.06230 \text{ mol } (CH_3)_2 NH_2Cl$$

Thus, in 545 mL, the concentration of $(CH_3)_2NH$ is 0.09116 M and that of $(CH_3)_2NH_2Cl$ is 0.1143 M.

In order to use the Henderson–Hasselbalch equation, we need the pK_a of the conjugate acid, $(CH_3)_2NH_2^+$:

$$pK_w = pK_a + pK_b$$

$$14.00 = pK_a + (-\log(5.4 \times 10^{-4}))$$

$$14.00 = pK_a + 3.2676$$

So the pK_a of $(CH_3)_2NH_2^+$ is 10.7324.

Inserting the pK_a and the acid/base concentrations into the Henderson–Hasselbach equation, we have:

$$pH = 10.7324 + \log \frac{(0.09116)}{(0.1143)}$$

The pH of the buffer solution is 10.634. Our final answer should have two significant figures (from the K_b value), so the pH is reported as 10.63.

---

vi. To determine the pH of a buffer after the addition of a strong acid or strong base, two steps should be followed: (1) a stoichiometry problem to neutralize (i.e., remove) the strong acid/base and (2) an equilibrium problem to find the final pH of the solution.

EXAMPLE:

Calculate the change in pH that occurs in the hypochlorous acid buffer from the first example upon the addition of 0.0012 mol of HCl.

The strong acid will react stoichiometrically with the base component of the buffer solution, OCl^-:

$$HCl(aq) + ClO^-(aq) \rightarrow HClO(aq) + Cl^-(aq)$$

Thus, we start with a stoichiometry problem:

$$0.0012 \text{ mol HCl added} \times \frac{1 \text{ mol ClO}^-}{1 \text{ mol HCl}} = 0.0012 \text{ mol ClO}^- \text{ consumed}$$

$$0.0012 \text{ mol ClO}^- \text{ consumed} = 0.0012 \text{ mol HClO produced}$$

The new initial concentration of hypochlorite ions in the buffer solution is the difference between the initial amount and the amount that reacted over the solution volume:

$$\frac{0.1397 \text{ mol ClO}^- - 0.0012 \text{ mol ClO}^-}{1.23 \text{ L solution}} = 0.1126 \text{ M ClO}^-$$

Similarly, the new initial concentration of hypochlorous acid in the buffer solution is:

$$\frac{0.129 \text{ mol HClO} + 0.0012 \text{ mol HClO}}{1.23 \text{ L solution}} = 0.106 \text{ M HClO}$$

Now we will use an ICE table to calculate the pH as in the above example:

$$HClO(aq) + H_2O(l) \leftrightharpoons ClO^-(aq) + H_3O^+(aq)$$

Reaction Condition	[HClO] (M)	[H$_2$O] (M)	[ClO$^-$] (M)	[H$_3$O$^+$] (M)
Initial	0.106	—	0.1126	0
Change	$-x$	—	$+x$	$+x$
Equilibrium	$0.106 - x$	—	$0.1126 + x$	x

We will now write the base ionization expression for the reaction:

$$K_a = \frac{[ClO^-][H_3O^+]}{[HClO]}$$

$$2.9 \times 10^{-8} = \frac{(0.1126 + x)(x)}{(0.106 + x)}$$

Again, we can assume that the "x is small" approximation is valid:

$$2.9 \times 10^{-8} = \frac{(0.1126)(x)}{(0.106)}$$

We find that $x = 2.730 \times 10^{-8}$, which is equal to the hydronium ion concentration at equilibrium. The pH is then equal to 7.56.

Notice that the pH of the solution with the added HCl is the same as in the original solution.

---

3. Buffer Effectiveness: Buffer Range and Buffer Capacity

    a. The buffer capacity refers to the amount of acid or base that a buffer can neutralize before its pH changes appreciably.

        i. A buffer system is most effective when the concentration of the weak acid (or weak base) and the concentration of the conjugate base (or conjugate acid) are (a) kept large and (b) approximately equal to each other.

        ii. When the concentrations of the weak acid (or weak base) and conjugate base (or conjugate acid) are exactly equal at equilibrium, the pH of the solution will be equal to the pK_a.

    b. The buffer range is the pH range over which a buffer effectively neutralizes added acids or bases.

        i. An effective range for a buffer system is a pH = p$K_a \pm 1$; this is where the concentrations of the weak conjugate pairs are not more than a factor of 10 different from one another.

EXAMPLE:

Select the most appropriate acid for preparing a buffer solution with a pH of 4.97 using the K_a values listed in Appendix II of your book.

We want to find an acid that has a pK_a value that is very close to the desired pH. We can more easily compare the desired [H$_3$O$^+$] to the K_a values listed in the Appendix. The [H$_3$O$^+$] value is equal to 10^{-pH}, which is 1.07×10^{-5}. Looking at the K_a values in the appendix, we find that propanoic acid has a K_a value of 1.3×10^{-5}, which is closest to the value that we are looking for.

---

4. Titrations and pH Curves

    a. In an acid–base titration, an acid (or base) solution of known concentration (called the titrant) is added to a base (or acid) of unknown concentration (called the analyte).

        i. The equivalence point of an acid–base titration is the point at which the moles of acid are stoichiometrically equal to the moles of base.

            1. The equivalence point can be monitored using an acid–base indicator or a pH probe.

        ii. A titration curve is a plot of pH of the analyte versus the volume of titrant added.

        iii. Each stage of a titration involves an acid–base reaction. The pH is determined by the prevailing species in solution.

    b. The titration of a strong acid with a strong base is essentially a stoichiometry problem.

      ▶ This is because the strong acid and strong base react completely—there is no significant reverse reaction.

    c. The titration of a weak acid (or base) with a strong base (or acid) is a two-part problem in which the acid–base stoichiometry is first accounted for and then the equilibrium reaction of the weak acid (or base) is considered.

      ▶ This two-step problem-solving approach is for convenience. It is mathematically simpler to remove the strong base in the calculation and then allow the equilibrium to be reestablished. You should realize that, in reality, the equilibrium is established immediately—the reaction does not go all the way in one direction and then go back.

EXAMPLE:

Lactic acid, $HC_3H_5O_3$, has a K_a value of 1.4×10^{-4}. Consider titrating 25.0 mL of a 0.187 M lactic acid solution with a 0.167 M KOH solution. Calculate the initial pH, the pH after the addition of 15.2 mL of KOH, the pH at the equivalence point, and the pH after the addition of 40.5 mL of KOH.

This is a titration of a weak acid with a strong base. Thus, all parts of this problem will require that we have the acid-ionization equation and corresponding K_a expression:

$$HC_3H_5O_3(aq) + H_2O(l) \leftrightarrows H_3O^+(aq) + C_3H_5O_3^-(aq)$$

$$K_a = \frac{[H_3O^+][C_3H_5O_3^-]}{[HC_3H_5O_3]}$$

Now we will approach each point in the titration as a separate problem:

1. The initial pH (before the addition of KOH):

This is an equilibrium problem similar to those from Chapter 15. We will use an ICE table and the acid-ionization constant as before:

Reaction Condition	$[HC_3H_5O_3]$ (M)	$[H_2O]$ (M)	$[C_3H_5O_3^-]$ (M)	$[H_3O^+]$ (M)
Initial	0.187	—	0	0
Change	$-x$	—	$+x$	$+x$
Equilibrium	$0.187 - x$	—	x	x

$$1.4 \times 10^{-4} = \frac{(x)(x)}{(0.187 - x)}$$

Since the acid ionization constant and the initial acid molarity differ by a factor of 10^3, we can use the "x is small" approximation:

$$1.4 \times 10^{-4} = \frac{(x)(x)}{(0.187)}$$

$$x = 0.005117 = [H_3O^+]$$

The pH of the solution is then $-\log(0.005117) = 2.29$, before any of the strong base has been added.

2. The pH after the addition of 15.2 mL of KOH:

The neutralization reaction will have a 1:1 mole ratio of acid to base:

$$HC_3H_5O_3(aq) + OH^-(aq) \rightarrow C_3H_5O_3^-(aq) + H_2O(l)$$

To begin, we can calculate the moles of acid and base initially present in the solution.

$$15.2 \text{ mL KOH} \times \frac{1 \text{ L}}{1000 \text{ mL}} \times \frac{0.167 \text{ mol KOH}}{1 \text{ L KOH}} = 0.002538 \text{ mol KOH}$$

$$25.0 \text{ mL HC}_3\text{H}_5\text{O}_3 \times \frac{1 \text{ L}}{1000 \text{ mL}} \times \frac{0.187 \text{ mol HC}_3\text{H}_5\text{O}_3}{1 \text{ L HC}_3\text{H}_5\text{O}_3} = 0.004675 \text{ mol HC}_3\text{H}_5\text{O}_3$$

We see that the base is the limiting reagent. All of the KOH will react to form a stoichiometric amount of $C_3H_5O_3^-$ (0.002538 mol). The lactic acid that remains will be the difference between the moles that are initially present (0.004675 mol) and those that reacted with the base (0.002538 mol); so 0.002137 mol of lactic acid are present after KOH is added and consumed.

In order to use the ICE table to determine the pH of the solution, we need to convert the number of moles present into concentrations. The total volume is the volume of acid plus the volume of base, or 40.2 mL. The new concentrations of acid and conjugate base are:

$$\frac{0.002538 \text{ mol C}_3\text{H}_5\text{O}_3^-}{0.0402 \text{ L solution}} = 0.06313 \text{ M C}_3\text{H}_5\text{O}_3^-$$

$$\frac{0.002137 \text{ mol HC}_3\text{H}_5\text{O}_3}{0.0402 \text{ L solution}} = 0.05316 \text{ M HC}_3\text{H}_5\text{O}_3$$

Now we will use an ICE table to calculate equilibrium concentrations:

Reaction Condition	$[HC_3H_5O_3]$ (M)	$[H_2O]$ (M)	$[C_3H_5O_3^-]$ (M)	$[H_3O^+]$ (M)
Initial	0.05316	—	0.06313	0
Change	$-x$	—	$+x$	$+x$
Equilibrium	$0.05316 - x$	—	$0.06313 + x$	x

$$1.4 \times 10^{-4} = \frac{(0.06313 + x)(x)}{(0.05316 - x)}$$

We will use the "x is small" approximation:

$$1.4 \times 10^{-4} = \frac{(0.06313)(x)}{(0.05316)}$$

$$x = 0.0001179 = [H_3O^+]$$

Since x is less than 5% of the original concentrations of acid and conjugate base, the approximation is valid. The pH of the solution is then $-\log(0.0001179) = 3.93$ after the addition of 15.2 mL of KOH.

3. The pH at the equivalence point:

   Recall that at the equivalence point, the moles of acid and base are equivalent. From the previous problem, we know that there are 0.04675 mol of lactic acid initially in the solution. With this information, we can use the concentration of the KOH solution to calculate the volume of KOH added:

   $$0.004675 \text{ mol HC}_3\text{H}_5\text{O}_3 \times \frac{1 \text{ mol KOH}}{1 \text{ mol HC}_3\text{H}_5\text{O}_3} \times \frac{1 \text{ L KOH solution}}{0.167 \text{ mol KOH}} \times \frac{1000 \text{ mL}}{1 \text{ L}} = 27.99 \text{ mL KOH solution}$$

   All of the weak acid and strong base will be consumed to produce an equimolar amount of the conjugate base, $C_3H_5O_3^-$. The initial concentration of the conjugate base is then:

   $$\frac{0.004675 \text{ mol C}_3\text{H}_5\text{O}_3^-}{0.05299 \text{ L solution}} = 0.08822 \text{ M C}_3\text{H}_5\text{O}_3^-$$

   In order to calculate the equilibrium concentrations, we need to consider the behavior of this weak base in water:

   $$C_3H_5O_3^-(aq) + H_2O(l) \leftrightarrows HC_3H_5O_3(aq) + OH^-(aq)$$

   This is the base-ionization equation, so we can write the equilibrium constant expression:

   $$K_b = \frac{[OH^-][HC_3H_5O_3]}{[C_3H_5O_3^-]}$$

   Since we are given the K_a for lactic acid, we can solve for the K_b of its conjugate base using K_w:

   $$K_w = K_a K_b$$

   The value of K_b for lactate is 7.14×10^{-11}. Using the base-ionization equation and the K_b value, we solve for x, $[OH^-]$:

Reaction Condition	$[C_3H_5O_3^-]$ (M)	$[H_2O]$ (M)	$[HC_3H_5O_3]$ (M)	$[OH^-]$ (M)
Initial	0.08822	—	0	0
Change	$-x$	—	$+x$	$+x$
Equilibrium	$0.08822 - x$	—	x	x

   $$7.14 \times 10^{-11} = \frac{(x)(x)}{(0.08822 - x)}$$

   Using the "x is small" approximation, we find x is 2.510×10^{-6} M. This is the concentration of hydroxide ions in the solution, so the pOH is 5.60 and the pH is 8.40 at the equivalence point. This value makes sense because we expect that the presence of a weak base will cause the pH to be greater than 7.

4. The pH after the addition of 40.5 mL of KOH:

   At this point in the titration, all of the weak acid has been converted to a weak base and excess strong base is in solution. We will determine the amount of KOH that remains in solution using the stoichiometry of the reaction as in the previous two parts of this reaction.

   $$40.5 \text{ mL KOH} \times \frac{1 \text{ L}}{1000 \text{ mL}} \times \frac{0.167 \text{ mol KOH}}{1 \text{ L KOH}} = 0.006764 \text{ mol KOH}$$

   From part 3, we know that 0.004675 mol of KOH will be consumed by the weak acid. This means that after the addition of KOH to $HC_3H_5O_3(aq)$, there are 0.002089 mol of KOH remaining in solution.

The concentration of hydroxide ions from the strong base is one thousand times more than the concentration of hydroxide ions produced when the conjugate base ionizes in water (from part 3). This means that the pOH of the solution will be dominated by the excess KOH. Using the new total volume of (25.0 + 40.5 mL), we can calculate the pOH:

$$\text{pOH} = -\log\left(\frac{0.002089 \text{ mol}}{0.0655 \text{ L}}\right)$$

The pOH is then 1.50, and the pH of the solution is 12.50.

---

    d.   The titration of a weak base with a strong acid involves the same type of calculation but employs the base ionization equation instead of the acid ionization equation.

    e.   When a titration experiment involves a polyprotic acid, multiple equivalence points must be considered.

    f.   A pH indicator is a weak acid that changes color at the titration's equivalence point.

        i.   In water, the indicator (HInd) dissociates as a weak acid:

$$\text{HInd}(aq) + \text{H}_2\text{O}(l) \rightleftharpoons \text{H}_3\text{O}^+(aq) + \text{Ind}^-(aq)$$

        ii.   The endpoint of a titration is when the indicator color change occurs. The equivalence point is when the moles of acid are equal to the moles of base.

        iii.   A shift in equilibrium results in a change in the solution color.

            1.   If the solution pH is greater than the pK_a of the indicator, then the color of the solution will be dominated by the color of Ind^-.

            2.   If the solution pH is less than the pK_a of the indicator, then the color of the solution will be dominated by the color of HInd.

          ▶   You can think about Le Chatelier's principle in predicting the dominant species at a given pH. A high pH means that there is a low concentration of H_3O^+ and the reaction shifts forward to make more H_3O^+ and more of the Ind^-. A low pH means that there is a high concentration of H_3O^+ and the reaction shifts in the reverse direction to make more HInd.

        iv.   The pK_a of the indicator should be within 1 unit of the pH at the equivalence point in order to ensure that the endpoint is indicative of the equivalence point.

5.   Solubility Equilibria and the Solubility-Product Constant

    a.   The equilibrium expression for a reaction in which a "slightly soluble" ionic solid dissolves in water is called a solubility-product constant, K_{sp}:

$$a\text{XY}(s) \rightleftharpoons b\text{X}^+(aq) + c\text{Y}^-(aq)$$

$$K_{sp} = [\text{X}^+]^a[\text{Y}^-]^b$$

    b.   The K_{sp} value is a measure of the solubility of a solid in water (at a specified temperature, typically 25 °C).

        i.   The molar solubility is the quantity (in mol) of a solid that will dissolve in 1 L of water.

EXAMPLE:

Calculate the solubility of calcium sulfate given that the K_{sp} value is 7.10×10^{-5}.

The solubility equation for calcium sulfate is:

$$CaSO_4(s) \leftrightarrows Ca^{2+}(aq) + SO_4^{2-}(aq)$$

The solubility-product constant expression is:

$$K_{sp} = [Ca^{2+}][SO_4^{2-}]$$

We can set up an ICE table for this reaction using S to represent the solubility:

Reaction Condition	[CaSO₄] (M)	[Ca²⁺] (M)	[SO₄²⁻] (M)
Initial	—	0	0
Change	—	+s	+s
Equilibrium	—	s	s

Including the K_{sp} value and the equilibrium concentrations into the K_{sp} expression:

$$7.10 \times 10^{-5} = S^2$$

$$S = \sqrt{7.10 \times 10^{-5}}$$

So the solubility of $CaSO_4$ in water is 0.00843 M. Notice that this is the concentration of Ca^{2+} and SO_4^{2+} in solution and also the number of moles of the solid $CaSO_4$ that will dissolve in 1 L of solution.

EXAMPLE:

The solubility of $PbBr_2$ is 0.01058 M at 25 °C. Calculate the solubility product constant value for $PbBr_2$.

We will start by writing the solubility equation, the solubility product constant expression, and then setting up an ICE table:

$$PbBr_2(s) \leftrightarrows Pb^{2+}(aq) + 2Br^-(aq)$$

$$K_{sp} = [Pb^{2+}][Br^-]^2$$

Reaction Condition	[PbBr₂] (M)	[Pb²⁺] (M)	[Br⁻] (M)
Initial	—	0	0
Change	—	+s	+2s
Equilibrium	—	s	2s

In this case, we have two bromide ions produced for each $PbBr_2$ molecule that dissolves in solution. The equilibrium concentration of bromide ions is then twice the solubility:

$$K_{sp} = (s)(2s)^2$$

$$K_{sp} = (0.01058)(0.02116)^2$$

$$K_{sp} = 4.737 \times 10^{-6}$$

c. The addition of a common ion causes the solubility of an ionic solid to decrease.

▶ According to Le Chatelier's principle, the solubility equilibrium shifts in the direction that minimizes the change—in this case, the change is the addition of products (ions), so the shift is toward reactants (solid salt).

EXAMPLE:

Calculate the solubility of $PbBr_2$ in a 0.100 M NaBr solution.

We will use the same equilibrium equation and the same solubility product constant as in the previous example. This time, however, we will set up the ICE table with an initial concentration of bromide ions:

Reaction Condition	$[PbBr_2]$ (M)	$[Pb^{2+}]$ (M)	$[Br^-]$ (M)
Initial	—	0	0.100
Change	—	$+S$	$0.100 + 2S$
Equilibrium	—	S	$0.100 + 2S$

We know the solubility constant expression and the solubility product constant value, so:

$$K_{sp} = [Pb^{2+}][Br^-]^2$$

$$4.737 \times 10^{-6} = (S)(0.100 + 2S)^2$$

Since the initial concentration of bromide ions and the solubility product constant value are five orders of magnitude different, we can use the "x is small" (or in this case, "S is small") approximation:

$$4.737 \times 10^{-6} = (S)(0.100)^2$$

$$S = 4.737 \times 10^{-4}$$

So the solubility of $PbBr_2$ in a 0.100 M NaBr solution is 4.737×10^{-4} M.

d. The solubility of an ionic compound may be changed when the solution pH is changed.

i. The way in which the solubility changes depends on how the ionic compound interacts with the added acid or base.

EXAMPLE:

Which of these will be more soluble in an acidic solution than in a neutral solution: $Al(OH)_3$, $BaSO_4$, or $Hg_2(CN)_2$?

For all of these salts, we need to consider the reaction that would occur upon the addition of an acid:

$$Al(OH)_3(s) + 3H_3O^+(aq) \rightarrow 6H_2O(l) + Al^{3+}(aq)$$

When water is formed, it will not ionize to reform OH^- to any appreciable extent. The solubility of $Al(OH)_3$ will be increased in an acidic solution because little OH^- is available to react with Al^{3+}.

$$BaSO_4(s) + 2H_3O^+(aq) \rightarrow H_2SO_4(aq) + H_2O(l) + Ba^{2+}(aq)$$

The sulfuric acid that forms is a strong acid that will dissociate completely to give hydronium ions and sulfate ions. The barium ions will combine with the sulfate ions to reform barium sulfate. The solubility of $BaSO_4$ will not, therefore, be more soluble in an acidic solution.

$$Hg_2(CN)_2(s) + 2H_3O^+(aq) \rightarrow 2HCN(aq) + 2H_2O(l) + Hg_2^{2+}(aq)$$

A weak acid, HCN, is formed in this reaction. HCN will dissociate partially to give hydronium ions and cyanide ions. The solubility will increase in this example, but it will not increase as much as it did in the first example where water was formed.

---

6. Precipitation

    a. In order to determine whether or not a precipitate will form under a given set of conditions, we compare the reaction quotient to the solubility product constant of a reaction.

        i. If $Q < K_{sp}$, then the solution is unsaturated and additional solid will dissolve.

        ii. If $Q = K_{sp}$, then the solution is saturated—it is at equilibrium.

        iii. If $Q > K_{sp}$, then the solution is supersaturated and a precipitate will form.

EXAMPLE:

Will a solution that is prepared by dissolving 1.04 g of NaOH and 2.34 g of $Pb(NO_3)_2$ in 1.42 L of water be unsaturated, saturated, or supersaturated?

We need to first consider the chemical reaction that will occur between these two substances:

$$2NaOH(aq) + Pb(NO_3)_2(aq) \leftrightarrows Pb(OH)_2(s) + 2NaNO_3(aq)$$

This reaction shows that we are concerned with the solubility of lead(II) hydroxide:

$$Pb(OH)_2(s) \leftrightarrows Pb^{2+}(aq) + 2OH^-(aq) \quad K_{sp} = 1.43 \times 10^{-20}$$

We will now calculate the concentrations of lead ions and hydroxide ions in the solution:

$$1.04 \text{ g NaOH} \times \frac{1 \text{ mol NaOH}}{39.998 \text{ g NaOH}} \times \frac{1 \text{ mol OH}^-}{1 \text{ mol NaOH}} = 0.02600 \text{ mol OH}^-$$

This amount of hydroxide ions is dissolved in 1.42 L of water, so the $[OH^-] = 0.01831$ M.

$$2.34 \text{ g Pb(NO}_3)_2 \times \frac{1 \text{ mol Pb(NO}_3)_2}{331.22 \text{ g Pb(NO}_3)_2} \times \frac{1 \text{ mol Pb}^{2+}}{1 \text{ mol Pb(NO}_3)_2} = 0.007065 \text{ mol Pb}^{2+}$$

This amount of lead ions is dissolved in 1.42 L of water, so the $[Pb^{2+}] = 0.004975$ M.

We will use the initial concentrations to calculate Q for this reaction:

$$Q = [Pb^{2+}][OH^-]^2$$

$$Q = (0.004975)(0.01831)^2$$

$$Q = 1.668 \times 10^{-6}$$

Since $Q > K_{sp}$, the solution is supersaturated and a precipitate will form in order to establish equilibrium.

---

    b. Selective precipitation is a method for separating out cations in a solution based on differences in their solubilities.

EXAMPLE:

A solution contains an equimolar amount of Fe^{2+}, Cu^{2+}, and Co^{2+}. If the total ion concentration in the solution is 4.52 M, what is the minimum number of grams of Na_2CO_3 that must be added to 1.00 L of the solution in order for a precipitate to form? What is the formula of the precipitate that forms?

First, we need to find the K_{sp} values for the metal carbonate solids that will form. From Appendix II:

$$K_{sp} = 3.07 \times 10^{-11} \text{ for FeCO}_3$$

$$K_{sp} = 2.4 \times 10^{-10} \text{ for CuCO}_3$$

$$K_{sp} = 1.0 \times 10^{-10} \text{ for CoCO}_3$$

The smaller the K_{sp} value is, the less the solid will dissolve. Comparing the K_{sp} values for the carbonate salts, we see that iron(II) carbonate is the least soluble, so $FeCO_3$ will precipitate out of solution first.

Next, we need to determine the concentration of iron ions in the solution. Since there are three ions present in equimolar amounts in 1.0 L and the total ion concentration is 4.52 M, each metal cation will have a concentration of 1.5067 M. We will now calculate the concentration of carbonate ions that must be added to the solution in order for the iron(II) carbonate to precipitate out:

$$K_{sp} = [Fe^{2+}][CO_3^{2-}]$$

$$3.07 \times 10^{-11} = (1.5067)[CO_3^{2+}]$$

$$[CO_3^{2+}] = 2.038 \times 10^{-11} \text{ M}$$

Now we will convert this into grams of sodium carbonate:

$$\frac{2.038 \times 10^{-11} \text{ mol CO}_3^{2-}}{1 \text{ L solution}} \times 1.00 \text{ L solution} \times \frac{1 \text{ mol Na}_2\text{CO}_3}{1 \text{ mol CO}_3^{2-}} \times \frac{105.99 \text{ g Na}_2\text{CO}_3}{1 \text{ mol Na}_2\text{CO}_3} = 2.16 \times 10^{-9} \text{ g Na}_2\text{CO}_3$$

---

7. Complex Ion Equilibria

    a. In solution, metal cations, acting as Lewis acids, are hydrated by water molecules, acting as Lewis bases.

    b. A complex ion is a compound that contains a central metal cation bound to one or more ligands.

        i.   Ligands are Lewis bases that can be charged or neutral.

        ii. Hydrated metal ions, $[M(H_2O)_x]^{y+}$, are usually written without water ligands, $M^{y+}(aq)$, for simplicity.

    c. In a solution, another ligand can displace water in a complex ion formation equilibrium:

$$M^+(aq) + XL(aq) \leftrightharpoons [ML_x]^+(aq)$$

$$K_f = \frac{[ML_x^+]}{[M^+][L]^X}$$

    where K_f is called the formation constant.

EXAMPLE:

What is the concentration of $CdBr_4^{2-}$ that forms when a 452 mL solution of 0.143 M $Cd(C_2H_3O_2)_2$ is combined with 596 mL of a 0.218 M solution of KBr? The formation constant for $CdBr_4^{2-}$ is 5.3×10^3.

We will first calculate the number of moles of Cd^{2+} and Br^- initially in the solution:

$$452 \text{ mL solution} \times \frac{1 \text{ L}}{1000 \text{ mL}} \times \frac{0.143 \text{ mol Cd(C}_2\text{H}_3\text{O}_2)_2}{1 \text{ L solution}} \times \frac{1 \text{ mol Cd}^{2+}}{1 \text{ mol Cd(C}_2\text{H}_3\text{O}_2)_2} = 0.06464 \text{ mol Cd}^{2+}$$

$$596 \text{ mL solution} \times \frac{1 \text{ L}}{1000 \text{ mL}} \times \frac{0.218 \text{ mol KBr}}{1 \text{ L solution}} \times \frac{1 \text{ mol Br}^-}{1 \text{ mol KBr}} = 0.1299 \text{ mol Br}^-$$

We will assume that the reaction goes to completion because the formation constant is so large. Given the 4:1 Br:Cd stoichiometry of our complex ion $CdBr_4^{2-}$, we note that bromide ions will limit the formation of the complex ion (Br^- is the limiting reagent), so:

$$0.1299 \text{ mol } Br^- \times \frac{1 \text{ mol } Cd^{2+}}{4 \text{ mol } Br^-} = 0.03248 \text{ mol } Cd^{2+}$$

The reaction stoichiometry shows us that the number of moles of Cd^{2+} used equals the moles of $CdBr_4^{2-}$ produced, so we are able to calculate the concentrations of all species in solution after the reaction occurs:

$$[Cd^{2+}] = \frac{(0.06464 - 0.03248) \text{ mol } Cd^{2+}}{1.048 \text{ L}} = 0.03069 \text{ M } Cd^{2+}$$

$$[Br^-] = \frac{0 \text{ mol } Br^-}{1.048 \text{ L}} = 0 \text{ M } Br^-$$

$$[CdBr_4^{2-}] = \frac{0.03248 \text{ mol } CdBr_4^{2-}}{1.048 \text{ L}} = 0.03099 \text{ M } CdBr_4^{2-}$$

We will now allow the reaction to proceed in the backward direction in order to reach equilibrium. We will write the balanced equilibrium reaction, the equilibrium constant expression (which is the inverse of the formation constant expression), and an ICE table:

$$CdBr_4^{2-}(aq) \rightleftharpoons Cd^{2+}(aq) + 4Br^-(aq)$$

$$K_c = \frac{1}{K_f} = \frac{[Cd^{2+}][Br^-]^4}{[CdBr_4^{2-}]}$$

Reaction Condition	$[CdBr_4^{2-}]$ (M)	$[Cd^{2+}]$ (M)	$[Br^-]$ (M)
Initial	0.03099	0.03069	0
Change	$-x$	$+x$	$+4x$
Equilibrium	$0.03099 - x$	$0.03069 + x$	$4x$

$$1.887 \times 10^{-4} = \frac{(0.03069 + x)(4x)^4}{(0.03099 - x)}$$

We will use successive approximation to solve for x:

$$1.887 \times 10^{-4} = \frac{(0.03069)(4x)^4}{(0.03099)}$$

$$x = 0.02937$$

$$1.887 \times 10^{-4} = \frac{(0.06006)(4x)^4}{(0.00162)}$$

$$x = 0.01187$$

$$1.887 \times 10^{-4} = \frac{(0.04256)(4x)^4}{(0.01912)}$$

$$x = 0.02399$$

$$1.887 \times 10^{-4} = \frac{(0.05468)(4x)^4}{(0.00700)}$$

$$x = 0.01753$$

$$1.887 \times 10^{-4} = \frac{(0.04822)(4x)^4}{(0.01346)}$$

$$x = 0.02130$$

$$1.887 \times 10^{-4} = \frac{(0.05199)(4x)^4}{(0.00969)}$$

$$x = 0.01925$$

$$1.887 \times 10^{-4} = \frac{(0.04994)(4x)^4}{(0.01174)}$$

$$x = 0.02040$$

$$1.887 \times 10^{-4} = \frac{(0.05109)(4x)^4}{(0.01059)}$$

$$x = 0.01977$$

which agrees with the previous solution (to two sig. figs.). We have the amount of $CdBr_4^{2-}$ that dissociates, but we are asked to solve for the equilibrium concentration of $CdBr_4^{2-}$:

$$[CdBr_4^{2-}] = (0.03099 - 0.01977) \text{ M} = 0.0112 \text{ M}$$

Our answer should have two significant figures, so the equilibrium concentration is reported as 0.011 M.

---

    d.   The solubility of a salt increases when complex ion formation is favorable.

EXAMPLE:

Which of the following insoluble salts will become more soluble upon the addition of NaCN: AgBr, $Al(OH)_3$, $FeBr_2$, or $PbBr_2$?

The salts that will become more soluble upon the addition of cyanide ions are those that form a complex ion with it. The formation of the complex ion increases solubility by removing the metal cation from the solution.

Looking in Appendix II, we see that Ag^+ and Fe^{2+} form complex ions with cyanide ions. This means that the solubilities of $FeBr_2$ and AgBr will both increase when NaCN is added to solutions of them.

---

    e.   Some metal hydroxide complex ions are amphoteric; these complex ions will be soluble at high and low pH but insoluble at neutral pH.

        i.   A metal hydroxide that is soluble in an acidic solution reacts according to a neutralization reaction that gives hydronium ions and the metal cation:

$$M(OH)_x(aq) + XH_2O(l) \leftrightarrows XH_3O^+(aq) + M^{x+}(aq)$$

    ii. A metal hydroxide that is soluble in a basic solution reacts with the hydroxide ion to give a complex ion:

$$M(OH)_x(aq) + YOH^-(aq) \leftrightharpoons M(OH)_{x+y}{}^{y-}(aq)$$

    iii. Only hydroxides of Al^{3+}, Cr^{3+}, Zn^{2+}, Pb^{2+}, and Sn^{2+} are amphoteric; all other metal hydroxides will be soluble only in acidic solutions.

Fill in the Blank:

1. A titration curve is a plot of _____ versus _____ in a titration experiment.

2. A complex ion contains a central metal atom that acts as a(n) _____ bound to one or more ligands that act as _____.

3. Buffer solutions contain significant amounts of a weak acid and _____ or weak bases and _____.

4. The equilibrium constant expression for a reaction that represents the dissolution of a sparingly soluble salt is called the _____.

5. At the _____ of a titration, the moles of acid and base are stoichiometrically equal.

6. The solubility of an ionic compound _____ in the presence of a Lewis base that can complex to the cation.

7. The pH of a buffer system is approximately equal to the acid _____ value.

8. The _____ of a salt is the number of moles of the salt that dissolve in 1 L of solution.

9. When the change in concentration to achieve equilibrium concentrations is small, the pH of a buffer solution can be calculated using the _____ equation.

10. The effective range for a buffering system is _____.

11. The solubility of a(n) _____ metal hydroxide increases when either acids or bases are added.

12. A(n) _____ will have multiple equivalence points when titrated with a strong base.

13. When two cations both form salts with a given anion, the cation with the _____ value of K_{sp} will precipitate out of solution first upon the addition of the anion.

14. The _____ is the amount of acid or base that can be added to a buffer without destroying its effectiveness.

15. A solution is supersaturated when the reaction quotient is _____ the K_{sp} value.

16. The _____ of a titration is where the indicator color changes.

17. The relative concentrations of the acid and conjugate base in a buffer system should not differ by more than _____.

18. The equilibrium constant for the formation of a complex ion is called the _____.

19. The solubility of an ionic substance _____ when dissolved in a solution containing a common ion.

Problems:

1. 1.0 L of a buffer is prepared using 24.2 g of acetic acid.

   a. What quantity (in g) of sodium acetate must be used in order for the concentrations of the acid and conjugate base to be equal at equilibrium?

   b. What will the pH of the buffer system be when 24.2 g of sodium acetate are added to the original solution? Assume that no volume change occurs upon the addition.

   c. What will the pH of the buffer system be when the amount of sodium acetate added to the acetic acid solution is equal to the number of moles of acetic acid originally in the solution? Assume that no volume change occurs upon the addition.

   d. Explain the difference between your answers in parts b and c, and determine which solution is a better buffer.

   e. Calculate the pH of the buffer system that you prepared in part a, upon the addition of 50.0 mL of a 0.024 M NaOH solution.

   f. What volume of the 0.024 M NaOH solution is required to exceed the buffering capacity of the solution that you prepared in part a? *Hint*: Consider how you would make the concentrations of the buffer components differ by a factor of 10 or more.

2. Buffers are used extensively in chemistry and biology because they serve to minimize changes in pH.

   a. Calculate the change in pH (ΔpH) that occurs when 10.0 mL of a 0.10 M HCl solution is added to 250.0 mL of pure water.

   b. Calculate the change in pH (ΔpH) that occurs when 10.0 mL of a 0.10 M HCl solution is added to 250.0 mL of a buffer solution that is 0.115 M in benzoic acid and 0.105 M in sodium benzoate. (The K_a of benzoic acid is 6.5×10^{-5}.)

   c. Explain the difference in the calculated ΔpH values from parts a and b.

3. Consider the titration of 25.0 mL of a 0.143 M solution of NaOH with a 0.143 M solution of HCl.

   a. Without doing any calculations, what volume of HCl will be required to reach the equivalence point?

   b. Will this titration experiment have a buffer region? Why or why not?

   c. Calculate the pH of the solution at the beginning of the experiment (before the addition of any HCl).

   d. Calculate the pH of the solution after 15.0 mL of the HCl solution has been added.

   e. Calculate the pH of the solution at the equivalence point.

   f. Calculate the pH of the solution after 45.0 mL of the HCl solution has been added.

4. Consider titrating 15.0 mL of a 0.245 M solution of hydrazine, H_2NNH_2, with a 0.453 M HCl solution. The K_b value for hydrazine is 1.3×10^{-6}.

   a. Without doing any calculations, determine whether the pH of the solution at the equivalence point is greater than 7, equal to 7, or less than 7.

   b. What volume of HCl needs to be added to the solution in order to have a buffer system with the maximum capacity for added acid or base?

   c. What is the pH of the solution before any HCl has been added?

   d. What is the pH of the solution after the addition of 3.0 mL of HCl?

   e. What is the pH of the solution at the equivalence point?

   f. What is the pH of the solution when 5.0 mL more of the HCl solution is added after the equivalence point has been reached?

g. What is the pH of the solution when 10.0 mL more of the HCl solution is added after the equivalence point has been reached?

h. What is the pH of the solution when 15.0 mL more of the HCl solution is added after the equivalence point has been reached?

5. The pH at half equivalence is an important reference point for an acid.

a. Calculate the pH at half equivalence when 50.0 mL of 0.0252 M NH_3 is titrated with a 0.0385 M HCl solution.

b. Based on your answer in part a, what do you think the significance of the pH at half equivalence is for chemists?

6. Consider titrating 20.0 mL of a 0.150 M solution of iodoacetic acid (ICH_2CO_2H) with a 0.100 M solution of sodium hydroxide. The pK_a of iodoacetic acid is 3.18.

a. What is the pH of the solution after 10.0 mL of NaOH has been added to the acid?

b. How will the pH of the solution that you calculated in part a change upon the addition of 5.0 mL of a 0.100 M solution of HCl?

c. How will the pH of the solution that you calculated in part a change upon the addition of 10.0 mL of a 0.100 M solution of HCl?

d. Comment on any differences between the values that you calculated in parts b and c.

e. What is the pH of the solution at the equivalence point?

f. Which of the following indicators is best for the titration of iodoacetic acid with sodium hydroxide?

i.   Thymol blue ($pK_a \approx 1.9$)

ii. Methyl red ($pK_a \approx 5.1$)

iii. Bromothymol blue ($pK_a \approx 6.8$)

iv. Phenol red ($pK_a \approx 7.9$)

v.   Phenolphthalein ($pK_a \approx 9.1$)

g. Sketch the titration curve for this experiment, being sure to indicate the buffer region(s), the initial pH, the pH at the equivalence point, the pH at the end of the experiment, and appropriate labels for the axes.

7. 14.3 g of K_2CO_3 are dissolved in enough water to make 1.0 L of solution.

a. What is the solubility of $Mg(NO_3)_2$ in this solution?

b. What is the maximum amount of $Mg(NO_3)_2$ that you can dissolve in this solution without forming a precipitate?

c. Will a precipitate form upon the addition of HCl (after the addition of the $Mg(NO_3)_2$)? If so, what is the precipitate? If not, why not?

d. Will a precipitate form upon the addition of NaOH? If so, what is the precipitate? If not, why not?

8. Na_2CO_3 is added to a solution containing Ca^{2+}, Mg^{2+}, and Fe^{2+}. In what order will these cations precipitate out of solution? Explain your answer.

9. Silver bromide is a sparingly soluble salt.

a. What is the solubility of silver bromide in water?

b. What is the solubility of silver bromide in a 0.0015 M solution of HCl?

c. What is the solubility of silver bromide in a 0.0150 M solution of HBr?

d. What is the solubility of silver bromide in a 0.250 M solution of NaCN? (*Hint:* Ag forms a complex ion with the cyanide ion.)

10. Copper(II) hydroxide is a sparingly soluble salt with a K_{sp} of 2.2×10^{-20}.

    a. Calculate the solubility of copper(II) hydroxide in water.

    b. How do you expect the solubility of copper(II) hydroxide to be affected (increase, decrease, or stay the same) by the addition of HCl? Explain.

    c. How do you expect the solubility of copper(II) hydroxide to be affected (increase, decrease, or stay the same) by the addition of NaOH? Explain.

    d. Copper forms many complex ions including one with ammonia. Given that the K_f of $[Cu(NH_3)_4]^{2+}$ is 1.7×10^{13}, calculate the solubility of copper(II) hydroxide in a 0.250 M solution of ammonia.

Concept Questions:

1. pH indicators are weak acids, as discussed in this chapter. Explain why their influence on the pH of a solution is never accounted for in a titration experiment.

2. Which species (HInd or Ind⁻) will be the dominant form of the indicator just before and just after the equivalence point when titrating a weak acid with a strong base? Which species will dominate when titrating a weak base with a strong acid?

3. The equivalence point for a titration can be estimated based on the pK_a of the weak acid used. Explain how and why this can be done. Estimate the pH at the equivalence point for the titration of pyruvic acid with sodium hydroxide.

4. In many of the calculations described in this chapter, we suspend the universe. For example, when calculating the pH of a titration experiment at equivalence, we split the problem into two parts: a reaction between the strong base and weak acid that goes to completion and then an equilibrium problem; in reality, these reactions are constantly occurring in both directions.

    a. Explain why this is a reasonable approach to the problem.

    b. Is this actually what occurs (on a molecular level) in the experiment?

    c. If this is NOT what actually occurs, explain why the problem isn't solved according to the physical reality of the situation.

5. The region of a titration curve in which a weak acid has been titrated with an amount of strong base that is less than the amount needed to reach the equivalence point is called the buffer region. On a titration curve, this region looks very flat, i.e., the pH doesn't change much as base is continually added. The region after the equilibrium point also looks flat on a titration curve, but this is not a buffer region. Explain why this region is not a buffer and why, despite this, it looks flat.

Chapter 18: Free Energy and Thermodynamics

Key Learning Outcomes:

-   Predict the Sign of Entropy Change

-   Calculate ΔS for a Change of State

-   Calculate Entropy Changes in the Surroundings

-   Calculate Gibbs Free Energy Changes and Predicting Spontaneity from ΔH and ΔS

-   Calculate Standard Entropy Changes (ΔS°_{rxn})

-   Calculate the Standard Change in Free Energy for a Reaction Using $\Delta G^{\circ}_{rxn} = \Delta H^{\circ}_{rxn} - T\Delta S^{\circ}_{rxn}$

-   Calculate ΔG°_{rxn} from Standard Free Energies of Formation

-   Calculate ΔG°_{rxn} for a Stepwise Reaction

-   Calculate ΔG°_{rxn} under Nonstandard Conditions

-   Relate the Equilibrium Constant and ΔG°_{rxn}

Chapter Summary:

In this chapter, we will explore the fundamental concepts that dictate whether or not a chemical change will occur spontaneously. We will first define spontaneous change and then use the second law of thermodynamics as its predictor. Entropy will be explained as a measure of a system's ability to disperse or arrange itself energetically. We will use the Boltzmann equation to relate the probability of a given arrangement to a system's entropy and then apply these ideas to common chemical systems. We will identify the entropy change of the universe as the predictor of spontaneous change and then define it in terms of system properties. Gibbs free energy will then be introduced as a thermodynamic function that determines spontaneity and is equal to the maximum (nonexpansion) work available from a system. We will learn how to use the third law of thermodynamics in defining the zero of entropy and investigate how to use this information to calculate the entropy change of a chemical reaction. Standard free energy changes will be calculated using three different methods, and then nonstandard free energy changes will be explored. Finally, you will learn about the relationship between the equilibrium constant value and the thermodynamic properties of a chemical system.

Chapter Outline:

1. Nature's Heat Tax: You Can't Win and You Can't Break Even

    a. The observation that energy spreads out is a description of the second law of thermodynamics.

    b. With each step (or transaction) in a process, heat is lost to the surroundings and efficiency decreases.

2. Spontaneous and Nonspontaneous Processes

    a. A spontaneous process is one that occurs without any ongoing outside intervention.

    i. A spontaneous chemical process occurs in order to minimize a system's chemical potential energy, just as a mechanical process occurs in a direction that minimizes its mechanical potential energy.

    b. Thermodynamics studies the direction and extent of a reaction, not the speed of a chemical reaction.

    i. Thermodynamics is unaffected by catalysts, which increase the rate or speed of a reaction but do not make the reaction occur to any greater (or lesser) extent.

c. Nonspontaneous processes can be made spontaneous through the application of energy from an external source.

3. Entropy and the Second Law of Thermodynamics

a. Enthalpy is not the criterion for spontaneous change; although most spontaneous processes are exothermic, there are some endothermic processes that occur spontaneously.

b. Entropy, S, is a thermodynamic function that is related to the number of energetically equivalent ways to arrange a system:

$$S = k \ln W$$

where k is the Boltzmann constant (1.38×10^{-23} J/K) and W is the multiplicity or number of energetically equivalent ways to arrange a system.

   i. The quantity W is the number of microstates that give the same energetic macrostate.

   ii. As the number of energetically equivalent ways to arrange a system increases, the entropy of the system increases.

c. The second law of thermodynamics states that a chemical process will proceed in a direction that increases the entropy of the universe:

$$\Delta S_{universe} > 0 \text{ for a spontaneous process}$$

d. Entropy has units of J/K and is a state function:

$$\Delta S = S_{final} - S_{initial}$$

e. The multiplicity (W) is a measure of probability; the more ways that a system can be arranged in a given state, the more likely the system is to be found in that state.

f. Entropy is maximized when energy dispersion is maximized.

   i. Heat flows from a hot body to a cold body as a result of energy dispersion.

4. Entropy Associated with State Changes

a. The entropy of a system always increases (ΔS is (+)) when:

   i. A solid changes into a liquid.

   ii. A solid changes into a gas.

   iii. A liquid changes into a gas.

   iv. A chemical reaction occurs in which the moles of gas increase ($\Delta n_{gas} > 0$).

b. A change in entropy that occurs for a phase change is one that occurs when heat is exchanged at a constant temperature (T):

$$\Delta S^{\circ}_{rxn} = \frac{q_{rev}}{T}$$

Entropy is a measure of the energy dispersal per unit temperature.

c. A reversible process is one that reverses direction upon an infinitesimally small change in some property (such as a temperature or pressure change).

   i. Reversible processes are in a constant state of equilibrium and represent idealized conditions.

5. Heat Transfer and Entropy Changes of the Surroundings

a. The entropy change of the universe is the sum of the entropy changes of the system (ΔS_{sys}) and its surroundings (ΔS_{surr}):

$$\Delta S_{universe} = \Delta S_{surr} + \Delta S_{sys}$$

b. Since the entropy change of the universe must be positive for a spontaneous process, it follows that:

$$\Delta S_{surr} + \Delta S_{sys} > 0$$

c. This means that the change in entropy of the surroundings must be greater than the negative change in entropy of the system for a spontaneous process:

$$\Delta S_{surr} > -\Delta S_{sys}$$

    i. For an exothermic process ($\Delta S_{surr} > 0$), the entropy of the surroundings must increase more than the entropy of the system decreases (if it decreases).

    ii. For an endothermic process ($\Delta S_{surr} < 0$), the entropy of the system must increase more than the entropy of the surroundings decreases.

d. The spontaneity of a process is temperature dependent since entropy measures heat dispersion per unit temperature.

e. Since entropy is temperature dependent and related to the enthalpy change of a chemical reaction, we can relate the entropy change of the surroundings to the temperature and enthalpy change:

$$\Delta S_{surr} = \frac{-\Delta H_{sys}}{T}$$

EXAMPLE:

Calculate the entropy change of the surroundings and determine the sign of ΔS_{sys} when the following reactions are carried out.

a. $4Fe(s) + 3O_2(g) \rightarrow 2Fe_2O_3(s)$ $\Delta H = -1648$ kJ/mol at 25 °C

The change in entropy of the surroundings is:

$$\Delta S_{surr} = \frac{-(-1648 \text{ kJ/mol})}{(298 \text{ K})} = +5.53 \text{ kJ/mol} \cdot \text{K}$$

The entropy change of the system is negative since the change in moles of gas molecules is negative.

b. $2H_2O(l) + O_2(g) \rightarrow 2H_2O_2(l)$ $\Delta H = +196$ kJ/mol at 25 °C

$$\Delta S_{surr} = \frac{-(+196 \text{ kJ/mol})}{(298 \text{ K})} = -0.658 \text{ kJ/(mol} \cdot \text{K)} = -658 \text{ J/(mol} \cdot \text{K)}$$

The entropy change of the system is negative since the change in moles of gas molecules is negative.

c. $2F_2(g) + Si(s) \rightarrow SiF_4(g)$ $\Delta H = -1615$ kJ/mol at 25 °C

$$\Delta S_{surr} = \frac{-(-1615 \text{ kJ/mol})}{(298 \text{ K})} = +5.42 \text{ kJ/mol} \cdot \text{K}$$

The entropy change of the system is negative since the change in moles of gas molecules is negative.

6. Gibbs Free Energy

a. We can rewrite the change in entropy of the universe (the Second Law) using system terms by replacing the entropy change of the surroundings with the enthalpy change divided by the temperature:

$$\Delta S_{universe} = \frac{-\Delta H_{sys}}{T} + \Delta S_{sys}$$

b. Rearranging, we get:

$$T\Delta S_{universe} = -\Delta H_{sys} + T\Delta S_{sys}$$

c. We introduce a new thermodynamic function, Gibbs free energy (G):

$$G = H - TS$$

d. The Gibbs free energy change at constant temperature is:

$$\Delta G = \Delta H - T\Delta S$$

e. We can see that the expressions for the entropy change of the universe and the Gibbs free energy of a system are the same. This means that if the entropy change of the universe is a predictor of spontaneity, the Gibbs free energy of a system is as well.

f. The free energy is often referred to as the chemical potential and is used to predict if a chemical process is spontaneous.

g. When $\Delta G < 0$, the process occurs spontaneously in the forward direction.

    i. When the entropy change (ΔS) is positive and the enthalpy change (ΔH) is negative, the free energy change will always be negative.

    ii. When the entropy change (ΔS) is positive and the enthalpy change (ΔH) is positive, the free energy change will be negative at high temperatures.

        1. This is called an entropically driven reaction.

    iii. When the entropy change is negative (ΔS) and the enthalpy change (ΔH) is negative, the free energy change will be negative at low temperatures.

        1. This is called an enthalpically driven reaction.

    iv. When the entropy change is negative and the enthalpy change is positive, the free energy change will never be negative.

▶ The graphic below can be used to remember how enthalpy and entropy relate to spontaneity.

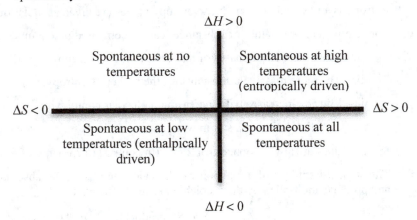

EXAMPLE:

Calculate the free energy change for the following processes, and determine whether or not the process occurs spontaneously at 25 °C. If the reaction is nonspontaneous at 25 °C, at what temperature, if any, will it become spontaneous?

a. $HCl(s) + NH_3(g) \rightarrow NH_4Cl(s)$ $\Delta H = -176.2$ kJ/mol, $\Delta S = -285.1$ J/(mol·K)

This reaction is enthalpy favored but entropy disfavored; we expect it to be spontaneous at low temperatures. In order to add the enthalpy and entropy terms, both need to be in units of kJ or J. We will convert the entropy into units of kJ/(mol·K) and then calculate the free energy change at 298 K:

$$\Delta G = -176.2 \text{ kJ/mol} - [(298 \text{ K})(-0.2851 \text{ kJ})/(\text{mol·K})] = -91.3 \text{ kJ/mol}$$

So the reaction is spontaneous at 25 °C.

b. $2C_4H_{10}(l) + 13O_2(g) \rightarrow 8CO_2(g) + 10H_2O(g)$ $\qquad \Delta H = -5271 \text{ kJ/mol}, \Delta S = +468.8 \text{ J/(mol·K)}$

This reaction is enthalpy and entropy favored; we expect it to be spontaneous at all temperatures. To confirm, we can calculate ΔG, first converting entropy to units of kJ/(mol·K) for the entropy change:

$$\Delta G = -5271 \text{ kJ/mol} - [(298 \text{ K})(+0.4688 \text{ kJ})/(\text{mol·K})] = -5411 \text{ kJ/mol}$$

This reaction is also spontaneous at 25 °C.

---

7. Entropy Changes in Chemical Reactions: ΔS°_{rxn}

   a. The standard entropy change of a reaction (ΔS°_{rxn}) is the change in entropy for a process in which all species are in their standard states.

   b. The standard state is defined based on the phase of the substance.

      i. The standard state for a gas is the pure gas at a pressure of exactly 1 atm.

      ii. The standard state for a liquid or solid is the pure substance in its most stable form at a pressure of 1 atm and at the temperature of interest (often taken to be 25 °C).

      iii. The standard state for a substance in solution is a concentration of 1 M.

   c. The third law of thermodynamics states that the entropy of a perfect crystal at 0 K is zero.

      i. A perfect crystal has only one configuration, so $W = 1$.

      ii. The third law provides a zero of entropy so that we can calculate absolute entropy values of all substances.

   d. Entropy is an extensive property, so entropy values are given as molar quantities.

   e. The relative entropy values of substances can be compared in a number of different cases.

      i. More disordered phases of matter have higher entropy.

      ii. The more massive a substance is, the higher its entropy.

      iii. Entropy increases with increasing molecular complexity.

      iv. The dissolution of a salt in water increases the system entropy.

      ▶ S° values are always nonzero at 25 °C (in contrast to enthalpy of formation values).

   f. The standard entropy of a reaction can be calculated using the absolute entropy values of reactants and products multiplied by their stoichiometric coefficients:

$$\Delta S^{\circ}_{rxn} = \sum n_p S^{\circ}(\text{products}) - \sum n_r S^{\circ}(\text{reactants})$$

      ▶ Note that this looks just like Hess's law from Chapter 10.

EXAMPLE:

Calculate the standard entropy change for the following reactions using the values given in Appendix II of your textbook.

a. $C(s, \text{graphite}) + O_2(g) \rightarrow CO_2(g)$

The standard entropy change for the reaction is given by the expression:

$$\Delta S^{\circ}_{rxn} = \sum n_p S^{\circ}(\text{products}) - \sum n_r S^{\circ}(\text{reactants})$$

From Appendix II:

C(*s*, graphite)	$S^{\circ} = 5.7$ J/(mol·K)
$O_2(g)$	$S^{\circ} = 205.2$ J/(mol·K)
$CO_2(g)$	$S^{\circ} = 213.8$ J/(mol·K)

So the enthalpy change for the reaction is:

$$\Delta S^{\circ}_{rxn} = 1 \text{ mol } CO_2 \times S^{\circ}(CO_2(g)) - \{1 \text{ mol } O_2 \times S^{\circ}(O_2(g)) + 1 \text{ mol } C \times S^{\circ}(C(s, \text{graphite}))\}$$

$$\Delta S^{\circ}_{rxn} = 213.8 \text{ J/(mol·K)} - \{205.2 \text{ J/(mol·K)} + 5.7 \text{ J/(mol·K)}\} = +2.9 \text{ J/K}$$

The entropy change is very small, which makes sense because the change in moles of gas is zero.

b. $NaCl(s) \xrightarrow{\;H_2O\;} Na^+(aq) + Cl^-(aq)$

From Appendix II:

$Cl^-(aq)$	$S^{\circ} = 56.6$ J/(mol·K)
$Na^+(aq)$	$S^{\circ} = 58.45$ J/(mol·K)
$NaCl(s)$	$S^{\circ} = 72.1$ J/(mol·K)

So the enthalpy change for the reaction is:

$$\Delta S^{\circ}_{rxn} = \{1 \text{ mol } Cl^- \times S^{\circ}(Cl^-(aq)) + 1 \text{ mol } Na^+ \times S^{\circ}(Na^+(aq))\} - 1 \text{ mol } NaCl \times S^{\circ}(NaCl(s))$$

$$\Delta S^{\circ}_{rxn} = \{56.6 \text{ J/(mol·K)} + 58.45 \text{ J/(mol·K)}\} - 72.1 \text{ J/(mol·K)} = +43.0 \text{ J/K}$$

This reaction describes the dissolution of salt in water, which should have a positive entropy change in agreement with our calculated value.

c. $2H_2(g) + O_2(g) \rightarrow 2H_2O(g)$

From Appendix II:

$H_2O(g)$	$S^{\circ} = 188.8$ J/(mol·K)
$H_2(g)$	$S^{\circ} = 130.7$ J/(mol·K)
$O_2(g)$	$S^{\circ} = 205.2$ J/(mol·K)

So the enthalpy change for the reaction is:

$$\Delta S^{\circ}_{rxn} = \{2 \text{ mol } H_2O(g) \times S^{\circ}(H_2O(g))\} - \{2 \text{ mol } H_2 \times S^{\circ}(H_2(g)) + 1 \text{ mol } O_2 \times S^{\circ}(O_2(g))\}$$

$$\Delta S^{\circ}_{rxn} = \{2 \times 188.8 \text{ J/(mol·K)}\} - \{2 \times 130.7 \text{ J/(mol·K)} + 205.2 \text{ J/(mol·K)}\} = -89.0 \text{ J/K}$$

In this reaction the number of moles of gas decreases, so it is reasonable to expect the entropy change to be negative.

EXAMPLE:

Determine the sign of the entropy change in the following processes:

a. $PCl_5(s) \rightarrow PCl_3(s) + Cl_2(g)$

The formation of a gas upon the decomposition of a solid results in a positive entropy change; that is, entropy increases, so ΔS is (+).

b. $Hg(l) \rightarrow Hg(g)$

A phase transition in which a liquid becomes a gas has a positive entropy change; that is, entropy increases, so ΔS is (+).

c. $C_2H_5OH(g) \rightarrow C_2H_5OH(l)$

The condensation of gaseous ethanol to liquid ethanol has a negative entropy change; that is, entropy decreases, so ΔS is (−).

8. Free Energy Changes in Chemical Reactions: Calculating ΔG°_{rxn}

   a. There are three ways to calculate a ΔG°_{rxn} value:

      i. $\Delta G^\circ_{rxn} = \Delta H^\circ_{rxn} - T\Delta S^\circ_{rxn}$

         ▶ This approach can be applied to reactions at any temperature, with the assumption that ΔH°_{rxn} and ΔS°_{rxn} do not change.

EXAMPLE:

Calculate the standard free energy change for the reaction of aluminum oxide with hydrogen to produce solid aluminum and water vapor at 45 °C:

$$Al_2O_3(s) + 3H_2(g) \rightarrow 2Al(s) + 3H_2O(g)$$

We can look up the standard enthalpy and entropy values in Appendix II:

$Al_2O_3(s)$	$\Delta H^\circ_f = -1675.7$ kJ/mol	$S^\circ = 50.9$ J/(mol·K)
$H_2(g)$	$\Delta H^\circ_f = 0$ kJ/mol	$S^\circ = 130.7$ J/(mol·K)
$Al(s)$	$\Delta H^\circ_f = 0$ kJ/mol	$S^\circ = 28.32$ J/(mol·K)
$H_2O(g)$	$\Delta H^\circ_f = -241.8$ kJ/mol	$S^\circ = 188.8$ J/(mol·K)

We will now calculate the value of ΔH°_{rxn} and ΔS°_{rxn} from these values:

$$\Delta H^\circ_{rxn} = \{2 \times [0 \text{ kJ/mol} + 3 \times (-241.8 \text{ kJ/mol})]\} - \{-1675.7 \text{ kJ/mol} + [3 \times 0 \text{ kJ/mol}]\} = +950.3 \text{ kJ/mol}$$

$$\Delta S^\circ_{rxn} = \{2 \times [28.32 \text{ J/(mol·K)}] + [3 \times 188.8 \text{ J/(mol·K)}]\} - \{50.9 \text{ J/(mol·K)} + [3 \times 130.7 \text{ J/(mol·K)}]\}$$

$$= +180.0 \text{ J/(mol·K)} = +0.1800 \text{ kJ/(mol·K)}$$

Using these values and the temperature, 318 K, we can calculate the standard free energy change:

$$\Delta G^\circ_{rxn} = +950.3 \text{ kJ/mol} - [(318 \text{ K})(+0.1800 \text{ kJ})/(\text{mol·K})]$$

$$\Delta G^\circ_{rxn} = +893.1 \text{ kJ/mol}$$

This reaction is nonspontaneous at room temperature because the value of the free energy change is positive.

ii. $\Delta G^{\circ}_{rxn} = \sum n_p G^{\circ}(products) - \sum n_r G^{\circ}(reactants)$

▶ Again, note that this looks just like the calculation of a reaction enthalpy in Chapter 10.

EXAMPLE:

Using the free energy of formation values in Appendix II, calculate the standard free energy change for the formation of rust at 25 °C:

$$4Fe(s) + 3O_2(g) \rightarrow 2Fe_2O_3(s)$$

From Appendix II:

  $Fe_2O_3(s)$ $\Delta G^{\circ}_f = -742.2$ kJ/mol

  $O_2(g)$ $\Delta G^{\circ}_f = 0$ kJ/mol

  $Fe(s)$ $\Delta G^{\circ}_f = 0$ kJ/mol

The standard free energy of this reaction is:

$$\Delta G^{\circ}_f = \{2 \times [-742.2 \text{ kJ/mol}]\} - \{[4 \times 0 \text{ kJ/mol}] + [3 \times 0 \text{ kJ/mol}]\} = -1484.4 \text{ kJ/mol}$$

---

iii. Using a procedure similar to Hess's law for calculating ΔH°_{rxn}; we can use a series of reactions that add up to the overall reaction to calculate ΔG°_{rxn}.

EXAMPLE:

Calculate the value of ΔG°_{rxn} for the synthesis of boron oxide from its elements:

$$4B(s) + 3O_2(g) \rightarrow 2B_2O_3(s)$$

Given the following ΔG°_{rxn} values:

1. $B_2O_3(s) + 3H_2O(g) \rightarrow 3O_2(g) + B_2H_6(g)$ $\Delta G^{\circ}_{rxn} = +1967.7$ kJ/mol

2. $H_2O(l) \rightarrow H_2O(g)$ $\Delta G^{\circ}_{rxn} = -108.2$ kJ/mol

3. $2H_2(g) + O_2(g) \rightarrow 2H_2O(l)$ $\Delta G^{\circ}_{rxn} = -240.8$ kJ/mol

4. $2B(s) + 3H_2(g) \rightarrow 2B_2H_6(g)$ $\Delta G^{\circ}_{rxn} = +87.6$ kJ/mol

We first need to arrange these equations so that they will add up to give the correct overall equation.

Reaction 1 will be reversed and multiplied by 2 to give the $2B_2O_3(s)$ molecules as products; this means that the ΔG°_{rxn} of the reaction will be multiplied by 2 and the sign will be reversed:

1. $6O_2(g) + 2B_2H_6(g) \rightarrow 2B_2O_3(s) + 6H_2O(g)$ $\Delta G^{\circ}_{rxn} = -2 \times (+1967.7 \text{ kJ/mol}) = -3935.4$ kJ/mol

Reaction 2 will be reversed and multiplied by 6 so that the gaseous water in the first reaction is replaced by liquid water:

2. $6H_2O(g) \rightarrow 6H_2O(l)$ $\Delta G^{\circ}_{rxn} = -6 \times (-108.2 \text{ kJ/mol}) = +649.2$ kJ/mol

Reaction 3 will be multiplied by 3 and reversed to cancel out the formation of liquid water:

3. $6H_2O(l) \rightarrow 6H_2(g) + 3O_2(g)$ $\Delta G^{\circ}_{rxn} = -3 \times (-240.8 \text{ kJ/mol}) = +722.4$ kJ/mol

Finally, we multiply reaction 4 by 2 to give the correct number of boron atoms as a reactant:

4. $4B(s) + 6H_2(g) \rightarrow 4B_2H_6(g)$ $\Delta G^{\circ}_{rxn} = 2 \times (+87.6 \text{ kJ/mol}) = +175.2 \text{ kJ/mol}$

Adding these reactions and their corresponding ΔG°_{rxn} values gives:

$$4B(s) + 3O_2(g) \rightarrow 2B_2O_3(s) \qquad \Delta G^{\circ}_{rxn} = -2388.6 \text{ kJ/mol}$$

---

    b.   Free energy is a measure of the maximum amount of work available from a system.

         i.   Entropy measures the amount of energy that is lost as heat (divided by temperature), while free energy measures the energy that remains to do work.

         ii.   Reversible reactions are reactions that occur infinitesimally slowly; in a reversible reaction, the free energy can be drawn out of the system in infinitesimally small amounts that exactly match the energy produced.

         iii.   All real reactions are irreversible, meaning that the real energy that can be extracted as work is less than the theoretical amount (ΔG).

         iv.   Thus, ΔG° values also represent the minimum amount of energy required to drive a nonspontaneous reaction forward.

9.   Free Energy Changes for Nonstandard States: The Relationship between ΔG°_{rxn} and ΔG_{rxn}

    a.   Ordinarily, reactions are not carried out at standard conditions.

    b.   The relationship between the standard free energy change and the free energy change for an ordinary, nonstandard state system is:

$$\Delta G_{rxn} = \Delta G^{\circ}_{rxn} + RT \ln Q$$

where R is the gas constant, T is the absolute temperature, and Q is the reaction quotient, with all gas-phase species expressed as partial pressures and all aqueous species expressed in molarity.

         i.   At standard conditions, $Q = 1$ and $\Delta G_{rxn} = \Delta G^{\circ}_{rxn}$.

         ii.   At equilibrium, $Q = K$ and $\Delta G_{rxn} = 0$.

            ▶   Note that at equilibrium, ΔG°_{rxn} does NOT equal zero; only ΔG_{rxn} does. This is a common mistake.

EXAMPLE:

Calculate the free energy change for the combustion of methane, $CH_4(g)$, at 25 °C given the following initial reaction conditions and the standard free energy change, $\Delta G^{\circ}_{rxn} = -801.1$ kJ/mol.

a. $P_{CH_4} = 0.142$ atm, $P_{O_2} = 0.715$ atm, $P_{H_2O} = 0.574$ atm, and $P_{CO_2} = 0.283$ atm

    First, we can write the equilibrium reaction:

$$CH_4(g) + 2O_2(g) \leftrightarrows CO_2(g) + 2H_2O(g) \qquad \Delta G^{\circ}_{rxn} = -801.1 \text{ kJ/mol}$$

    The reaction quotient expression for this reaction is:

$$Q = \frac{P_{CO_2} P_{H_2O}^2}{P_{CH_4} P_{O_2}^2}$$

Now we can calculate the value of Q for the conditions given:

$$Q = \frac{(0.283 \text{ atm})(0.574 \text{ atm})^2}{(0.142 \text{ atm})(0.715 \text{ atm})^2} = 1.284$$

Next, we can calculate the value of ΔG_{rxn}:

$$\Delta G_{rxn} = -801.1 \text{ kJ/mol} + (8.3145 \text{ J/(mol·K)})(298 \text{ K}) \ln(1.284) \times \frac{1 \text{ kJ}}{1000 \text{ J}} = -800.5 \text{ kJ/mol}$$

Since the value of ΔG_{rxn} is less than zero, the reaction will proceed spontaneously in the forward direction.

b. $P_{CH_4} = 0.00082$ atm, $P_{O_2} = 0.000051$ atm, $P_{H_2O} = 42.10$ atm, and $P_{CO_2} = 29.53$ atm

We will recalculate Q under the new conditions:

$$Q = \frac{(29.53 \text{ atm})(42.10 \text{ atm})^2}{(0.00082 \text{ atm})(0.000051 \text{ atm})^2} = 2.454 \times 10^{16}$$

Substituting into the expression for ΔG_{rxn}:

$$\Delta G_{rxn} = -801.1 \text{ kJ/mol} + (8.3145 \text{ J/(mol·K)})(298 \text{ K}) \ln(2.454 \times 10^{16}) \times \frac{1 \text{ kJ}}{1000 \text{ J}} = -707.6 \text{ kJ/mol}$$

The reaction is spontaneous in the forward direction under this second set of conditions.

---

10. Free Energy and Equilibrium: Relating ΔG_{rxn}^o to the Equilibrium Constant (K)

   a. The standard free energy change predicts the direction of spontaneous change when all species (reactants and products) are combined under standard conditions.

   b. The equilibrium constant indicates whether a reaction is product-favored or reactant-favored.

   c. The standard free energy change and the equilibrium constant are related:

$$\Delta G_{rxn}^o = -RT \ln K$$

   i. When the reaction favors products ($K > 1$), the reaction will proceed spontaneously in the forward direction from standard conditions ($\Delta G_{rxn}^o < 0$).

   ii. When the reaction favors reactants ($K < 1$), the reaction will not proceed spontaneously in the forward direction from standard conditions ($\Delta G_{rxn}^o > 0$).

   iii. When the reaction reaches equilibrium at a position where the ratio of products to reactants is 1 ($K = 1$), the reaction is at equilibrium under standard conditions ($\Delta G_{rxn}^o = 0$).

EXAMPLE:

Calculate the equilibrium constant for the combustion of methane at 25 °C, and determine if the reaction is product-favored or reactant-favored at this temperature.

We were given $\Delta G_{rxn}^o = -801.1$ kJ/mol in the previous example, so we can calculate the equilibrium constant using the equation:

$$\Delta G_{rxn}^o = -RT \ln K$$

We need to rearrange to solve for K:

$$K = e^{-\Delta G^{\circ}_{rxn}/RT}$$

Substituting the given values:

$$K = e^{-(-801100 \text{ J/mol})/(8.3145 \text{ J/K}\cdot\text{mol})(298 \text{ K})}$$

The equilibrium constant is so large that most calculators will overflow with this input! This is consistent with the calculations that we did in the previous problem where we saw that even at very high partial pressures of the products, the reaction would continue to proceed in the forward direction spontaneously.

---

d. The temperature dependence of the equilibrium constant can be used to determine thermodynamic quantities by combining the expression for ΔG°_{rxn} in terms of the equilibrium constant and the expression for ΔG°_{rxn} in terms of the enthalpy and entropy changes:

$$\Delta G^{\circ}_{rxn} = \Delta H^{\circ}_{rxn} - T\Delta S^{\circ}_{rxn}$$

$$-RT \ln K = \Delta H^{\circ}_{rxn} - T\Delta S^{\circ}_{rxn}$$

e. This expression can be arranged to give the equation of a line, plotting $\ln K$ versus $1/T$:

$$\ln K = \frac{-\Delta H^{\circ}_{rxn}}{R}\left(\frac{1}{T}\right) + \frac{\Delta S^{\circ}_{rxn}}{R}$$

where the slope is equal to $-\Delta H^{\circ}_{rxn}/R$ and the y-intercept is equal to $\Delta S^{\circ}_{rxn}/R$.

Fill in the Blank:

1. The change in _____ is proportional to the negative of the entropy change of the universe.

2. The entropy of a sample of matter _____ when it goes from the liquid phase to the solid phase.

3. A(n) _____ is a process that cannot occur without ongoing outside intervention.

4. The free energy of formation of pure elements in their standard states is _____.

5. _____ is a measure of the number of energetically equivalent ways that a system can be prepared.

6. The entropy of a perfect crystal at 0 K is _____.

7. The value of the equilibrium constant is _____ when the standard free energy change is a positive number.

8. Entropy is a(n) _____, meaning that the change in entropy can be calculated by finding the difference between the final entropy and the initial entropy of a system.

9. In a reversible chemical reaction, the change in magnitude of the free energy is equal to the _____.

10. A reaction that is nonspontaneous at high temperatures but spontaneous at low temperatures has a(n) _____ entropy change and a(n) _____ enthalpy change.

11. In a spontaneous process, a chemical system proceeds in a direction that _____ the entropy of the universe.

12. The entropy of He is _____ than the entropy of Ne because of _____.

13. When a chemical reaction is multiplied by some factor, the ΔG_{rxn} is _____.

14. _____ is the study of the extent of a reaction, while _____ is the study of the speed of a reaction.

15. In a spontaneous exothermic reaction, the change in enthalpy of the system is _____ while the change in entropy of the surroundings is _____.

16. The free energy of a reaction under standard conditions can be calculated by _____.

17. A spontaneous change occurs when the change in free energy is _____.

Problems:

1. Which of the following substances would you expect to have the highest entropy at room temperature?

   a. $BF_3(s)$ b. $COF_2(s)$ c. $C(s)$ d. $BF_3(l)$ e. $COF_2(l)$

   Explain the reasoning behind your choice.

2. Without doing any calculations, predict the signs of ΔH°_{rxn}, ΔS°_{rxn}, and ΔG°_{rxn} for the dissolution of barium sulfate in water at standard conditions:

$$BaSO_4(s) \leftrightharpoons Ba^{2+}(aq) + SO_4^{2-}(aq)$$

3. Methane and hexane are both compounds that contain only carbon and hydrogen (hydrocarbons).

   Methane: $T_{bp} = 112\ °C$ $\Delta H_{vap} = 8.20\ kJ/mol$

   Hexane: $T_{bp} = 342\ °C$ $\Delta H_{vap} = 28.90\ kJ/mol$

   a. Calculate the entropy change of the surroundings that occur upon the vaporization of methane and hexane.

   b. Provide an explanation for the difference in entropy changes that you calculated.

4. Consider the reactions shown below.

   $2H_2(g) + O_2(g) \rightarrow 2H_2O(l)$

   $C(graphite) + O_2(g) \rightarrow CO_2(g)$

   $CH_4(g) + 2O_2(g) \rightarrow CO_2(g) + 2H_2O(l)$

   a. Using the values given in Appendix II of your textbook, calculate ΔH°_{rxn} and ΔS°_{rxn} for each of the reactions.

   b. Using the values that you calculated in part a, calculate ΔG°_{rxn} for each reaction at 25 °C.

   c. Calculate the value of ΔG°_{rxn} for each reaction at 25 °C using the values of ΔG°_f given in Appendix II.

   d. Calculate the value of ΔG°_{rxn} for the reaction of graphite and hydrogen gas to produce methane at 25 °C.

   e. Calculate the equilibrium constant of the reaction using the value of ΔG°_{rxn} that you calculated in part d.

   f. Over what temperature range will the reaction in part d be spontaneous?

5. Consider the reactions shown below.

   $2O(g) \rightarrow O_2(g)$

   $H_2O(l) \rightarrow H_2O(g)$

   $2H(g) + O(g) \rightarrow H_2O(g)$

   $C(graphite) + 2O(g) \rightarrow CO_2(g)$

   $C(graphite) + O_2(g) \rightarrow CO_2(g)$

   $C(graphite) + 2H_2(g) \rightarrow CH_4(g)$

   $2H(g) \rightarrow H_2(g)$

a. Without doing any calculations, predict the sign of the entropy change for each of the reactions given.

b. Using the values given in Appendix II of your textbook, calculate ΔH°_{rxn} and ΔS°_{rxn} for each of the reactions.

c. Using the values that you calculated in part a, calculate ΔG°_{rxn} for each reaction at 25 °C.

d. Calculate the value of ΔG°_{rxn} for each reaction at 25 °C using the values of ΔG°_f given in Appendix II.

e. Calculate the value of ΔG°_{rxn} for the reaction combustion of methane at 25 °C.

f. Calculate the equilibrium constant for the combustion of methane at 25 °C, using the value of ΔG°_{rxn} that you calculated in part e.

g. Over what temperature range will the reaction in part d be spontaneous?

6. Under what conditions will a reaction be spontaneous at all temperatures?

7. Urea is a common fertilizer that is synthesized industrially by reacting carbon dioxide and ammonia:

$$CO_2(g) + 2NH_3(g) \leftrightarrows CO(NH_2)_2(s) + H_2O(l)$$

a. Calculate the standard free energy change for this reaction using the values given in Appendix II of your textbook.

b. If 50.0 g of urea are placed into a closed 1.0 L container at 25 °C, what will the partial pressure of CO_2 be at equilibrium?

c. What percentage of urea has decomposed at equilibrium?

d. At what temperature is the decomposition of urea spontaneous?

e. Suggest two conditions that would aid in the efficient synthesis of urea using the above chemical reaction, and explain how the equilibrium position will change when these conditions are imposed. *Note:* You should not add or remove material in answering this question.

8. Hydrogen gas and iodine vapor react to produce hydrogen iodide gas:

$$H_2(g) + I_2(g) \leftrightarrows 2HI(g)$$

a. Calculate the standard free energy of the reaction (ΔG°_{rxn}) using the free energy of formation values given in your textbook.

b. Calculate the equilibrium constant for the reaction at 25 °C.

c. Calculate the free energy change, ΔG_{rxn}, at 25 °C when the partial pressures of all gases are 0.33 atm.

d. Calculate the free energy change, ΔG_{rxn}, at 25 °C when the partial pressures of H_2 and I_2 gases are 0.33 atm while the partial pressure of HI is 0.033 atm.

e. Which situation, the one in part c or the one in part d, is closer to equilibrium? Use the values of ΔG_{rxn} that you calculated to support your answer.

9. An important reaction in atmospheric chemistry is that between nitrogen monoxide and ozone to form nitrogen dioxide and oxygen gas:

$$NO(g) + O_3(g) \leftrightarrows NO_2(g) + O_2(g)$$

a. Calculate the standard free energy change for the reaction using the values given in Appendix II in your textbook.

b. Based on your answer in part a, is the reaction product-favored, reactant-favored, or neither?

c. Calculate the equilibrium constant for the reaction.

d. Calculate the standard entropy change for this reaction using the values given in Appendix II in your textbook.

    e.   Explain the value of the standard entropy change that you calculated in part d.

    f.   Do you think that this reaction is entropically or enthalpically driven? Justify your answer.

10. Structural isomers are compounds that have the same molecular formula but different structures. Butane and isobutane (C_4H_{10}) are examples of structural isomers.

    a.   Calculate the mole percent of butane in a mixture of these isomers at 25 °C given that the standard free energy of formation for isobutane is −18.0 kJ/mol and the standard energy of formation for butane is −15.9 kJ/mol.

    b.   Based on your answer in part a, which species do you believe is more stable at room temperature?

Concept Questions:

1. The precipitation of AgCl upon the combination of aqueous solutions of NaCl and $AgNO_3$ seems to be a violation of the second law of thermodynamics because solids are more ordered than ions in solution. Explain why this is not, in fact, a violation of the second law.

2. Another statement of the second law of thermodynamics is that absolute zero can never be reached. Explain this statement using what you have learned in this chapter.

3. Explain why water vapor condenses at low temperatures but not at high temperatures based on the size of ΔS_{surr} for the process.

4. Use the second law of thermodynamics to explain why a gas will spontaneously fill its container.

5. For each of the following pairs, determine which has the greater absolute entropy. Explain.

    a. $C(s)$ or $Ag(s)$

    b. $Br_2(g)$ at 298 K or $Br_2(g)$ at 398 K

    c. $F_2(g)$ or $2F(g)$

6. We metabolize sugars in our bodies through a combustion reaction much like the burning of gasoline in a car motor. Our body carries out this reaction in many steps, whereas a car carries it out in a single step. In terms of the reversibility of a chemical reaction, explain why more useful work can be obtained by the body than by a combustion engine.

7. The value of ΔG°_{rxn} is proportional to the natural log of the equilibrium constant. This means that when the equilibrium constant is equal to 1, the ΔG°_{rxn} is 0. Explain what this means physically.

Consider the burning of solid glucose in gaseous oxygen to produce gaseous carbon dioxide and water vapor. What are the signs of ΔH°, ΔG°, and ΔS° for this reaction?

$$C_6H_{12}O_6(s) + 6O_2(g) \leftrightarrows 6CO_2(g) + 6H_2O(g)$$

8. The standard entropy change for the vaporization of water is +79.2 J/(mol·K), which is smaller than the standard entropy change for many other liquids. Consider the structure of water to explain why this is so. *Hint:* Methanol also has a relatively small entropy change associated with vaporization.

9. A substance's heat capacity is related to its entropy. Explain why this makes physical sense using the concept of entropy as a measure of energy dispersion.

Chapter 19: Electrochemistry

Key Learning Outcomes:

- Use the Half-Reaction Method of Balancing Aqueous Redox Equations in Acidic Solution

- Balance Redox Reactions Occurring in Basic Solution

- Calculate Standard Potentials for Electrochemical Cells from Standard Electrode Potentials of the Half-Reactions

- Predict Spontaneous Redox Reactions and Sketching Electrochemical Cells

- Relate ΔG^o and E^o_{cell}

- Relate E^o_{cell} and K

- Calculate E^o_{cell} under Nonstandard Conditions

- Predict the Products of Electrolysis Reactions

- Use Stoichiometry to Calculate the Quantity of Reactants Consumed or Products Produced in an Electrolytic Cell

Chapter Summary:

This chapter focuses on a common and useful reaction type: oxidation–reduction (also called redox) reactions. You will begin by learning how to balance these reactions in both acidic and basic media using the half-reaction method. Next, you will be introduced to two types of electrochemical cells—galvanic and electrolytic—which house spontaneous and nonspontaneous redox reactions, respectively. You will learn how to construct a galvanic cell and how to use a shorthand notation called cell diagrams to convey the overall chemistry in a galvanic cell. You will become familiar with how to calculate the cell potential using a table of standard electrode potentials. You will then relate the free energy concept from the previous chapter to the cell potential for reactions that are at standard and nonstandard conditions. The concept of nonstandard conditions will be exploited in your exploration of concentration cells. The concept of galvanic cells will be extended to the operation of batteries. Next, you will learn how external voltage can be used to drive nonspontaneous reactions in electrolytic cells and how reaction stoichiometry and Faraday's laws can be applied to calculate current, charge, or mass. Finally, you will discover that corrosion is simply an undesirable redox reaction.

Chapter Outline:

1. Lightning and Batteries

2. Balancing Oxidation–Reduction Equations

    a. Oxidation is the loss of electrons, which results in an increase in oxidation state.

    b. Reduction is the gain of electrons, which results in a decrease in oxidation state.

    c. In a redox reaction, both mass and charge must be balanced.

    d. Redox reactions can be balanced in a systematic way using the half-reaction method:

        i. Assign oxidation states to all species.

        ii. Break up the reaction into two half-reactions: one for oxidation and one for reduction.

        iii. Balance the half-reactions for mass.

            1. First, balance all atoms except O and H.

            2. Oxygen is balanced by adding water, H_2O.

3. Hydrogen is balanced by adding hydrogen ions, H^+.

4. For a reaction carried out in a basic solution, add enough hydroxide ions, OH^-, to both sides to neutralize all hydrogen ions.

iv. Add electrons to balance charge.

v. Multiply each half-reaction by the smallest whole number that results in the half-reactions having the same number of electrons.

vi. Add half-reactions to get the overall reaction.

▶ After completing all steps, check that the reaction balances for mass and overall charge, and be sure that all electrons have canceled out.

EXAMPLE:

Balance the redox reaction between dichromate ions and nitrous acid:

$$Cr_2O_7^{2-}(aq) + HNO_2(aq) \rightarrow Cr^{3+}(aq) + NO_3^-(aq)$$

We will follow the steps given above.

First, we will determine the oxidation state on all atoms:

Reactants:

Cr: +6

O (in $Cr_2O_7^{2-}$): –2

H: +1

N: +3

O (in HNO_2): –2

Products:

Cr: +3

N: +5

O: –2

We see that chromium is being reduced (each Cr has gained 3 electrons) and nitrogen is being oxidized (each N has lost 2 electrons). Note that this is consistent with the fact that the reactant has a nitrogen atom bonded to two oxygen atoms and the product has a nitrogen atom bonded to three oxygen atoms.

Second, we will separate the reaction into two half-reactions:

Oxidation: $HNO_2(aq) \rightarrow NO_3^-(aq)$

Reduction: $Cr_2O_7^{2-}(aq) \rightarrow Cr^{3+}(aq)$

Third, we will balance mass; we balance the N and Cr atoms first, and then the oxygen atoms are balanced by adding water and the hydrogen atoms are balanced by adding acid:

Oxidation: $HNO_2(aq) + H_2O(l) \rightarrow NO_3^-(aq) + 3H^+(aq)$

Reduction: $Cr_2O_7^{2-}(aq) + 14H^+(aq) \rightarrow 2Cr^{3+}(aq) + 7H_2O(l)$

Fourth, we will add electrons to balance the charge:

Oxidation: $HNO_2(aq) + H_2O(l) \rightarrow NO_3^-(aq) + 3H^+(aq) + 2e^-$

Reduction: $Cr_2O_7^{2-}(aq) + 14H^+(aq) + 6e^- \rightarrow 2Cr^{3+}(aq) + 7H_2O(l)$

Fifth, we will make the number of electrons equal by multiplying the oxidation half-reaction by 3:

Oxidation: $3HNO_2(aq) + 3H_2O(l) \rightarrow 3NO_3^-(aq) + 9H^+(aq) + 6e^-$

Reduction: $Cr_2O_7^{2-}(aq) + 14H^+(aq) + 6e^- \rightarrow 2Cr^{3+}(aq) + 7H_2O(l)$

Finally, we will add the half-reactions and cancel out species that are on both sides of the reaction:

$3HNO_2(aq) + 3H_2O(l) + Cr_2O_7^{2-}(aq) + 14H^+(aq) + 6e^- \rightarrow 3NO_3^-(aq) + 9H^+(aq) + 6e^- + 2Cr^{3+}(aq) + 7H_2O(l)$

Overall reaction: $3HNO_2(aq) + Cr_2O_7^{2-}(aq) + 5H^+(aq) \rightarrow 3NO_3^-(aq) + 2Cr^{3+}(aq) + 4H_2O(l)$

To check: Mass balance (8H, 3N, 13O, 2Cr on each side) and charge balance (+3 on each side) is achieved.

---

3. Voltaic (or Galvanic) Cells: Generating Electricity from Spontaneous Chemical Reactions

   a. Electrical current is the flow of electrical charge.

      i. An example is the flow of electrons through a wire or ions migrating in a solution.

   b. In a redox reaction, electrons can be transferred directly (when both species are combined in a single solution) or made to flow through an external circuit when the reaction is housed in an electrochemical cell.

      i. When the oxidation and reduction reactions are separated from each other and electrons are allowed to flow through a wire, electrical work is possible.

         1. An electrochemical cell generates electricity in this way.

      ii. A voltaic (or galvanic) cell produces electrical current from a spontaneous chemical reaction.

      iii. Electrolytic cells use electrical current to drive nonspontaneous reactions.

   c. A half-cell is a cell in which half of the reaction (oxidation or reduction) takes place. It is composed of an electrolyte and an electrode.

   d. An electrode is a conductive surface through which electrons can enter or leave half-cells.

   e. The rate of current flow is measured in amperes (A). 1 ampere is equal to one coulomb per second.

      i. 1 A corresponds to the flow of 6.242×10^{18} electrons per second.

   f. Electrical current, I, flows due to a potential energy difference, V, and is measured in volts. 1 volt is equal to one joule per coulomb.

      i. A potential difference of 1 V means that a charge of 1 C experiences a potential energy difference of 1 J between electrodes.

   g. Electromotive force (emf) is the force that drives electron motion due to a potential energy difference.

   h. The difference between the potential energy of each half-reaction is the cell potential (E_{cell}) or the cell emf.

      i. The cell potential is a measure of the relative tendencies of reactants to undergo oxidation or reduction.

   i. The cell potential at standard conditions (1 M aqueous solution and 1 atm pressure for gases) is called the standard cell potential (E^o_{cell}) or standard condition emf.

   j. The anode is the electrode where oxidation occurs and electrons are released; the anode has a negative charge in a galvanic cell.

   k. The cathode is the electrode where reduction occurs and electrons are taken up; the cathode has a positive charge in a galvanic cell.

      i. Electrons flow from the negative terminal (anode) to the positive terminal (cathode).

l.   A salt bridge connects the anode and cathode half-cells.

   i.   The salt bridge contains a strong electrolyte capped with a semipermeable membrane that allows ions to flow in order to ensure charge neutrality.

      1.   Anions travel toward the anode in order to balance the loss of electrons (the buildup of positive charge).

      2.   Cations travel toward the cathode in order to balance the gain of electrons (the buildup of negative charge).

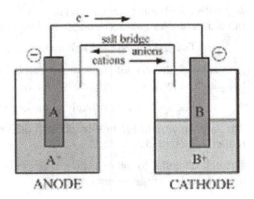

m.   Cell diagrams or line notation is a shorthand way of writing cells.

   i.   In a cell diagram, the components of the oxidation half-cell (electrode and electrolyte) are always written on the left, and the components of the reduction half-cell (electrolyte and electrode) are always written on the right.

      1.   The oxidation and reduction half-cells are separated by double vertical lines indicating the salt bridge.

   ii.   Phases are separated by a single vertical line.

      1.   When there is more than one species in the same phase, the species are separated by a comma.

      2.   If all species are in the aqueous phase, an inert electrode (such as solid platinum) is used to provide a surface on which electron transfer can occur.

   iii.   Phases are written in the same order as seen in the chemical reaction.

EXAMPLE:

Draw the cell diagram for the reaction that was balanced in the last example.

$$3HNO_2(aq) + Cr_2O_7^{2-}(aq) + 5H^+(aq) \rightarrow 3NO_3^-(aq) + 2Cr^{3+}(aq) + 4H_2O(l)$$

The balanced half-cell reactions are:

Oxidation: $3HNO_2(aq) + 3H_2O(l) \rightarrow 3NO_3^-(aq) + 9H^+(aq) + 6e^-$

Reduction: $Cr_2O_7^{2-}(aq) + 14H^+(aq) + 6e^- \rightarrow 2Cr^{3+}(aq) + 7H_2O(l)$

We write oxidation on the left and reduction on the right.

$$Pt(s) \,|\, HNO_2(aq), NO_3^-(aq), H^+(aq) \,||\, Cr_2O_7^{2-}(aq), H^+(aq), Cr^{3+}(aq) \,|\, Pt(s)$$

Notice that we have used platinum electrodes on both sides because all species are aqueous.

---

4. Standard Electrode Potentials

   a. Half-cell potentials cannot be measured directly.

      i. We assign the standard hydrogen electrode (SHE) as the zero of potential and measure the potential of all other half-cells against this:

$$2H^+(aq) + 2e^- \rightarrow H_2(g) \qquad E^o_{cell} = 0.00 \text{ V}$$

   b. The more negative the electrode potential, the greater the potential of an electron at that electrode.

   c. A positive half-cell potential means that the species under consideration has a greater tendency than the standard hydrogen electrode to undergo reduction.

      i. The species is less likely to undergo oxidation than the SHE.

   d. A negative half-cell potential means that the species under consideration has a greater tendency than the standard hydrogen electrode to undergo oxidation.

      i. The species is less likely to undergo reduction than the SHE.

   e. The cell potentials for half-reactions combined with the SHE are tabulated in a table of standard reduction potentials.

   f. The cell potential for an oxidation reaction is found by reversing the reduction reaction and taking the negative value of the reduction potential.

$$E^o_{oxidation} = -E^o_{reduction}$$

   g. The cell potential is found by adding the cell potentials for the two half-reactions as compared with the SHE.

$$E^o_{cell} = E^o_{oxidation} + E^o_{reduction}$$

   h. Cell potentials are intensive properties and are therefore not multiplied by factors when changing the stoichiometric coefficients in reactions.

      i. The cell potential is a measure of potential energy differences per unit charge and therefore does not depend on the amount of material.

EXAMPLE:

What is the cell potential for the redox reaction shown below if all species are initially at a concentration of 1 M?

$$Fe^{3+}(aq) + Mg(s) \rightarrow Fe^{2+}(aq) + Mg^{2+}(aq)$$

The half-reactions with the cell potentials from Table 18.1 in your textbook are:

Oxidation:	$Mg(s) \rightarrow Mg^{2+}(aq) + 2e^-$	$E^o_{oxidation} = -E^o_{reduction} = 2.37 \text{ V}$
Reduction:	$Fe^{3+}(aq) + e^- \rightarrow Fe^{2+}(aq)$	$E^o_{reduction} = 0.77 \text{ V}$

Notice that the oxidation reaction is the reverse of the reduction reaction listed in the table; the half-cell potential is therefore the reverse of the value given in the table.

The cell potential when these two reactions are combined is the sum of the half-cell potentials, so:

$$E^o_{cell} = 2.37 \text{ V} + 0.77 \text{ V} = 3.14 \text{ V}$$

---

   i. The table of standard reduction potentials can be used to predict the direction of redox reactions at standard conditions.

      i. Substances that are at the top of the table have a strong tendency to be reduced; they are excellent oxidizing agents.

ii. Substances that are at the bottom of the table have a strong tendency to be oxidized; they are excellent reducing agents.

iii. Any half-reaction will be spontaneous when paired with a half-reaction that is below it on the table, provided it is reversed to be an oxidation.

j. Metals can be dissolved in acidic solutions to produce hydrogen gas if the metal is below the SHE on the table of standard reduction potentials.

i. Nitric acid is a stronger oxidizing agent than other acids because its nitrogen reduces more readily than its acidic hydrogen.

EXAMPLE:

Predict whether or not the following reactions will be spontaneous at standard conditions:

a. $Cu^+(aq) + Ag(s) \rightarrow Ag^+(aq) + Cu(s)$

We look on the standard reduction table and find that the reduction of Cu^+ is below Ag^+; this means that this reaction will not be spontaneous as written and the reverse reaction will be spontaneous.

b. $2H^+(aq) + Zn(s) \rightarrow Zn^{2+}(aq) + H_2(g)$

On the standard reduction table, the reduction H^+ is featured above that of Zn^{2+}, meaning that the hydrogen ion is more likely to be reduced. This reaction will be spontaneous as written.

c. $2K^+(aq) + OH^-(aq) + H_2(g) \rightarrow 2H_2O(l) + 2K(s)$

On the standard reduction table, the reduction of K^+ is below that of H_2O, meaning that the potassium ion is less likely to be reduced. This reaction will not be spontaneous as written, and the reverse reaction will be spontaneous.

---

5. Cell Potential, Free Energy, and the Equilibrium Constant

a. The cell potential is its potential energy difference per unit charge:

$$E^o_{cell} = \frac{\text{Potential Difference (J)}}{\text{Charge (C)}}$$

i. The potential difference is the maximum (useful) work available from the system, so it is equal to the free energy.

ii. The charge is equal to the number of moles of electrons (n) times the charge of 1 mole of electrons (F, Faraday's constant = 96,485 C/mol e⁻).

b. Rearranging, we obtain the relationship between the standard free energy of reaction and the standard cell potential for an electrochemical cell:

$$\Delta G^o = -n \cdot F \, E^o_{cell}$$

c. A process is spontaneous at standard conditions if it has a cell potential that is positive, a free energy change that is negative, and an equilibrium constant greater than 1.

d. A process is nonspontaneous at standard conditions if it has a cell potential that is negative, a free energy change that is positive, and an equilibrium constant less than 1.

EXAMPLE:

Calculate the standard free energy change for the reaction of copper ions with silver metal at standard conditions:

$$Cu^+(aq) + Ag(s) \rightarrow Ag^+(aq) + Cu(s)$$

Using the table of standard reduction potentials, we can calculate the standard cell potential:

$$E^o_{cell} = E^o_{oxidation} + E^o_{reduction} = +0.52\text{ V} + -0.80\text{ V} = -0.28\text{ V}$$

This reaction is already balanced, and there is one electron transferred, so $n = 1$ and the standard free energy change is:

$$\Delta G^o = -n \cdot F \cdot E^o_{cell} = -1 \cdot \left(\frac{96,485\text{ C}}{\text{mol e-}}\right) \cdot \left(-0.28\frac{\text{J}}{\text{C}}\right) = +2.7 \times 10^4\text{ J}$$

Notice that the standard free energy change is a positive number, in agreement with our prediction that this reaction is not spontaneous as written.

e. We can relate the cell potential to the equilibrium constant:

$$E^o_{cell} = \frac{RT}{nF}\ln K = \frac{0.0592\text{ V}}{n}\log(K) \text{ (at 25 °C)}$$

EXAMPLE:

Use the cell potential to calculate the equilibrium constant for the reaction in the above example:

$$Cu^+(aq) + Ag(s) \rightarrow Ag^+(aq) + Cu(s) \qquad E^o_{cell} = -0.28\text{ V}$$

We need to rearrange the equation given for the cell potential in terms of the equilibrium constant in order to solve for the equilibrium constant:

$$E^o_{cell}\frac{n}{0.0592\text{ V}} = \log(K)$$

$$K = 10^{E^o_{cell}\frac{n}{0.0592\text{V}}} = 10^{(-0.28\text{ V})\frac{1}{0.0592\text{V}}} = 1.9 \times 10^{-5}$$

This answer is in agreement with our prediction that this reaction is nonspontaneous at standard conditions since the equilibrium constant is much less than 1 (it favors reactant formation).

6. Cell Potential and Concentration

a. Using the relationship between the free energy at standard and nonstandard conditions, we can find an expression for the nonstandard cell potential, called the Nernst equation:

$$E_{cell} = E^o_{cell} - \frac{0.0592\text{ V}}{n}\log(Q)$$

EXAMPLE:

Calculate the potential of a cell that is prepared using a 1.45 M solution of magnesium nitrate and a 2.32 M solution of silver nitrate. In this cell, magnesium is the anode and silver is the cathode.

We will begin by writing the half-reactions along with their cell potentials:

$$Ag^+(aq) + e^- \rightarrow Ag(s) \qquad E^o_{reduction} = 0.80\text{ V}$$

$$Mg(s) \rightarrow Mg^{2+}(aq) + 2e^- \qquad E^o_{oxidation} = -E^o_{reduction} = 2.37\text{ V}$$

We will begin by calculating the cell potential at standard conditions:

$$E^o_{cell} = E^o_{oxidation} + E^o_{reduction} = 2.37 \text{ V} + 0.80 \text{ V} = 3.17 \text{ V}$$

The balanced chemical reaction will be:

$$2Ag^+(aq) + Mg(s) \rightarrow 2Ag(s) + Mg^{2+}(aq)$$

Now we can use the Nernst equation to calculate the cell potential, using $n = 2$ and $Q = [Mg^{2+}]/[Ag^+]^2$:

$$E_{cell} = E^o_{cell} - \frac{0.0592 \text{ V}}{n} \log(Q) = 3.17 \text{ V} - \frac{0.0592 \text{ V}}{2} \log\left(\frac{1.45}{2.32^2}\right) = 3.19 \text{ V}$$

The cell potential is almost the same as the standard cell potential since the concentrations of both solutions are almost equal initially.

---

    b.   A concentration cell is a galvanic cell comprised of two equivalent half-cells that differ only in the electrolyte concentrations.

        i.   In this case, the standard cell potential is zero and the cell potential is:

$$E_{cell} = -\frac{0.0592 \text{ V}}{n} \log(Q)$$

        ii.   Electrons flow in order to equalize the concentrations of all species on both sides of the cell. The dilute cell (anode) will increase its concentration of dissolved ions, and the concentrated cell (cathode) will decrease its concentration of dissolved ions.

EXAMPLE:

Calculate the cell potential of a galvanic cell prepared using a 0.54 M solution of $Cu(NO_3)_2$ on one side of the cell and a 1.43 M solution of $Cu(NO_3)_2$ on the other side. State the direction in which electrons will flow.

We can predict that electrons will flow from the low concentration side of the cell to the high concentration side of the cell in order to equalize the concentrations.

$$Cu^{2+}(aq, conc) + e^- \rightarrow Cu(s) \qquad E^o_{reduction} = 0.34 \text{ V}$$

$$Cu(s) \rightarrow Cu^{2+}(aq, dil) + 2e^- \qquad E^o_{oxidation} = -E^o_{reduction} = -0.34 \text{ V}$$

Overall reaction: $Cu^{2+}(aq, conc) \rightarrow Cu^{2+}(aq, dil) \qquad E^o_{cell} = E^o_{oxidation} + E^o_{reduction} = 0.0 \text{ V}$

There are two electrons transferred in this reaction, so the cell potential will be:

$$E_{cell} = -\frac{0.0592 \text{ V}}{n} \log(Q) = -\frac{0.0592 \text{ V}}{2} \log\left(\frac{0.54}{1.43}\right) = 0.013 \text{ V}$$

---

    7.   Batteries: Using Chemistry to Generate Electricity

        a.   In a battery, redox reactions are used to generate electricity that can be used to do work.

            i.   Such cells are often characterized as primary (single use), secondary (rechargeable), or flow through.

        b.   A dry-cell battery does not contain significant quantities of liquids.

            i.   An inexpensive battery uses a zinc case as the anode and an ammonium chloride/manganese oxide paste as the cathode:

                Anode: $Zn(s) \rightarrow Zn^{2+}(aq) + 2e^-$

                Cathode: $2MnO_2(s) + 2NH_4^+(aq) + 2e^- \rightarrow Mn_2O_3(s) + 2NH_3(aq) + H_2O(l)$

ii. An alkaline battery is the most common disposable battery:

Anode: $Zn(s) + 2OH^-(aq) \rightarrow Zn(OH)_2(s) + 2e^-$

Cathode: $MnO_2(s) + 2H_2O(l) + 2e^- \rightarrow MnO(OH)(s) + 2OH^-(aq)$

c. Most car batteries are lead-storage batteries, which can be run in the reverse direction with the application of current to reverse the reaction:

Anode: $Pb(s) + HSO_4^-(aq) \rightarrow PbSO_4(s) + H^+(aq) + 2e^-$

Cathode: $PbO_2(s) + HSO_4^-(aq) + 3H^+(aq) + 2e^- \rightarrow PbSO_4(s) + 2H_2O(l)$

d. Nickel–cadmium batteries:

Anode: $Cd(s) + 2OH^-(aq) \rightarrow Cd(OH)_2(s) + 2e^-$

Cathode: $2NiO(OH)(s) + 2H_2O(l) + 2e^- \rightarrow 2Ni(OH)_2(s) + 2OH^-(aq)$

e. Nickel–metal hydride (NiMH) batteries:

Anode: $MH(s) + OH^-(aq) \rightarrow M(s) + H_2O(l) + e^-$

Cathode: $NiO(OH)(s) + H_2O(l) + e^- \rightarrow Ni(OH)_2(s) + OH^-(aq)$

f. Lithium ion batteries operate in ways that differ from the other examples and are beyond the scope of this book.

g. Fuel cells are a different type of battery because they require a constant supply of fuel in order to continually generate current. The hydrogen–oxygen fuel cell is:

Anode: $2H_2(g) + 4OH^-(aq) \rightarrow 4H_2O(l) + 4e^-$

Cathode: $O_2(g) + 2H_2O(l) + 4e^- \rightarrow 4OH^-(aq)$

8. Electrolysis: Driving Nonspontaneous Chemical Reactions with Electricity

a. Electrolysis is the driving of a nonspontaneous redox reaction forward with the application of an external power source.

i. Electrons are drawn from the anode so that the charge labels are the opposite of what they are in galvanic cells.

b. In the electrolysis of a pure molten salt, the nonmetal anion is oxidized and the metal cation is reduced.

i. In a mixture, the cation that is most easily reduced will be reduced first and the anion that is most easily oxidized will be oxidized first.

c. The electrolysis of aqueous solutions is more complex because water itself can be oxidized or reduced when a voltage is applied.

i. Cations of active metals cannot be reduced in aqueous solutions because water will be reduced first.

EXAMPLE:

Predict the reaction that will occur at the anode and cathode for the electrolysis of a mixture of molten $CuCl_2$ and $FeCl_3$.

The oxidation reaction is:

$$2Cl^-(l) \rightarrow Cl_2(g) + 2e^-$$

So this reaction takes place at the anode.

The reduction reaction is one of the following:

$$Cu^{2+}(l) + 2e^- \rightarrow Cu(s)$$

$$Fe^{3+}(l) + 3e^- \rightarrow Fe(s)$$

Looking at the table of standard reduction potentials, we see that Cu^{2+} is above Fe^{3+} on the table. This means that Cu^{2+} is more likely to be reduced and its reduction reaction will be the one that occurs preferentially at the cathode.

---

    d.   Overvoltage is the additional voltage, measured experimentally, that must be applied in order to drive a nonspontaneous reaction forward.

        i.   The predicted reaction may not occur because of overvoltage.

    e.   In determining the quantities of metal produced or consumed at the cathode or anode, respectively, we must consider the stoichiometry of the half-reactions.

    f.   The current applied for a given time period will give the overall charge transferred:

$$i, \text{ current (C/s)} \times \text{time } (s) = Q, \text{ charge (C)}$$

$$1 \text{ C/s} = 1 \text{ Ampere (A)}$$

EXAMPLE:

What current (in Amperes) is needed in order to plate out 5.0 g of solid silver from an aqueous solution of Ag(I) in ten minutes?

The half-reaction that is taking place is:

$$Ag^+(aq) + e^- \rightarrow Ag(s)$$

We can use the stoichiometric relationship between Ag(s) and electrons to calculate the charge transferred, where $Q = nF$:

$$5.0 \text{ g Ag} \times \frac{1 \text{ mol Ag}}{107.87 \text{ g Ag}} \times \frac{1 \text{ mol e}^-}{1 \text{ mol Ag}} \times \frac{96,485 \text{ C}}{1 \text{ mol e}^-} = 4.472 \times 10^3 \text{ C}$$

This is the total charge that must be transferred in ten minutes. The current is the charge per second, so we can continue the problem:

$$\frac{4.472 \times 10^3 \text{ C}}{10.0 \text{ min}} \times \frac{1 \text{ min}}{60 \text{ s}} = 7.5 \frac{\text{C}}{\text{s}}$$

The units of current are C/s or A, so the current needed is 7.5 A.

---

    9.   Corrosion: Undesirable Redox Reactions

        a.   Corrosion is the gradual, undesirable oxidation of metals via reaction with oxygen:

$$M(s) \rightarrow M^{n+}(aq) + ne^-$$

$$O_2(g) + 2H_2O(l) + 4e^- \rightarrow 4OH^-(aq) \qquad E^o_{reduction} = +0.40 \text{ V}$$

        b.   Oxidation occurs to a more significant extent in an acidic solution:

$$O_2(g) + 4H^+(l) + 4e^- \rightarrow 2H_2O(l) \qquad E^o_{reduction} = +1.23 \text{ V}$$

        c.   Some oxides are very strong and form a barrier that resists further oxidation.

        d.   Iron oxides form a very brittle, porous iron oxide (rust).

            i.   Rust can be prevented by using a sacrificial anode that will be more easily oxidized than iron.

Fill in the Blank:

1. The _____ relates the cell potential at nonstandard conditions to the standard cell potential.

2. A(n) _____ cell uses electrical current to drive a nonspontaneous reaction.

3. Electromotive force is the _____.

4. Oxidation corresponds to a(n) _____ in oxidation state and a(n) _____ of electrons.

5. In a chemical reaction, both _____ and _____ must be balanced.

6. The quantity of charge that flows in an electrochemical cell is given by _____.

7. The generation of electricity by separating the two half-reactions and connecting them with a wire is done in a(n) _____ cell.

8. In an electrochemical cell, oxidation occurs at the _____.

9. The _____ is a measure of the difference in potential energy per unit charge and has units of _____.

10. A(n) _____ battery does not contain large amounts of liquid water.

11. A(n) _____ is needed in a galvanic cell to maintain charge neutrality.

12. _____ is the additional voltage required to drive a nonspontaneous reaction forward.

13. A half-cell with a greater tendency to undergo reduction than the standard reduction potential will have a(n) _____ cell potential.

14. Species at the top of the table of standard reduction potentials have a strong tendency to serve as _____ agents and will therefore be _____.

15. A metal whose reduction half-reaction lies _____ the reduction of H^+ to form H_2 on the table of reduction potentials will dissolve in acids.

16. A spontaneous reaction will have a negative free-energy change and a(n) _____ cell potential.

17. A(n) _____ is different from a standard battery because it needs to have a constant fuel supply.

18. The reduction potential of the standard hydrogen electrode is _____.

19. The electrode where oxidation occurs is called the _____, and the electrode where reduction occurs is called the _____.

20. In a(n) _____ cell, the standard cell potential is zero while the actual cell potential is nonzero.

21. The standard hydrogen electrode uses a _____ electrode.

22. In an aqueous solution undergoing electrolysis, the predicted reduction or oxidation reaction might be incorrect due to the electrolysis of _____.

Problems:

1. Balance the following redox reactions:

   a. $Al(s) + O_2(g) \rightarrow Al_2O_3(s)$

   b. $NO_3^-(aq) + Al(s) \rightarrow NH_3(aq) + Al(OH)_4^-(aq)$

   c. $Cl_2(g) \rightarrow Cl^-(aq) + ClO_3^-(aq)$

   d. $HNO_3(aq) + H_3AsO_3(aq) \rightarrow NO(g) + H_3AsO_4(aq)$

   e. $NO_2(g) + H_2(g) \rightarrow NH_3(g) + H_2O(l)$

   f. $Cu(s) + HNO_3(aq) \rightarrow Cu(NO_3)_2(aq) + NO(g) + H_2O(l)$

g. $Cr(OH)_3(s) + ClO_3^-(aq) \rightarrow CrO_4^{2-}(aq) + Cl^-(aq)$ (in a basic solution)

h. $Au^{3+}(aq) + I^-(aq) \rightarrow Au(s) + I_2(s)$

i. $FeO(s) + CO(g) \rightarrow Fe(s) + CO_2(g)$

j. $TiO_2(s) + Cu(s) \rightarrow Cu^{2+}(aq) + Ti(s) + H_2O(l)$

2. A 0.010 M $CrCl_3$ aqueous solution is placed in a beaker with a chromium-plated wire, and a 0.20 M aqueous solution of $CuSO_4$ is placed in a second beaker with a copper wire. The two wires are connected to a voltmeter.

   a. The cell, as described, will not have a voltage reading. What is missing, and why does its absence result in zero voltage?

   b. Sketch the complete cell. Label the anode, cathode, and direction of electron flow, anion flow, and cation flow.

   c. Write the half-reactions for this cell, and label them as oxidation and reduction.

   d. Write the balanced, overall redox reaction, and calculate the standard cell potential.

   e. Write the cell diagram for this cell.

   f. What is the standard free-energy change for this cell?

   g. What is the equilibrium constant for this reaction?

   h. Will the potential reading on the voltmeter be the same as the standard cell potential that you calculated in part d? Why or why not?

   i. If they are different, calculate the actual cell potential.

3. An electrochemical cell is prepared from a 0.10 M aqueous solution of nickel(II) nitrate with a nickel-plated electrode for one half-cell, and an aqueous solution of lead(II) nitrate of unknown concentration and a lead wire for the other half-cell. The voltage of this cell is measured to be 0.18 V.

   a. What is the concentration of the lead nitrate solution used?

   b. What will the concentration of the lead nitrate solution be at equilibrium?

   c. Without doing any calculations, how could you increase the cell voltage of this system?

4. Consider electroplating 10.0 g of chromium from an aqueous solution of $Cr(NO_3)_3$.

   a. How long (in minutes) will it take for you to carry out this process using a 10 mA current?

   b. What current will be required to electroplate the chromium in 6.4 hours?

   c. What current will be required to electroplate the chromium in one day?

   d. Will it require more or less current to carry out this process in a chromium nitrate solution that is twice as dilute? Explain your answer.

5. An electrochemical cell is prepared using TiO_2 in an acidic solution and a titanium electrode in one half-cell and $Cu^{2+}(aq)$ and a copper electrode in the other half-cell. The voltage reading for this cell is measured as a positive number.

   a. Does this reaction occur spontaneously? Explain how you can tell.

   b. Write the half-reactions, and label them as oxidation and reduction.

   c. Write the cell diagram for this reaction.

   d. What is the standard cell potential for this reaction? Given that the reduction potential for TiO_2 is -0.870 V and the standard reduction potential for Cu^{2+} is 0.340 V.

   e. What is the cell potential if $[Cu^{2+}] = 0.0560$ M, the pH of the titanium half-cell is 2.00, and the temperature is 25 °C?

   f. What will the cell potential be when the reaction reaches equilibrium?

6. An electrochemical cell is prepared with a silver electrode dipped in a silver nitrate aqueous solution as one half-cell and a cadmium electrode in a cadmium chloride aqueous solution as the other half-cell, with the two half-cells connected by a salt bridge.

   a. What is the spontaneous electrochemical reaction that occurs, and what is the maximum potential produced by this cell?

   b. What would be the effect on the potential of this cell if sodium sulfide were added to the Cd/Cd^{2+} half-cell and CdS is precipitated out of solution?

   c. What would be the effect on the potential of the cell if the size of the silver electrode were doubled?

7. A mixture of NaCl, KCl, and KBr is melted, and electrolysis is carried out on the molten salt solution.

   a. What reactions will occur at the anode and the cathode? Explain the reasoning behind your answers.

   b. Will the reactions be different if electrolysis is carried out on an aqueous solution of these salts? If not, why not? If so, explain why.

8. What reactions will occur at the anode and cathode when an aqueous solution of $Ca(OH)_2$ undergoes electrolysis?

9. Consider the following structures and standard reduction potentials:

| pyruvate | acetate | lactate |

pyruvate → lactate	$E^o_{cell} = -0.190$ V
acetate + CO_2 → pyruvate	$E^o_{cell} = -0.700$ V
$NAD^+ + H^+$ → NADH	$E^o_{cell} = -0.320$ V

NAD^+ is a biologically important nucleotide that is involved in a number of biochemical reactions. It can be produced in metabolism.

   a. Write a balanced chemical reaction for the conversion of lactate to acetate and CO_2.

   b. Calculate the standard cell potential and the standard free energy change for the conversion of lactate to pyruvate.

   c. Calculate the standard free energy change for the conversion of NAD^+ to NADH.

   d. Which of the reactions given is spontaneous, and which is nonspontaneous?

   e. The conversion of lactate to acetate and CO_2 is an important step in metabolism. The energy of this conversion can be coupled to the energy associated with the NADH reaction. When these reactions are carried out, will NAD^+ be consumed or produced?

   f. How much NAD^+ will be consumed or produced when 1 mole of lactate is converted to acetate and CO_2?

Concept Questions:

1. Consider four metals: A, B, C, and D. Experiments conducted on these metals gave the following results:

   • A and C react with hydrochloric acid to give hydrogen gas.

   • When C is added to solutions of the other metals, solid A, B, and D are formed.

   • Metal D spontaneously reduces metal B.

    a.   List the metals in order of increasing strength as a reducing agent.

    b.   A cell composed of which two half-cells will give the largest positive voltage? Be sure to indicate which reaction takes place at the anode and which reaction takes place at the cathode.

2.   What are the values of K_c, $\Delta G°$, and $E°_{cell}$ for a product-favored reaction?

3.   For certain reactions, the standard cell potential and the measured cell potential are the same.

    a.   What conditions are necessary in order for this to be true?

    b.   Is it possible for the standard cell potential to be zero for a reaction other than the standard hydrogen electrode? If not, why not? If so, when?

4.   Are standard reduction potentials (those listed on the table of thermodynamic values) absolute or relative values?

5.   All concentration cells have the same values of $\Delta G°$, $E°$, and K. Explain what the values are and why.

6.   Explain why the terminals of a battery are labeled in the opposite way as the electrodes of an electrolytic cell.

7.   In the last chapter, you learned that the free energy of a spontaneous reaction is the maximum useful work available for the system. Conversely, the free energy of a nonspontaneous reaction is the minimum amount of work required to drive it forward. Explain how this relates to the concept of overvoltage discussed in this chapter.

Chapter 20: Radioactivity and Nuclear Chemistry

<u>Key Learning Outcomes:</u>

- Write Nuclear Equations for Alpha Decay

- Write Nuclear Equations for Beta Decay, Positron Emission, and Electron Capture

- Predict the Type of Radioactive Decay

- Use Radioactive Decay Kinetics

- Use Radiocarbon Dating

- Use Uranium/Lead Dating to Estimate the Age of a Rock

- Determine the Mass Defect and Nuclear Binding Energy

<u>Chapter Summary:</u>

In this chapter, you will be introduced to radioactivity and nuclear chemistry. You will begin by learning about the various subatomic particles and how each of them is involved in particular types of nuclear reactions. You will see how the valley of stability is related to the ratio of protons and neutrons in a particular nuclide and how it can be used to predict whether or not a particular nuclide is radioactive. Next you will learn how to detect radioactivity and then the kinetics of radioactive decay. All radioactive decay processes obey simple first-order kinetics that can be used in radioactive dating experiments. Next you will learn about nuclear fission processes that can be used in nuclear bombs and for power generation. One of the most famous expressions of all time will then be discussed to relate the mass defect to the energy associated with nuclear reactions. Nuclear fusion will be introduced as a potential energy source that has yet to be utilized. The formation of new elements through transmutation will be described. Finally, the effects of radiation on life and the uses of radiation in medicine will be explored.

<u>Chapter Outline:</u>

1. Diagnosing Appendicitis

    a. Radioactivity is the spontaneous emission of subatomic particles or high-energy electromagnetic radiation by the nuclei of certain unstable atoms.

        i. Radioactive atoms are those that emit high-energy electromagnetic radiation.

2. The Discovery of Radioactivity

    a. Phosphorescence is the long-lived emission of light that sometimes follows the absorption of light.

    b. Radioactivity was discovered by Antoine-Henri Becquerel and Marie Curie in the late 1890s.

3. Types of Radioactivity

    a. In radioactivity discussions, a particular isotope of an element is called a nuclide.

    b. A nucleon is a subatomic particle found in the nucleus: protons or neutrons. The number of nucleons is given by the mass number of the atom.

    c. Subatomic particles are given symbols in accordance with the atomic symbols for the elements.

        i. Protons have the symbol $_1^1\text{p}$ to indicate that they have one proton and contribute to the mass number by a factor of 1.

        ii. Neutrons have the symbol $_0^1\text{n}$ to indicate that they have no protons and contribute to the mass number by a factor of 1.

iii. Electrons have the symbol $_{-1}^{0}e$ to indicate that they annihilate a proton and do not contribute to the mass number at all.

1. The symbol for the electron will become more obvious in the context of radioactive processes.

d. In a balanced nuclear reaction, the total mass number (A) and total charge (Z) of the reactants must equal those of the products.

e. Unstable nuclei undergo various types of radioactive decay: alpha decay, beta decay, gamma ray emission, positron emission, and electron capture.

f. Alpha (α) decay occurs when an unstable nucleus emits a particle with two protons and two neutrons.

i. An alpha particle is a helium nucleus: $_{2}^{4}\text{He}$

ii. In a nuclear equation, an alpha particle is emitted when a parent nuclide decays into a daughter nuclide:

$$_{92}^{238}\text{U} \rightarrow {}_{90}^{234}\text{Th} + {}_{2}^{4}\text{He}$$

iii. Alpha particles have high ionizing power (ability of radiation to ionize other molecules or atoms) because they have very high energy.

iv. Alpha radiation has very low penetrating ability (ability of radiation to penetrate matter) because alpha particles are so large.

EXAMPLE:

Predict the product of the alpha decay of americium-241.

We look on the periodic table and find that americium is element 95; the atomic symbol is then $_{95}^{241}\text{Am}$. We can determine the identity of the daughter nuclide by subtracting the helium-4 nucleus and identifying the symbol using the atomic number:

$$_{95}^{241}\text{Am} \rightarrow {}_{2}^{4}\text{He} + \underline{\hspace{1cm}}$$

$A = 241 - 4 = 237$; $Z = 95 - 2 = 93$. Thus, $_{95}^{241}\text{Am} \rightarrow {}_{2}^{4}\text{He} + {}_{93}^{237}\text{Np}$

g. Beta (β) decay occurs when an unstable nucleus emits an electron.

i. In radioactivity, a beta particle is emitted along with a proton when a neutron decays:

$$_{0}^{1}\text{n} \rightarrow {}_{1}^{1}\text{p} + {}_{-1}^{0}\text{e}$$

$$_{88}^{228}\text{Ra} \rightarrow {}_{89}^{228}\text{Ac} + {}_{-1}^{0}\text{e}$$

ii. Beta radiation has lower ionizing power and higher penetrating ability than does alpha decay because beta particles are smaller.

EXAMPLE:

Predict the product of the beta decay of hydrogen-3.

As in the last example, we will first determine the atomic number of hydrogen and write the atomic symbol as $_{1}^{3}\text{H}$. We will next determine the identity of the product by considering the loss of a beta particle in the nuclear reaction:

$$^{3}_{1}H \rightarrow \underline{\quad} + ^{0}_{-1}e$$

$A = 3 - 0 = 3; Z = 1 - (-1) = 2.$ Thus, $\qquad ^{3}_{1}H \rightarrow ^{3}_{2}He + ^{0}_{-1}e$

---

h. Gamma (γ) ray emission is the emission of electromagnetic radiation.

   i. Gamma rays are given the symbol: $^{0}_{0}\gamma$

   ii. Gamma rays are usually emitted along with other particles.

   iii. Gamma rays have very high penetrating power, which makes exposure to them very dangerous.

i. Positron emission occurs when a positron is emitted.

   i. A positron is the antiparticle of an electron: $^{0}_{+1}e$

   ii. When a positron and an electron collide, they annihilate each other and release energy as radiation.

   iii. A positron is formed with the decay of a proton to a neutron:

$$^{1}_{1}p \rightarrow ^{1}_{0}n + ^{0}_{+1}e$$

EXAMPLE:

Predict the product of the reaction when a positron is emitted from a carbon-11 nuclide.

Carbon has an atomic number of 6, so it has an atomic symbol of $^{11}_{6}C$. A positron is emitted when a proton splits to give a neutron and an emitted electron:

$$^{11}_{6}C \rightarrow \underline{\quad} + ^{0}_{+1}e$$

$A = 11 - 0 = 11; Z = 6 - 1 = 5.$ Thus, $\qquad ^{11}_{6}C \rightarrow ^{11}_{5}B + ^{0}_{+1}e$

---

j. Electron capture occurs when a nucleus assimilates an electron from an inner orbital of its electron cloud, thus converting a proton to a neutron:

$$^{1}_{1}p + ^{0}_{-1}e \rightarrow ^{1}_{0}n$$

EXAMPLE:

Predict the product of the electron capture reaction of beryllium-7.

Beryllium has an atomic number of 4, so it has an atomic symbol of $^{7}_{4}Be$. Electron capture essentially results in the conversion of a proton into a neutron:

$$^{7}_{4}Be + ^{0}_{-1}e \rightarrow \underline{\quad}$$

$A = 7 - 0 = 7; Z = 4 + (-1) = 3.$ Thus, $\qquad ^{7}_{4}Be + ^{0}_{-1}e \rightarrow ^{7}_{3}Li$

---

4. The Valley of Stability: Predicting the Type of Radioactivity

   a. The binding of protons in the nucleus is the result of the strong force, an attractive force that occurs only at short distances.

      i. Nucleus stability is a balance between the repulsive coulombic forces of the protons and the attraction of all nucleons via the strong force.

   b. A graph of the neutrons (N) versus protons (Z) for nuclides can be used to predict nuclear stability.

      i. From the graph, we can find the N/Z ratio of a particular isotope.

         1. For $Z \leq 20$, an N/Z ratio of approximately one is in the valley of stability and is therefore not likely to be radioactive. The ratio value increases (deviates from one) as the mass number increases, up to approximately 1.5.

         2. When the N/Z ratio is too high, neutrons are at a high-energy state and will be diminished via β decay.

         3. When the N/Z ratio is too low, the repulsions dominate and the number of protons will be diminished via positron emission, α decay, or electron capture.

         4. All nuclides with $Z > 83$ are radioactive and tend to undergo decay.

      ii. Stability can also be predicted by considering the average atomic mass (given on the periodic table), which represents a stable N/Z ratio.

EXAMPLE:

Predict which of the following is/are likely to decay via positron emission:

a. $^{210}_{84}\text{Po}$

Since the atomic number is greater than 83, this is an unstable nuclide. Positron emission can occur:

$$^{210}_{84}\text{Po} \rightarrow ^{210}_{83}\text{Bi} + ^{0}_{+1}\text{e}$$

It is more likely, however, that an alpha particle will be emitted to decrease the atomic number further:

$$^{210}_{84}\text{Po} \rightarrow ^{206}_{82}\text{Pb} + ^{4}_{2}\text{He}$$

b. $^{198}_{79}\text{Au}$

The ratio of neutrons to protons is $119/79 = 1.51$. We can see from Figure 21.5 in your textbook that this is above the valley of stability; the nuclide is neutron rich, so neutrons must be converted to protons. Positron emission will not accomplish this, so it will not likely occur.

c. $^{37}_{19}\text{K}$

The ratio of neutrons to protons is $18/19 = 0.95$. We can see that this is below the valley of stability; the nuclide is proton rich and can therefore decay via positron emission:

$$^{37}_{19}\text{K} \rightarrow ^{37}_{18}\text{Ar} + ^{0}_{+1}\text{e}$$

---

   c. Magic numbers are the numbers of protons and neutrons in nuclides with unique stability.

      i. Magic numbers result from the fact that nucleons occupy energy levels.

      ii. The majority of stable nuclides have an even number of protons and an even number of neutrons.

      iii. The least stable nuclides have an odd number of protons and an odd number of neutrons.

      iv. Atoms with very high atomic numbers (>83) usually decay in multiple steps.

5. Detecting Radioactivity

   a. Film-badge dosimeters use photographic film to monitor exposure to radiation.

    b. A Geiger-Müller counter uses a chamber filled with argon gas.

        i. The argon gas is ionized to give an electrical signal.

        ii. The counter will click with each radioactive particle that travels through the chamber.

    c. A scintillation counter contains a material that emits ultraviolet or visible light in response to excitation.

6. The Kinetics of Radioactive Decay and Radiometric Dating

    a. All elements above atomic number 83 have only radioactive isotopes, as well as Tc ($Z = 43$) and Pm ($Z = 61$).

    b. All radioactive elements decay via first-order kinetics.

    c. The rate of radioactive decay is equal to the number of radioactive nuclei (N) multiplied by the rate constant (k):

$$\text{Rate} = k \cdot N$$

    d. The half-life of a radioactive element is:

$$t_{1/2} = 0.693/k$$

        i. Recall that the half-life is the time required for half of a sample to decay.

    e. The integrated rate law is given in terms of the number of radioactive nuclei initially, N_0, as well as N, the number of radioactive nuclei after time, t, is elapsed:

$$\ln \frac{N_t}{N_0} = -kt$$

        i. The units of measure for N and N_0 can be in grams, atoms, or moles. The units for time are dependent upon the unit of measure for the rate constant.

    f. The integrated rate law can be expressed in terms of the rates:

$$\ln \frac{\text{rate}_t}{\text{rate}_0} = -kt$$

EXAMPLE:

Fluorine-21 decays to neon-21 with a half-life of 4.17 seconds. If you begin with a 14.0 g sample of fluorine-21, what quantity will remain after 1 minute?

We will calculate the rate constant using the half-life formula:

$$t_{1/2} = 0.693/k$$

$$k = 0.693/t_{1/2} = 0.693/(4.17 \text{ s}) = 0.166 \text{ s}^{-1}$$

We will now rearrange the integrated rate law to solve for the quantity of fluorine-21 after 1 minute (expressed in seconds to match the units of the half-life constant).

$$\ln \frac{N_t}{N_0} = -kt$$

$$N_t = N_0 e^{-kt} = 14.0 \text{ g} \cdot e^{-(0.166 \text{ s}^{-1})(60 \text{ s})} = 6.6 \times 10^{-4} \text{g}$$

---

    g. Radiometric dating is based on the variation of the rate at two different times.

        i. Radiocarbon dating uses the decay of carbon-14:

$$^{14}_{6}\text{C} \rightarrow {}^{14}_{7}\text{N} + {}^{0}_{-1}\text{e} \qquad t_{1/2} = 5730 \text{ years}$$

1. Since carbon is incorporated into all living organisms, the amount of C-14 is relatively constant while the organism is alive as it is constantly exchanging carbon with its environment. Upon death, the amount of C-14 decreases through beta decay. Thus, the age of an object can be determined by comparing the remaining C-14 fraction in the object to that expected from atmospheric C-14.

2. Radiocarbon dating is appropriate for objects that are less than 50,000 years old and have incorporated carbon during their life cycle.

ii. Objects that are very old or were never living can be dated using uranium/lead decay.

EXAMPLE:

Zircon is a mineral that contains large amounts of uranium but no lead when formed and is found throughout the earth. A certain metamorphic rock is found to have 1.52 g of lead-206 for every 1.00 g of uranium-238. Assuming all of the Pb-206 was formed from radioactive decay of U-238, how old is this sample? The half-life of U-238 is 4.5 billion years.

We will calculate the rate constant for the decay:

$$t_{1/2} = 0.693/k$$

$$k = 0.693/t_{1/2} = 0.693/(4.5 \times 10^9 \text{ years}) = 1.54 \times 10^{-10} \text{ yr}^{-1}$$

To determine the initial mass of uranium-238, we will carry out a conversion factor problem:

$$1.52 \text{ g Pb} \times \frac{1 \text{ mol Pb}}{206 \text{ g Pb}} \times \frac{1 \text{ mol U}}{1 \text{ mol Pb}} \times \frac{238 \text{ g U}}{1 \text{ mol U}} = 1.76 \text{ g U}$$

This is the quantity of uranium-238 that has decayed, so the original quantity was $1.00 + 1.76 = 2.76$ g.

Solving for time in the integrated rate law and using mass values for N_t/N_0:

$$\ln \frac{N_t}{N_0} = -kt$$

$$t = -\frac{1}{k} \ln \frac{N_t}{N_0} = -\frac{1}{1.54 \times 10^{-10} \text{ yr}^{-1}} \ln \frac{1.76}{2.76} = 2.9 \times 10^9 \text{ yr}$$

---

7. The Discovery of Fission: The Atomic Bomb and Nuclear Power

   a. Nuclear fission is the splitting of a uranium or other heavy atom and is accompanied by the release of large quantities of energy.

   b. When uranium-235 is bombarded with (slow-moving) neutrons, it splits into lighter atoms, neutrons, and energy:

   $$^{235}_{92}\text{U} + ^1_0\text{n} \rightarrow ^{140}_{56}\text{Ba} + ^{93}_{36}\text{Kr} + 3\,^1_0\text{n} + \text{energy}$$

      i. U-235 is a fissile material, whereas the more abundant isotope, U-238, is fissionable; it is split only by fast-moving neutrons.

   c. A chain reaction can result if there is enough uranium-235 since each nuclear reaction emits more neutrons, which can initiate additional reactions.

      i. The critical mass of uranium is the minimum amount of uranium-235 needed to produce a self-sustaining chain reaction.

   d. In a bomb, the energy is released rapidly; the energy can also be released slowly and used in a nuclear power plant.

8. Converting Mass to Energy: Mass Defect and Nuclear Binding Energy

    a. Matter can be converted to energy in a nuclear reaction:

$$E = mc^2$$

    b. The mass defect is the difference in mass between the individual nucleons and the whole nucleus; the energy corresponding to the mass defect is called the nuclear binding energy.

        i. One atomic mass unit is equal to 931.5 MeV (megaelectron volts).

        ii. The binding energy is usually reported per nucleon, where a nucleon is a particle in the nucleus (protons and neutrons).

    c. The higher the nuclear binding energy per nucleon, the more stable is a nuclide.

        i. Fe-56 has the highest nuclear binding energy per nucleon and therefore represents the most stable nucleus possible.

EXAMPLE:

What is the binding energy of a Cu-63 nucleus if the actual mass of a copper-63 nucleus is 62.91367 amu?

First, we need to calculate the mass defect. Copper-63 has 29 protons and 34 neutrons. The mass defect is then the difference between the actual mass and the mass of 29 hydrogen atoms and 34 neutrons:

$$\text{mass defect} = 29(1.00783 \text{ amu}) + 34(1.00866 \text{ amu}) - 62.91367 \text{ amu} = 0.60784 \text{ amu}$$

The nuclear binding energy is then calculated by multiplying the mass defect by the energy per amu:

$$\text{nuclear binding energy} = 0.60784 \text{ amu} \times \frac{931.5 \text{ MeV}}{\text{amu}} = 566.2 \text{ MeV}$$

Since there are 63 nucleons in a copper-63 atom:

$$\text{nuclear binding energy per nucleon} = \frac{566.2 \text{ MeV}}{63 \text{ nucleons}} = 8.987 \text{ MeV/nucleon}$$

9. Nuclear Fusion: The Power of the Sun

    a. Fusion is the combination of two light nuclei to form a heavier one.

        i. Energy is released when the binding energy of the newly formed species (product) is higher than that of the reactants.

    b. Fusion requires extremely high temperatures and materials that can withstand those temperatures.

    c. Fusion has not yet been shown to be a viable energy source.

10. Nuclear Transmutation and Transuranium Elements

    a. The process of changing one element into another is called transmutation.

        i. High-energy bombardment can result in transmutation.

        ii. Elements that are not found in nature can be made.

    b. A linear accelerator accelerates charged particles through an evacuated tube using alternating voltages.

    c. A cyclotron accelerates charged particles in a circle using alternating voltages and a strong magnetic field.

11. The Effects of Radiation on Life

   a. Acute radiation damage occurs from exposure to high levels of radiation for a short period of time.

      i. Acute radiation kills a large number of cells at one time.

   b. An increased cancer risk results from low levels of radiation over an extended period of time.

      i. Damage to DNA can result in cancer.

   c. Genetic defects result when the DNA of reproductive cells is damaged and then passed on to offspring.

   d. Radioactivity is measured in units of curies:

   $$1 \text{ curie (Ci)} = 3.7 \times 10^{10} \text{ decay events per second}$$

   e. A more useful measure of radioactivity takes the penetrating power of the radiation into account and is called the absorbed dose:

   $$1 \text{ gray (Gy)} = 1 \text{ J per kg of body tissue (*SI unit)}$$

   $$1 \text{ rad} = 0.01 \text{ J per kg of body tissue}$$

   f. We can also take into account the effect of different types of radiation on biological cells using the biological effectiveness factor (i.e., organ radiosensitivity) to calculate the dose equivalent:

   $$\text{dose in Gy} \times \text{biological effectiveness factor} = \text{dose in Sv (*Sievert, SI unit)}$$

   $$\text{dose in rads} \times \text{biological effectiveness factor} = \text{dose in rems}$$

      i. The RBE factor is ~1 for X-rays, γ rays, and beta particles; 3–10 for neutrons and protons, and 10–20 for α particles.

      ii. Death is likely for exposures above 600 rem (6 Sv).

12. Radioactivity in Medicine and Other Applications

   a. Radiotracers are radioactive nuclides attached to a compound or included in a mixture to track movement in the body.

   b. Positron emission tomography (PET) is a medical technique in which a positron-emitting nuclide such as ^{18}F is attached to a metabolically active substance and given to a patient. An ensuing positron-electron annihilation event emits two photons that can be detected by an external scanner, which correlates the number of such incidents to the density of molecules in a particular region of the body.

   c. Radiation can be used to treat cancer cells, kill microorganisms on food, clean medical devices, and make harmful insects infertile.

Fill in the Blank:

1. Radiocarbon dating measures the rate of the decay of carbon-14 to _____.

2. _____ occurs when a particle composed of two protons and two neutrons is emitted.

3. _____ is the combination of two light nuclei to form a heavier one.

4. In a(n) _____ a charged particle is accelerated by alternating voltages and kept in circular motion using a magnetic field.

5. In _____, a heavy atom is split into two or more smaller atoms.

6. In a nuclear reaction, the original atom is called the _____ and the product atom is called the _____.

7. A nucleus is held together by the _____.

8. Nuclides that lie above the valley of stability have too many _____ and will tend to undergo _____.

9. Emission of an electron occurs in _____.

10. A particular isotope of an element is called a(n) _____ in nuclear chemistry.

11. A(n) _____ is a radioactivity detector that contains argon gas that is ionized when exposed to a radioactive particle.

12. The difference in mass between the products and reactants in a nuclear reaction is called the _____. This mass is converted into _____ when the _____ of the reactants is larger than that of the products.

13. When an electron and a(n) _____ collide, they annihilate each other.

Problems:

1. Write balanced nuclear equations for the following:

    a. $^{210}_{84}$Po emits a beta particle.

    b. $^{222}_{88}$Ra emits an alpha particle.

    c. $^{234}_{90}$Th emits a beta particle.

    d. $^{138}_{57}$La emits a positron.

    e. $^{185}_{79}$Au emits an alpha particle.

    f. $^{40}_{19}$K undergoes electron capture.

2. Predict decay pathways for the following radioactive elements. In cases where sequential decay occurs, indicate this.

    a. technetium-98

    b. potassium-40

    c. einsteinium-252

    d. promethium-145

    e. thallium-201

    f. magnesium-22

    g. molybdenum-99

3. Neon-19 decays to fluorine-19 with a half-life of 17.22 s.

    a. Write the nuclear reaction for the decay of neon-19 to fluorine-19.

    b. If 13.2 g of neon-19 is placed in a container, what percentage of the neon-19 will remain after 2 hours? How many fluorine-19 atoms will be in the container?

    c. If 13.2 g of neon-19 is placed in a container, how long will it take for 6.6 g of it to decompose? *Note:* You should be able to answer this question without doing any calculations.

    d. If 13.2 g of neon-19 is placed in a container, how long will it take for 2/3 of it to decompose?

4. Iodine-131 has been in the news a lot because it is a by-product of the nuclear fission reaction that occurs in nuclear power plants.

    a. What type of decay process do you expect I-131 to undergo?

    b. Write the balanced nuclear reaction that corresponds to the decay process suggested in part a.

    c. I-131 is very dangerous in humans because it concentrates itself in the thyroid where it can undergo decay in close proximity to cells. A small dose can actually be more dangerous that a very large dose. Why do you think that this might be so?

5. A piece of wood is found buried in the ground. The rate of its decay is found to be 2.34 disintegrations per minute per gram of carbon.

    a. If living organisms have a decay rate of 15.3 disintegrations per minute per gram of carbon, how old is the wood?

    b. A much larger piece of wood is found nearby. If its decay rate is the same, will the age of the wood be older, younger, or the same? Explain your answer.

6. Calculate the mass defect and the binding energy per nucleon for the following:

    a. Chlorine-37, which has a mass of 36.965903 amu

    b. Gallium-71, which has a mass of 70.924701 amu

    c. Palladium-102, which has a mass of 103.904026 amu

    d. Holmium-165, which has a mass of 164.930332 amu

7. A neutron can decay into a proton and an electron. When this happens, energy is released.

    a. Calculate the amount of energy released.

    b. In this case, the energy that is released is turned into kinetic energy of the electron. How fast is the electron moving when it is emitted?

8. Some common household goods used to have considerable amounts of radioactive elements in them. For example, uranium oxides in glazes used in earthenware measured 20 millirad/hour on a Geiger-Müller counter. Calculate radiation absorbed by a 150–lb. person from earthenware over one year.

Concept Questions:

1. Which of the various types of radioactive decay is most dangerous and which is the least dangerous? Explain your answers.

2. Explain why it is necessary for neutrons to have energy levels when considering nuclide stability.

3. In this chapter, you learned that all radioactive decay processes follow first-order kinetics. Based on what you learned in Chapter 15 and your knowledge of nuclear processes, why do you think this is?

4. Carbon dating cannot be used to estimate the age of dinosaur bones. Explain why this is so.

5. Uranium-lead dating is usually performed on samples that contain zircon. Zircon incorporates uranium and thorium, but it strongly rejects lead. Explain why the rejection of lead in the sample is important for this type of dating.

Chapter 21: Organic Chemistry

Key Learning Outcomes:

- Write Structural Formulas for Hydrocarbons

- Name Alkanes

- Name Alkenes and Alkynes

- Write Addition Reactions

- Write Alcohol Reactions

Chapter Summary:

In this chapter, the branch of chemistry that focuses on carbon compounds will be explored. You will first learn why carbon's unique bonding properties lead to such a large variety of compounds. Then you will explore how to name and identify common reactions of the various organic compounds including hydrocarbons, alcohols, aldehydes, ketones, carboxylic acids, esters, ethers, and amines. Finally, you will be introduced to polymers, which will be important in the next chapter.

Chapter Outline:

1. Fragrances and Odors

    a. Most common smells are caused by organic molecules, which contain carbon and other elements such as H, N, O, and S.

    b. Organic chemistry is the study of carbon-containing compounds.

2. Carbon: Why It Is Unique

    a. Carbon has a strong tendency to form four covalent bonds since it has four valence electrons and an intermediate electronegativity value.

    b. Carbon can form double and triple bonds.

    c. Carbon bonds to itself to form a variety of structures including chains, branched chains, and rings; this is called catenation.

3. Hydrocarbons: Compounds Containing Only Carbon and Hydrogen

    a. Aliphatic hydrocarbons differ in the type of bonding that occurs between the carbon atoms. These hydrocarbons are classified as alicyclic (straight chain or branched) or cyclic.

        i.   Alkanes contain only carbon–carbon single bonds.

        ii. Alkenes contain one or more carbon–carbon double bonds.

        iii. Alkynes contain one or more carbon–carbon triple bonds.

    b. Aromatic compounds contain one or more benzene rings.

    c. There are five common ways to draw hydrocarbon molecules.

        i.   Structural formulas show all atoms and their connections to one another.

        ii. Condensed structures show the connections of carbons only.

        iii. Carbon skeletons show carbon–carbon bonds as lines where the end of each line implies a carbon atom; hydrogen atoms are implied based on the octet rule.

iv. Ball-and-stick models show atoms as spheres connected by sticks, which represent bonds.

v. The space-filling model shows how electrons fill space.

d. Single bonds rotate freely at room temperature so the arrangement of atoms does not matter, but only the connectivity matters.

e. Double and triple bonds are rigid, and therefore the spatial arrangement is important.

f. Structural isomers have the same formula but different atom connectivities.

EXAMPLE:
_____

Draw structural formulas and carbon skeleton formulas for the three structural isomers of pentane, C_5H_{12}.

We begin by sketching the carbon skeletons (without hydrogen atoms) of the isomers:

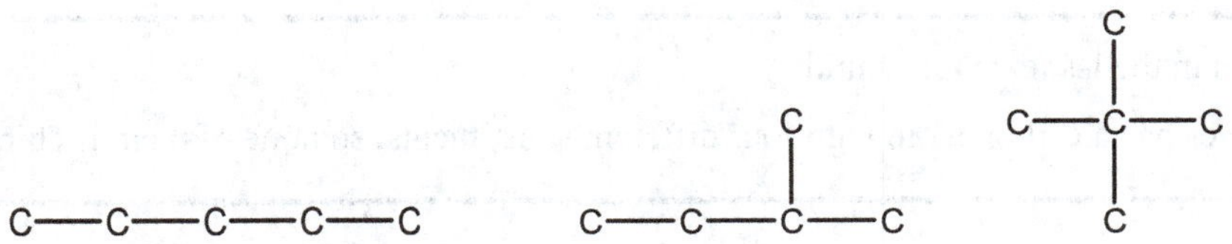

Now we will add the hydrogen atoms:

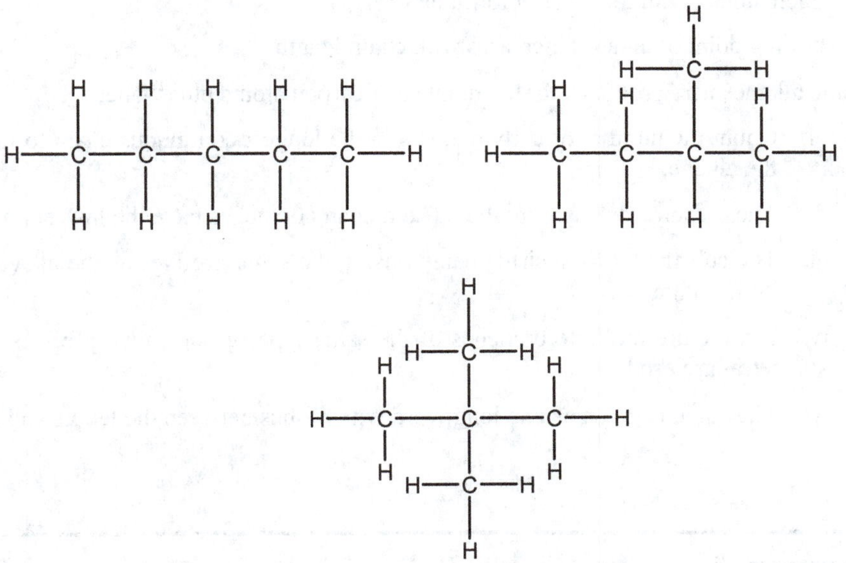

The carbon skeleton formulas are drawn by removing the hydrogen atoms and using lines to represent carbon–carbon bonds:

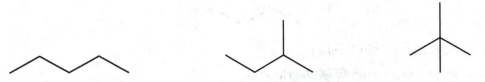

_____

g. Stereoisomers have the same atom connectivity but have a different spatial arrangement.

i. Geometric isomers are cis–trans isomers and will be discussed later.

ii. Optical isomers are nonsuperimposable mirror images.

1.   Optical isomers are also called enantiomers and are chiral.

2.   Any carbon atom attached to four different substituents is a chiral carbon center.

3.   Enantiomers have similar properties to each other except:

a.   They rotate polarized light in different directions.

i.   Right rotation is called dextrorotatory (d).

ii.   Left rotation is called levorotatory (l).

b.   They display different chemical properties in chiral environments.

i.   This is physiologically very important.

4.   An equimolar mix of two enantiomers is called a racemic mixture.

EXAMPLE:

Are any of the isomers in the last example chiral?

None of these molecules has a carbon atom with four different substituents, so none of them is chiral.

---

4.   Alkanes: Saturated Hydrocarbons

a.   Alkanes have only carbon–carbon single bonds; these alicyclic compounds are "saturated" with hydrogen atoms, with the general formula, C_nH_{2n+2}

b.   The boiling point of alkanes increases with chain length.

c.   Name alkanes using prefixes for the number of carbon atoms plus "-ane."

i.   Count the number of carbon atoms in the longest continuous chain to find the base name of the alkane.

ii.   The carbon atoms are numbered so that any branches have the lowest possible number.

iii.   The substituent (branch) is named using the same prefixes as the alkane name but with the "-yl" ending.

iv.   If there are multiple branches of the same type (group), then prefixes such as di-, tri- and tetra- are used.

v.   The name is given as one long word with dashes between the letters and any numbers.

EXAMPLE:

Name the alkanes shown in the first example.

The first alkane is the straight chain structure, $CH_3CH_2CH_2CH_2CH_3$:

Since there are five carbon atoms, this is called pentane.

The next structure is a branched alkane, $CH_3CH_2CH(CH_3)_2$:

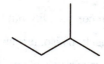

We see that the longest continuous chain is four carbon atoms long, so the base name is butane. There is a single substituent, which is a single carbon atom, so it is a methyl group. We number the carbon atoms to give the methyl group the smallest number. So the name of this molecule is 2-methylbutane.

The next isomer is also branched, $C(CH_3)_4$:

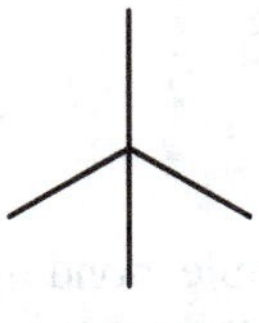

We see that the longest chain has three carbon atoms, so the base name is propane. There are two methyl groups attached to the main carbon chain; both are attached at carbon 2. The name is 2,2-dimethylpropane.

---

5. Alkenes and Alkynes

    a. Alkenes and alkynes are called unsaturated hydrocarbons.

        i. Alkenes have the general formula C_nH_{2n}.

        ii. Alkynes have the general formula C_nH_{2n-2}.

        iii. For each additional degree of unsaturation, two hydrogen atoms are removed.

    b. Alkenes are named in the same way as alkanes with a prefix plus "-ene."

        i. The base name is derived from the longest continuous chain containing the double bond.

        ii. The position of the double bond is given by a number preceding the name.

        iii. Numbering is done to give the alkene the smallest possible number.

EXAMPLE:

Name the following alkenes:

a.

The carbon chain has six carbon atoms, and there is a single double bond, so the base name is hexene. We number the carbons so that the double bond starts at the lowest number. The name of this compound is 2-hexene.

b.

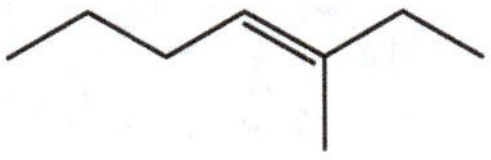

The longest carbon chain containing the double bond has seven carbon atoms, so the base name is heptene. The double bond and the methyl group are both on carbon 3, so the name is 3-methyl-3-heptene.

c.

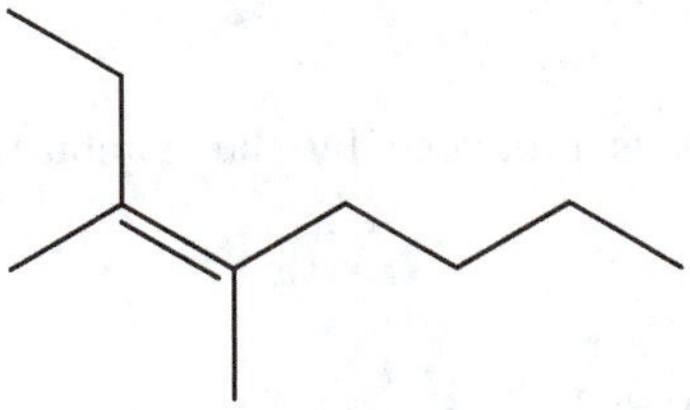

In this molecule, we see that there are eight carbon atoms in the longest carbon chain containing the double bond, so the base name is octene. There are two methyl groups on carbons 3 and 4. The double bond starts on carbon 3, so the name is 3,4-dimethyl-3-octene.

---

    c. Alkynes are named exactly as alkenes except they use the ending "-yne."

EXAMPLE:

Name the following alkynes:

a.

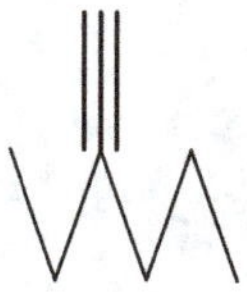

The longest carbon chain that contains the triple bond is five carbons long, so the base name is pentyne. There is a single ethyl substituent at carbon 2, so the name is 2-ethyl-1-pentyne.

b.

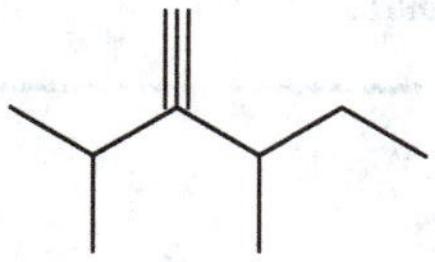

The longest carbon chain containing the triple bond again has five carbon atoms, so the base name is pentyne. There are two groups on this molecule: the isopropyl group at carbon 2 and the methyl group at carbon 3. The name of this compound is 2-isopropyl-3-methyl-1-pentyne.

---

    d.   In alkenes and alkynes there are both sigma and pi bonding so rotation is hindered.

        i.   Geometric (cis–trans) isomerism results.

EXAMPLE:

In the example for naming compounds with a double bond, we named the following compound incorrectly:

Give the full, correct name.

This molecule is a trans isomer because we see that the substituents (carbon chains) are on opposite sides of the double bond. The name should be *trans*-2-hexene.

The cis form of this molecule is:

---

6. Hydrocarbon Reactions

    a. Of all the energy produced in the United States, 85%–90% is produced by the combustion of hydrocarbon compounds.

    b. Alkanes commonly undergo halogen substitution:

$$CH_3CH_2CH_3 + Cl_2 \xrightarrow{\text{light or heat}} CH_3CH_2CH_2Cl + HCl$$

i. This reaction occurs via a reaction mechanism, which involves a radical species:

$$Cl_2 \xrightarrow{\text{light or heat}} 2Cl\cdot$$

$$Cl\cdot + R\text{–}H \longrightarrow R\cdot + HCl$$

$$R\cdot + Cl_2 \longrightarrow R\text{–}Cl + Cl\cdot$$

c. Alkenes and alkynes undergo similar chemical reactions.

    i. Halogens can add across a double or triple bond.

    ii. Hydrogen can add across a double or triple bond in order to increase the saturation.

        1. This is called a hydrogenation reaction.

    iii. When a polar reagent is added to an asymmetrical double bond, the positive end of the reagent (the less EN end) will add to the carbon atom with the most hydrogens; this is commonly called Markovnikov's rule.

EXAMPLE:

What is the product of the reaction between hydrochloric acid and 3-methyl-2-pentene?

We can draw the structure of 3-methyl-2-pentene:

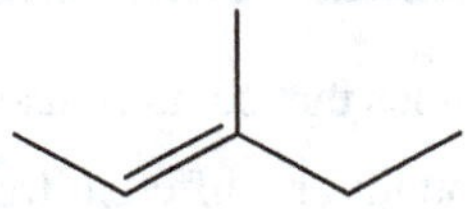

When HCl is added across the double bond, the chlorine atom will add to the more substituted carbon atom and the hydrogen will add to the carbon atom with a hydrogen atom attached to give:

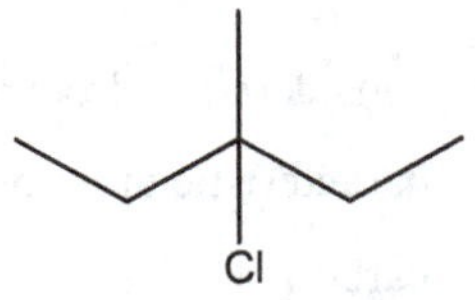

---

7. Aromatic Hydrocarbons

    a. Benzene is a six-carbon ring with six hydrogen atoms and is represented either by two resonance structures each with three double bonds alternating between carbon atoms or a single ring containing a circle.

        i. Benzene has a delocalized pi orbital system that causes it to be very stable.

    b. Compounds containing benzene are named using multiple schemes.

        i. If there is only one substituent, the compound is named using (substituent) benzene.

        ii. Some benzene-containing substances have common names including toluene, aniline, phenol, and styrene.

        iii. When benzene is the substituent, it is called phenyl.

        iv. Disubstituted benzene rings are numbered and listed alphabetically; when the substituents are identical, the prefix "di" is used.

1. Instead of numbering, we can use *o-*, *m-*, or *p-*.

   a. *Ortho* (*o-*) is equivalent to 1,2-.

   b. *Meta* (*m-*) is equivalent to 1,3-.

   c. *Para* (*p-*) is equivalent to 1,4-.

EXAMPLE:

Name the following organic compound:

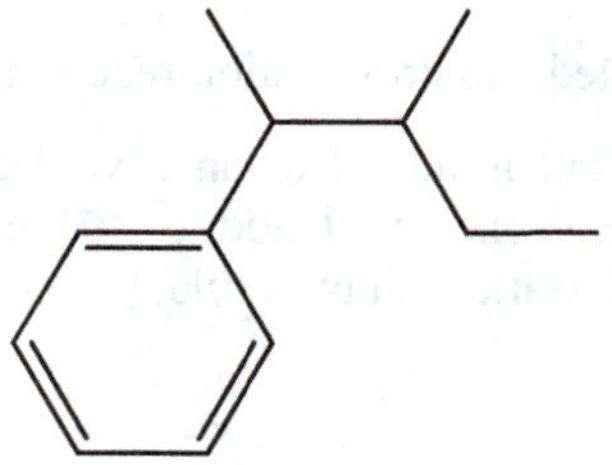

We will treat benzene as a substituent because the alkyl group is large. We see that the carbon chain has five carbon atoms, so the base name is pentane. There is one methyl group and one phenyl group. We number the carbon chain to give the smallest numbers for the substituents. So the name of this molecule is 3-methyl-2-phenylpentane.

---

   c. Polycyclic aromatics are compounds that contain fused benzene rings.

   d. The reactions of aromatic compounds are different from alkenes because of the pi system.

      i. Substitution of a single hydrogen atom occurs.

8. Functional Groups

   a. Other organic compounds can be considered hydrocarbons with a functional group attached.

      i. A functional group is a characteristic atom or group of atoms.

   b. "R" is used to represent the hydrocarbon.

9. Alcohols

   a. Alcohols contain the –OH (hydroxyl) group and are of the form R–OH.

   b. Alcohols are named using the base name of the longest carbon chain containing the OH group. The base name is combined with the ending "-ol" and a number that indicates the location of the OH group on the hydrocarbon chain.

EXAMPLE:

Name the following organic compound:

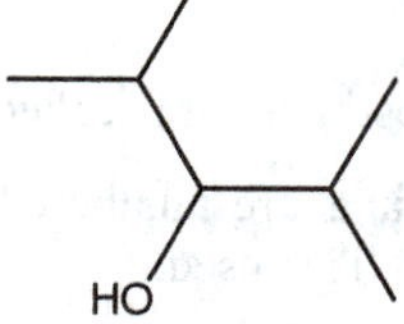

We look for the longest carbon chain containing the hydroxyl group and see that there are five carbon atoms so the base name is pentanol. There are two methyl groups, and they are attached to carbons 2 and 4. This molecule is 2,4-dimethyl-3-pentanol.

---

c. Reactions of alcohols fall into four main categories:

   i. Substitution reactions where the hydroxyl group is replaced:

   $$R-OH + HX \rightarrow R-X + H_2O$$

   ii. Elimination or dehydration reactions where an alkene is formed:

   $$RCH_2CH_2-OH \rightarrow RCH = CH_2 + H_2O$$

   iii. Oxidation reactions to give aldehydes or carboxylic acids:

   $$RCH_2OH \rightarrow RCOH \rightarrow RCO_2H$$

   iv. Reactions with active metals to give hydrogen gas:

   $$RCH_2OH + M \rightarrow RCOM + \tfrac{1}{2}H_2$$

EXAMPLE:

Predict the product that forms when 2,4-dimethyl-3-pentanol reacts with hydrochloric acid.

The hydroxyl group will be replaced with a chlorine atom to form 3-chloro-2,4-dimethylpentane:

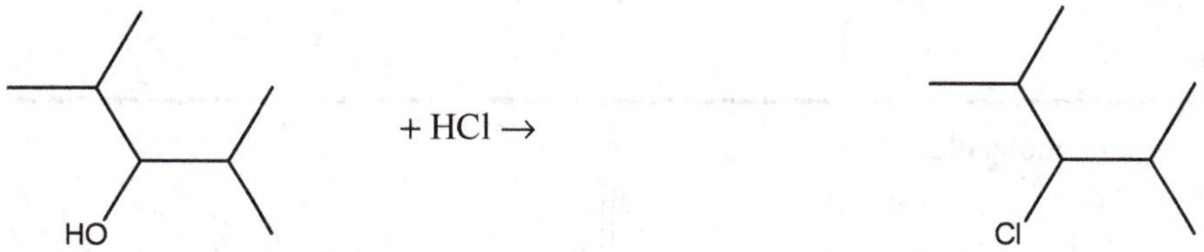

---

10. Aldehydes and Ketones

   a. Both aldehydes and ketones contain the carbonyl (C=O) group.

   b. Aldehydes have the general form RCOH and the structure:

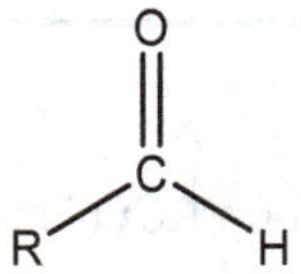

      i. Aldehydes are named by dropping the "-e" at the end of the alkane name and replacing it with "-al." The base name is derived from the longest chain that includes the carbonyl group, whose carbon is numbered 1.

      ii. Aldehydes are prepared by oxidizing alcohols and can be reduced to form alcohols.

EXAMPLE:

Name the following organic molecule:

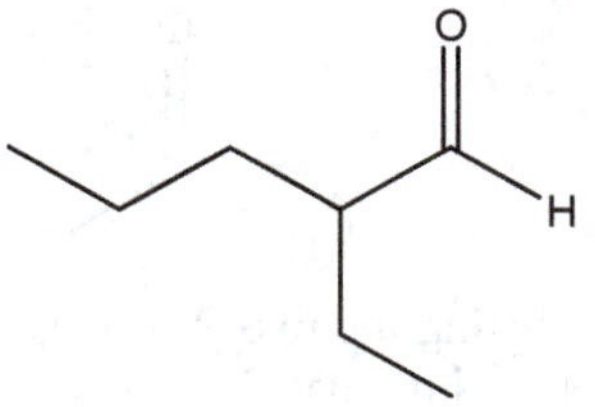

The longest carbon chain that contains the carbon of the aldehyde group has five carbon atoms, so the base name is pentanal. There is one substituent: an ethyl group. We number the carbon chain so that the aldehyde is at carbon 1, so the name is 2-ethyl-pentanal.

---

c. Ketones have the general form RCOR′ and the structure:

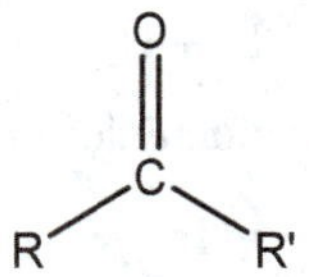

where R and R′ are the same or different alkyl groups.

  i. Ketones are named by dropping the "-e" at the end of the alkane name and replacing it with "-one."

d. Both aldehydes and ketones can undergo reduction reactions, where reagents add across the carbon–oxygen double bond (analogous to the reactions of alkenes).

e. Alcohols commonly undergo oxidation to become aldehydes and ketones, while ketones commonly undergo reduction.

EXAMPLE:

Name the following organic molecule:

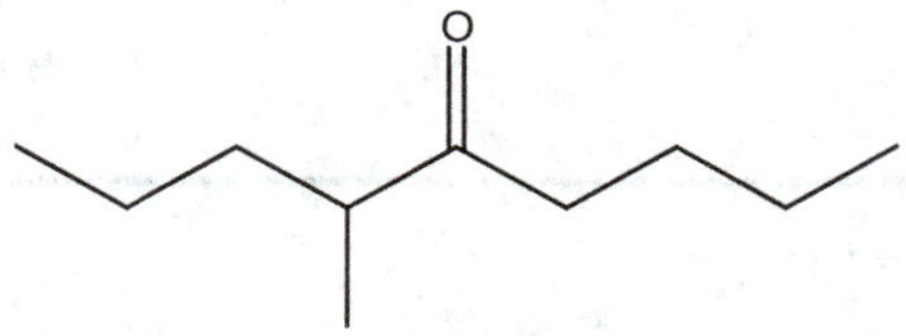

The longest carbon chain containing the carbonyl group has nine carbon atoms, and there is a single methyl substituent. We number the carbon chain to give the lowest numbering scheme, so this molecule is 4-methyl-5-nonanone.

---

11. Carboxylic Acids and Esters

a. Carboxylic acids have the general form RCO$_2$H and the structure:

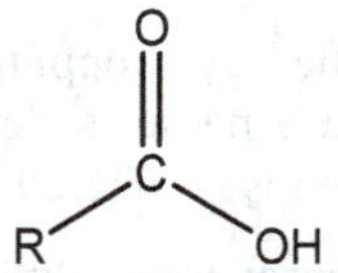

  i. Carboxylic acids are named by dropping the "-e" of the alkane name and replacing it with "-oic acid."

  ii. Carboxylic acids are weak acids.

b. Esters have the general form RCO$_2$R′ and the structure:

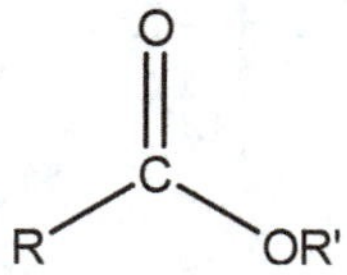

  i. Esters are named by listing the two R groups in alphabetical order and replacing the "-e" of the second alkyl group with "-oate."

c. Carboxylic acids react with alcohols to form esters in condensation reactions (elimination of water).

d. Condensation of two carboxylic acids at high heat results in the formation of an acid anhydride (without water).

EXAMPLE:
_____

Predict the ester that forms from the condensation reaction between butanoic acid and ethanol.

We will first write out the structures of each:

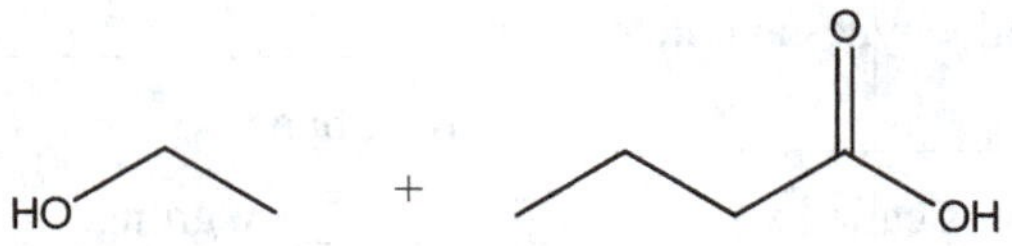

In the condensation reaction, we lose water and form an ester:

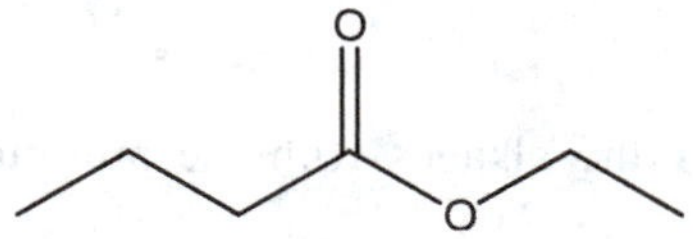

The product is ethyl butanoate.

_____

12. Ethers

a. Ethers have the general form ROR′ and the structure:

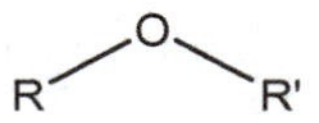

b. Ethers are named by listing the two alkyl groups in alphabetical order followed by "ether"; when the alkyl groups are the same, the prefix "di" is used.

13. Amines

a. Amines are derivatives of ammonia with one or more of the hydrogen atoms replaced by alkyl groups.

b. Amines are commonly identified by their bad smell.

c. Amines are named by listing the alkyl groups in alphabetical order followed by "amine."

d. Amines are weak bases and react accordingly.

e. Condensation reactions between amines and carboxylic acids are biologically very important and result in amide linkages and protein formation.

Fill in the Blank:

1. _____ are organic compounds with the general form of ROR′.

2. When a polar reagent is added to an unsymmetrical alkene, the electropositive end of the reagent adds to the carbon atom with _____.

3. The _____ of a molecule shows each atom and how the atoms are bonded to each other.

4. Aldehydes, ketones, carboxylic acids, and esters all contain the _____ group.

5. Molecules that have the same molecular formula but different structures are called _____.

6. Nonsuperimposable mirror images are called _____ and will cause polarized light to rotate in _____.

7. 1,4-dichlorobenzene is also called _____.

8. A(n) _____ is an equimolar mixture of enantiomers.

9. The base chain of an alkene is the longest continuous carbon chain that contains the _____.

10. Geometric isomers are also called _____.

11. _____ are molecules in which the atoms have the same connectivity but different spatial arrangement.

12. Alcohols contain the _____ functional group.

13. The most common reactions of aldehydes and ketones are _____.

14. Alkanes are often referred to as _____ hydrocarbons.

15. When benzene is a substituent, it is called a(n) _____ group.

Problems:

1. Draw all of the structural isomers for the alkanes with the molecular formula of C_7H_{16}. Name each of the molecules that you have drawn.

2. Draw all of the structural isomers for the alkenes with the molecular formula of C_5H_{10}. Name each of the molecules that you have drawn.

3. Identify the functional groups in each of the following compounds, and provide names for each.

a.

b.

c.

d.

e.

f.

g.

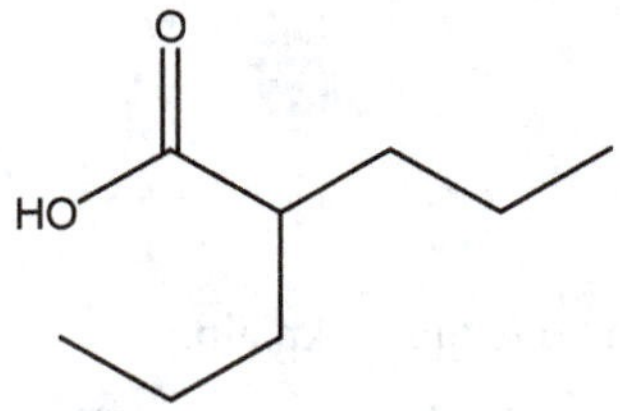

h.

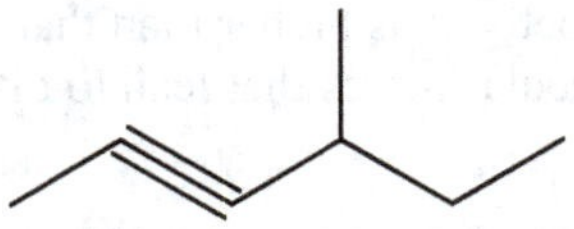

i.

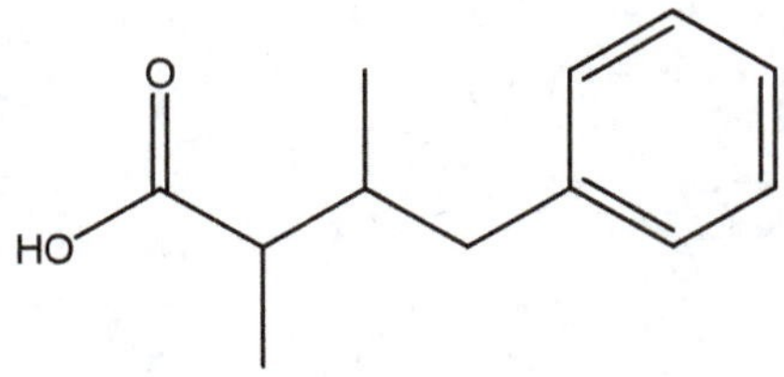

j.

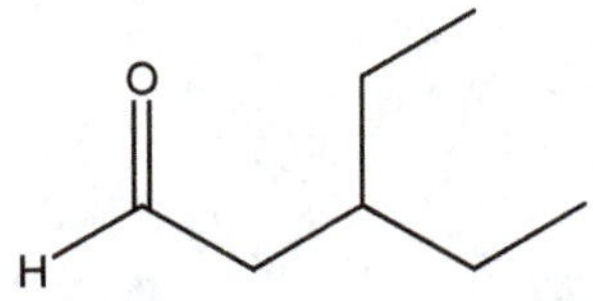

k.

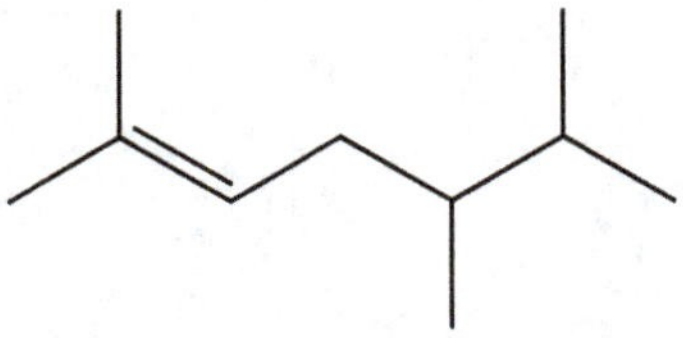

l.

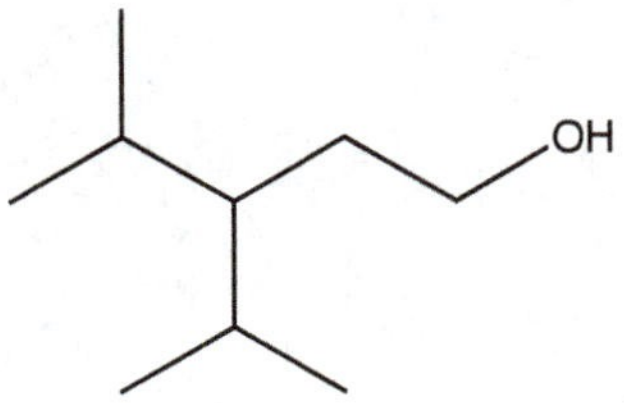

4. Identify which of the compounds in Question 1 have chiral carbons. Identify each of the chiral centers with a star.

5. Draw structures, and predict the products that form in the following reactions:

   a. F_2 reacts with benzene using iron(III) chloride as a catalyst

   b. Combustion of hexane

   c. Reaction of 2-chloro-1-propene with hydrochloric acid

   d. Condensation of butanoic acid with itself

   e. Reaction of ethyl amine with propanoic acid

   f. Reaction of hydrogen cyanide with ethanal

   g. Oxidation of 2-methyl-4-heptanol (write all oxidation products)

   h. Reduction of 2-methyl-pentanal

   i. Reaction of benzene with 2-chlorobutane

   j. Reaction of diethylamine with nitric acid

Concept Questions:

1. A container of three hydrocarbons (C_3H_8, $C_{20}H_{42}$, and $C_{40}H_{82}$) is heated.

   a. Which hydrocarbon will boil first? Explain.

   b. Only one of these compounds is solid at room temperature. Which compound is this? Explain.

2. In the discussion of cis–trans isomerism, you learned that isomers have different physical properties. For example, in Table 22.9 of your textbook, you see that the boiling point of *cis*-1,2-dichloroethene is higher than that of *trans*-1,2-dichloroethene. Why do you think that this is? *Hint:* Consider the intermolecular forces that lead to differences in boiling points.

3. Markovnikov's rule states that the more electronegative atom will add to the more substituted carbon atom of a double bond. Suggest a reason for this.

Chapter 22: Transition Metals and Coordination Compounds

Key Learning Outcomes:

- Write Electronic Configurations for Transition Metals and Their Ions

- Name Coordination Compounds

- Identify and Draw Geometric Isomers

- Recognize and Draw Optical Isomers

- Estimate Crystal Field Splitting Energy

- Determine the Number of Unpaired Electrons in Octahedral Complexes

Chapter Summary:

This chapter focuses on the properties and chemistry of transition metals. You will begin this exploration by looking at the electron configurations of transition metals and then move on to a discussion of periodic properties including atomic size, ionization energy, electronegativity, and oxidation state. Next you will learn basic definitions of transition metal bonding so that you can name these compounds in a systematic fashion. As you draw coordination compounds, you will see that they can have both structural isomers and stereoisomers, which you will learn how to identify. Your understanding of coordination compound bonding will be expanded with a discussion of crystal field theory, which explains the varied colors associated with transition metals as well as their magnetic properties. Finally, you will finish this chapter by learning about some common applications of coordination compounds.

Chapter Outline:

1. The Colors of Rubies and Emeralds

    a. Crystal field theory is a theory that explains the splitting of metal d orbitals.

2. Properties of Transition Metals

    a. Electrons are added to the $(n-1)d$ orbitals as you move from left to right across the periodic table.

        i. The ns and $(n-1)d$ orbitals are very close in energy, resulting in some exceptions to the normal order for orbital filling.

        ii. Transition metals form ions by losing the ns electrons before losing the $(n-1)d$ electrons.

    b. The trends in atomic radius differ from the main-group elements.

        i. There is little variation in atomic radius across a row because added electrons go into the $(n-1)d$ orbital while the number of outer-shell electrons (ns^2) remains constant.

        ii. There is a small increase in the atomic radius as you progress down the periodic table from the first to the second row of transition metals, but there is no meaningful change between the second and third row.

            1. The third transition row has additional electrons in the $(n-2)f$ subshell that are not effective at shielding. The fact that there is no change in atomic radius is called the lanthanide contraction.

    c. The trends in ionization energies also differ somewhat from the main-group elements.

        i. Ionization energies increase slowly across a row, but the increase is small compared to the main-group elements.

      ii. The ionization energies of the third row elements are generally greater than the second row elements because the nuclear charge increases substantially with added protons, while the size only increases a small amount with added electrons.

   d. The trends in electronegativity parallel those of ionization energies.

      i. Electronegativity values slowly increase across a row in the d-block.

      ii. The electronegativity generally increases down from the first to second row, but there is no change from the second to third row.

   e. Transition metals often exhibit multiple oxidation states.

      i. The d-block elements tend to lose their valence s-electrons, as well as a variable number of d-electrons (Groups 3 and 12 are exceptions).

         1. Elements close to the center of each row have the widest range of oxidation states.

         2. Elements in the second and third rows of the transition metals are more likely to reach higher oxidation states than those in the first row.

3. Coordination Compounds

   a. A complex ion consists of a central metal ion bound to one or more ligands.

      i. Ligands are Lewis bases that form coordinate covalent bonds with metal atoms or ions. (Lewis acids).

         1. Ligands may be ions (such as CN^-) or molecules (such as NH_3).

      ii. The primary valence is the oxidation state of the central metal atom.

      iii. The secondary valence is the number of ions or molecules that are directly bound to the central metal atom; the secondary valence is usually called the coordination number.

   b. A coordination compound is one in which a complex ion combines with one or more uncoordinated counterions (that are required to maintain charge neutrality).

      i. When writing the formula of a coordination compound:

         1. Cations are listed before anions.

         2. The coordination complex is placed in square brackets, and counterions are written outside of [ ].

EXAMPLE:

What is the primary valence and the coordination number of cobalt in $[Co(NH_3)_5Cl]Cl_2$?

With three negative ions present in this compound, the primary valence must be +3 to give an overall neutral compound. There are six ligands directly bound to the metal of the complex ion, so the coordination number is 6.

---

   c. Coordination compounds form when an empty orbital on the metal accepts an electron pair from a ligand.

      i. This type of bond is often referred to as a coordinate covalent bond.

      ii. Ligands that donate a single electron pair are called monodentate.

      iii. Some ligands can donate two or more electron pairs to the metal; these are called chelating agents, and the complex ion is called a chelate.

1. Bidentate ligands donate two electron pairs from different atoms; therefore, they must be large enough to form two different points of attachment to the central metal.

2. Ethylenediaminetetraacetate (EDTA) can donate up to six electron pairs to a central metal atom. EDTA is a polydentate ligand.

d. Linear, square planar, tetrahedral, and octahedral geometries are the most common for transition metals.

e. The rules for naming a complex ion or coordination compound are different from those for other ionic compounds (discussed in Chapter 3 of your book).

   i. First the ligands are named.

      1. Neutral ligands are named as molecules, except for water, ammonia, and carbon monoxide.

         a. Water, ammonia, and carbon monoxide ligands are named as aqua, ammine, and carbonyl, respectively.

      2. Anionic ligands are named using the ion name with a change of ending:

         a. "-ide" becomes "-o"   (e.g., fluoride becomes fluoro)

         b. "-ate" becomes "-ato"   (e.g., cyanate becomes cyanato)

         c. "-ite" becomes "-ito"   (e.g., nitrite becomes nitrito)

   ii. The ligands are listed in alphabetical order before the metal name.

   iii. A prefix (di-, tri-, tetra-, penta-, hexa-) is used to denote the number of each ligand type as in molecular compounds. The prefix does not affect the ordering of the ligand names.

      1. If a prefix is already used in the ligand name, the prefixes are altered to be "bis-" (instead of bi-), "tris-" (instead of tri-), or "tetrakis-" (instead of tetra-). The ligand name follows in parenthesis.

   iv. The metal is named.

      1. If the complex ion is a cation, the metal is named with the oxidation state written as a Roman neutral.

      2. If the complex ion is an anion, the ending of the metal is dropped and "-ate" is added followed by the oxidation state of the metal as a roman numeral; for example, cobalt(II) becomes cobaltate(II).

      3. If the symbol of the metal originates from a Latin name, then the Latin stem is used; for example, iron becomes ferrate.

   v. The entire name is written as a list of the ligands followed by the metal name in a single word.

      1. If it is a coordination compound, the cation and anion are listed as two words with the cation coming first (as in other ionic compounds).

EXAMPLE:

What is the name of the cobalt compound from the last example, $[Co(NH_3)_5Cl]Cl_2$?

First, we will name the ligands and determine the number of each. There are five ammine ligands, and there is one chloro ligand.

Next, we will determine the alphabetical listing of the ligands and the prefix for each. Ammine is listed first with the penta- prefix since there are five of them. Chloro comes after ammine, and since there is only one chloro ligand, no prefix is needed.

Next, we will name the metal as cobalt(III) since we determined that the oxidation state is +3 and the complex ion is a cation.

Thus, the name of this compound is: pentaamminechlorocobalt(III) chloride.

---

  f. When a formula is written for a complex ion, the symbol of the metal is written first, followed by the neutral molecules, and then the anions; when more than one anion or neutral molecule is present, they are listed in alphabetical order based on the chemical symbol.

EXAMPLE:

What is the formula of dichlorobis(ethylenediamine)platinum(IV) chloride?

The symbol for platinum is Pt, and the ligands are Cl and $H_2NCH_2CH_2NH_2$; there are two of each ligand present. Since ethylenediamine is neutral, it is listed first. So, the complex ion is $[Pt(H_2NCH_2CH_2NH_2)_2Cl_2]^{2+}$.

Two chloride counterions are necessary to maintain an overall neutral charge.

The formula for the coordination compound is therefore $[Pt(H_2NCH_2CH_2NH_2)_2Cl_2]Cl_2$.

---

 4. Structure and Isomerization

  a. There are two main types of isomerism for coordination compounds: structural isomers and stereoisomers.

   i. Structural isomers differ in the ways that atoms are connected to one another.

    1. Coordination isomers are those in which ligands and uncoordinated counterions exchange places.

    2. Linkage isomers occur when ligands can connect to metals in different ways. That is, the donor atom of at least one of the ligands is different.

     a. This requires ambidentate ligands such as NO_2^- versus ONO^-.

   ii. Stereoisomers differ in their spatial arrangement.

    1. Geometric isomers include:

     a. Cis–trans isomers of square planar complexes of the form (MA_2B_2) and octahedral complexes of the form (MA_4B_2); and,

     b. Fac–mer isomers, which occur in octahedral complexes of the form (MA_3B_3).

    2. Optical isomers are nonsuperimposable mirror images of one another.

     a. These molecules are chiral and are called enantiomers; they rotate polarized light in different directions.

EXAMPLE:

Does the following square planar molecule have a stereoisomer? If so, what kind?

$$\text{Cl} \longrightarrow \underset{\underset{\text{Cl}}{|}}{\overset{\overset{\text{NH}_3}{|}}{\text{Pt}}} \longrightarrow \text{NH}_3$$

We can see that there is a trans isomer of this complex:

$$\text{H}_3\text{N} \longrightarrow \underset{\underset{\text{Cl}}{|}}{\overset{\overset{\text{Cl}}{|}}{\text{Pt}}} \longrightarrow \text{NH}_3$$

We can also check to see if the mirror image of the original complex is superimposable:

$$\text{Cl} \longrightarrow \underset{\underset{\text{Cl}}{|}}{\overset{\overset{\text{NH}_3}{|}}{\text{Pt}}} \longrightarrow \text{NH}_3 \quad \vdots \quad \text{H}_3\text{N} \longrightarrow \underset{\underset{\text{Cl}}{|}}{\overset{\overset{\text{NH}_3}{|}}{\text{Pt}}} \longrightarrow \text{Cl}$$

We can see that the mirror image can be superimposed on the original molecule by rotating by 90°, so these are not enantiomers. Therefore, the original molecule has a geometric isomer but no optical isomer.

5. Bonding in Coordination Compounds

   a. Bonding in coordination compounds can be described in terms of valence bond theory in which the atomic orbitals of the metal are hybridized.

   b. Crystal field theory is a bonding model for transition metals that accounts for the colors and magnetism of coordination compounds.

      i. Crystal field theory focuses on the repulsions between the electron pairs of the ligands and the remaining d orbital electrons of the central metal ion.

         1. The degenerate d orbitals of the metal are split into higher- and lower-energy levels because of the ligand arrangement. The arrangement of ligands in the crystal is called the crystal field.

            a. Orbitals located in regions of space where ligands are located will increase in energy (as a result of maximum repulsion) more than those located in regions of space without ligands present.

            b. The difference in energy between the sets of orbitals is called the crystal field splitting.

         2. In strong-field complexes, the splitting is large; in weak-field complexes, the splitting is small.

      ii. The color of metal complexes is a result of the metal complex's absorption of some wavelengths of light and the reflection of others.

         1. The color of metal complexes results when all colors but one are absorbed; in this case, the metal will look like the reflected color.

2. The color of metal complexes can also result when only one color is absorbed and all others are reflected; the metal will look like the color that is complementary to the absorbed color.

iii. The energy of an absorbed photon is related to the splitting of the *d* orbitals.

EXAMPLE:

An unknown metal complex is green in aqueous solution. What is the splitting of the *d* orbitals in this complex (in kJ/mol)?

We first calculate the energy of the absorbed light. Since red is complementary to green on the color wheel, the wavelength of absorbed light is approximately 700 nm. Converting this to energy:

$$E = h\nu = \frac{hc}{\lambda} = \frac{(6.626 \times 10^{-34} J \cdot s)(3.00 \times 10^8 m/s)}{700 \times 10^{-9} m} = 2.84 \times 10^{-19} J = \text{crystal field splitting}$$

This is the energy per ion; the energy per mole of ions is:

$$2.84 \times 10^{-19} J/ion \times \frac{6.022 \times 10^{23} \text{ ions}}{1 \text{ mol}} \times \frac{1 \text{ kJ}}{1000 \text{ J}} = 171 \text{ kJ/mol}$$

iv. The magnitude of the crystal field splitting in a complex ion is dependent on the ligands that are attached to the central metal ion and the identity of the metal.

1. The ability of an ion to split the *d* orbitals is given in a spectrochemical series from highest splitting ability to lowest:

$$CN^- > NO_2^- > en > NH_3 > H_2O > OH^- > F^- > Cl^- > Br^- > I^-$$

2. The splitting increases as the charge on the metal increases.

v. The magnetic properties (i.e., diamagnetism vs. paramagnetism) of transition metal complexes are affected by *d* orbital splitting because once the lower energy *d* orbitals are half-filled, electrons can either be paired or occupy higher-energy orbitals.

1. If the splitting of the orbitals is weak, the electrons will occupy the higher-energy orbital, form high-spin complexes, and exhibit greater magnetism.

2. If the splitting of the orbitals is large, the electrons will pair up prior to occupying the higher-energy orbitals, form low-spin complexes, and exhibit less magnetism.

vi. The crystal field splitting is dependent on the geometry of the coordination compound.

1. For octahedral complexes, the z^2 and $x^2 - y^2$ orbitals are higher in energy than the *xy*, *yz*, and *xz* orbitals are.

2. For tetrahedral complexes, the *xy*, *xz*, and *yz* orbitals are higher in energy than the z^2 and $x^2 - y^2$ orbitals are.

3. For square planar complexes, the $x^2 - y^2$ orbital is highest in energy, the *xy* orbital is next highest in energy, the z^2 orbital is next, and the *xz* and *yz* orbitals are lowest in energy.

6. Applications of Coordination Compounds

a. Ligands selectively bind to metals, so they can be used to identify the presence of a given metal (similar to selective precipitation).

b. Coordination compounds are used as coloring agents because they absorb certain wavelengths of visible light.

    c.   Many metals are involved in biological systems.

          i.   In hemoglobin and in cytochrome c, an iron complex called a heme is connected to a protein.

              1.   Heme consists of an iron ion coordinated to a flat, polydentate ligand called a porphyrin.

              2.   In the blood vessels of the lung, oxygen molecules coordinate with the iron ion in hemoglobin. Hemoglobin carries oxygen to body tissues, where the coordinated oxygen is replaced by water molecules.

          ii.   In chlorophyll, a different porphyrin ring is coordinated to magnesium; chlorophyll is used by plants in photosynthesis.

         iii.   Carbonic anhydrase has a zinc ion bound to three nitrogen atoms from surrounding nucleic acids. The fourth site of the tetrahedral zinc complex can bind a water molecule.

              1.   Carbonic anhydrase catalyzes the reaction between carbon dioxide and water in respiration by making the water molecule more acidic than it normally is.

         iv.   Coordination compounds are also used in commercial drugs.

Fill in the Blank:

1. In _____ complexes, the splitting of d orbitals is very large, and in _____ complexes, the splitting of d orbitals is small.

2. The splitting of a ligand field _____ with increasing metal charge.

3. Complex ions contain a central metal atom bound to one or more _____.

4. _____ isomers occur when a coordinated ligand exchanges places with an uncoordinated ligand.

5. The list of ligands in order of their ability to split a ligand field is called the _____.

6. The _____ is the charge of the central atom, and the _____ is the number of ligands bound directly to the central atom.

7. Elements in the third row of the transition metals are approximately the same size as their counterparts in the second row due to _____.

8. _____ isomers result when the ligands bound to the metal have a different spatial arrangement. There are two types: _____ and _____.

9. _____ complexes exhibit magnetic properties due to their weak field ligand.

10. Ligands that have the ability to donate more than one electron pair to the central atom are called _____ or _____.

11. In an octahedral field, the _____ orbitals are higher in energy than the _____ orbitals.

12. In coordination compounds, the metal acts like a(n) _____ and the bound species act like _____.

Problems:

1. Provide names for the following compounds:

    a. $K_4[Fe(CN)_6]$

    b. $Pt(NH_3)_2Cl_4$

    c. $Cr(CO)_6$

    d. $[Cr(NH_3)_3(H_2O)_3]Cl_3$

    e. $[Co(NH_3)_4Cl_2]Cl$

    f. $[PtNH_3Cl_3]^-$

    g. $[Ni(en)_3]Cl_2$

    h. $[Co(NH_3)_4ClBr]$

2. Write formulas for the following compounds.

    a. Pentaamminebromocobalt(III) sulfate

    b. Pentaamminesulfatocobalt(III) ion

    c. Hexafluorocobalt(III) ion

    d. Diaquatetracyanocopper(VI) chloride

    e. Tris(ethylenediamine)cobalt(III) chloride

    f. Pentaamminechlorochromium(III) sulfate

    g. Diamminedichloroplatinum(II)

3. Determine the type of isomers each of the following are examples of:

    a. $[Co(NH_3)_4ClBr]Cl$ and $[Co(NH_3)_4Cl_2]Br$

    b. $[Cr(H_2O)_5CN]^{2+}$ and $[Cr(H_2O)_5NC]^{2+}$

    c. $[Cr(NH_3)_5(OSO_3)]Br$ and $[Cr(NH_3)_5Br]SO_4$

4. Draw all isomers and identify each type for the following complexes. If no isomers exist, state that.

    a. $[Co(NH_3)_4Cl_2]^+$

    b. $[Co(en)]^{2+}$

    c. $[PtBr_4]^{2-}$

    d. $[Co(en)_2Cl_2]^+$

    e. $Co(NH_3)_3Cl_3$

    f. $[Cr(H_2O)_5Cl]^{2+}$

5. Determine the number of unpaired electrons that you would expect to see for the following compounds:

    a. $[PtBr_4]^{2-}$

    b. $[Co(en)_3]^{3+}$

    c. $Cr(CO)_6$

    d. $[Co(H_2O)_6]^{2+}$

    e. $[Co(NH_3)_3(CN)_3]$

    f. $[CoCl_6]^{3+}$

6. Rank the cobalt compounds from question 5 in order of increasing magnetic properties.

7. Which of the complex cobalt ions from question 5 is/are most likely to be green in aqueous solution?

Concept Questions:

1. Isomers can have different colors. For example, $[Cr(NH_3)_5(OSO_3)]Br$ is a red complex, while $[Cr(NH_3)_5Br]SO_4$ is violet.

    a. Explain the difference in their colors.

    b. Do you think that all isomer types will have different colors? If so, explain. If not, which isomers will and which will not?

    c. Which compound do you expect to be more strongly attracted to a magnetic field? Explain.

    d. Which exerts a stronger ligand field, SO_4^{2-} or Br^-? Explain your answer.

    e. Calculate the ligand field splitting for each of the compounds in order to verify your answer from part d.

2. The $[Co(NH_3)_6]^{3+}$ ion has a yellow color, but when one of the NH_3 ligands is replaced by H_2O to give $[Co(NH_3)_5(H_2O)]^{3+}$, the color shifts to red.

    a. What is the name of $[Co(NH_3)_5(H_2O)]^{3+}$?

    b. Use the information given to determine which ligand induces a stronger ligand field, NH_3 or H_2O.

    c. Which complex is more magnetic? Explain.

3. Draw ligand field splitting diagrams for octahedral, square planar, and tetrahedral complexes, and explain the magnetic property differences that you expect to see as a result of these diagrams. In other words, do you expect one geometry to promote magnetism more than others? Explain your answer.

4. An experiment is carried out on two different transition metal complexes: MX_6 and MY_6. Both molecules are excited with electrical energy, and MX_6 appears red while MY_6 appears green.

    a. Which ligand, X or Y, exerts a stronger crystal field on the metal? Explain the reasoning behind your choice.

    b. Another experiment is done, and it is determined that only one of these complexes is magnetic. Which compound is it? Explain the reasoning behind your choice.

    c. How does the color of emitted light for each of these relate to the color of light that they will be when illuminated with white light? *Hint:* Think about the relationship between the emission in the original problem and absorption discussed in this chapter.

    d. Carbonyl groups are often used as tags to indicate the electron density of the metal. The metal donates electrons to the antibonding orbital of the carbonyl group, weakening the C–O bond. When one of the ligands (X or Y) is replaced with a carbonyl group, the energy of the carbonyl bond changes more in MX_5CO than in MY_5CO. Which of the two ligands, X or Y, donates more electron density to the metal? Explain the reasoning behind your choice.

Chapter E:

Fill in the Blank:

1. potential
2. law of conservation of energy
3. intensive
4. temperature; Kelvin
5. accuracy; precision

Problems:

1. a. 4
   b. 3
   c. 3
   d. 6
   e. ambiguous (3 or 4)
2. a. 52357
   b. 21
   c. 7.67×10^5
   d. 104
   e. -7.5×10^5
3. 4 g
4. a. 0.20
   b. 3.460×10^5
   c. 2.96×10^5
   d. 6.7580×10^{-2}
5. The average is 5.51 mL. This result is very accurate since it agrees exactly with the actual value when two significant figures are considered.
6. $6,000
7. 5.9×10^{12} miles
8. 76 days
9. a. 50 lb
   b. No, since the concrete is more dense than water (1000 kg/m^3), it will sink.
   c. No. Whether or not something floats in water depends only on whether its density is greater or less than water. Since density is an intensive property, the concrete will be more dense and therefore sink, no matter what size the pieces are.

Concept Questions:

1. a. No, the students did not use the same measuring device since the number of significant figures varies from two to three.
   b. We cannot comment on the accuracy since the actual value is not provided. The results are fairly precise, however, since they all agree up to two significant figures (which is the number of significant figures for the average based on the least precise value).
2. A simple example is gas mileage where we can find the distance traveled and divide it by the amount of gas in a full tank to determine the miles per gallon of a vehicle.
3. On a microscopic scale, the energy is released to the environment as heat. The temperature of the air around the fire increases, and the kinetic energy (as well as the internal energy) of the air molecules increases.

Chapter 1:

Fill in the Blank:

1. mole
2. atomic mass
3. carbon-12
4. law of definite proportions
5. electrons
6. electrons
7. created nor destroyed
8. nucleus
9. experiments
10. protons
11. natural abundance
12. Scientific laws
13. gases, liquids, and solids
14. homogeneous
15. ratio of small whole numbers
16. atoms
17. element
18. chemical
19. mass spectrometry
20. Cations
21. compound
22. kilogram

Problems:

1. a. mixture
   b. mixture
   c. mixture
   d. mixture
   e. pure substance (if an alloy, then mixture)
   f. mixture
2. a. heterogeneous mixture
   b. heterogeneous mixture (if filtered, then homogeneous)
   c. homogeneous mixture
   d. heterogeneous mixture
   e. homogeneous mixture
   f. homogeneous mixture
3. The first set of masses gives 2.91 mol C, 5.82 mol H, and 2.91 mol O. The second set of masses gives 0.0372 mol C, 0.0745 mol H, and 0.0373 mol O. Both of these give a ratio of moles C:H:O of 1:2:1.
4.

Atomic Symbol	Typical Ion Formed	Number of Electrons	Number of Protons	Number of Neutrons
^{40}Ca	Ca^{2+}	18	20	20
^{9}Be	Be^{2+}	2	4	5
^{79}Se	Se^{2-}	36	34	45
^{37}Cl	Cl^-	18	17	20

5. a. Na^+
   b. O^{2-}
   c. Mg^{2+}
   d. Al^{3+}
   e. S^{2-}
   f. Cl^-

6. 7 amu

7. 24.30 amu

8. The mass of 1 mole of Na-23 atoms is 23.19005459 g and 1 mole of Na-23 ions is 23.18950601 g, which differ in the ten-thousandths place. It is appropriate, therefore, to use the same mass for both as long as the mass doesn't determine the significant figures in the final result.

9. 5.86×10^{22}

10. 2.64×10^{8} g

Concept Questions:

1. As discussed in this chapter, theories are the backbone of science. Theories are used to explain the world around us and provide the best answer for the available experimental evidence. Throwing out all ideas that are only theories would result in, essentially, throwing out all of science.
2. a. This is a physical change since the species will be unchanged chemically and will just be partitioned into different areas of space.
   b. This is a homogeneous mixture as the particles of the gas will be in constant, random motion.
3. Carbon is arbitrarily defined to be exactly 12 amu, but protons and neutrons do not have a mass of exactly 1 amu and electrons do not have zero mass. These small deviations result in noninteger masses.
4. Silicon should have a molar mass of approximately 28 g/mol since most of the silicon atoms (92.3% of them) in a sample will have a mass of 28 amu.
5. 88.9 g of oxygen must be produced. The mass of products (hydrogen and oxygen) must equal the mass of reactants (water) according to the law of conservation of mass. The mass of the oxygen, therefore, is simply the mass of the water minus the mass of the hydrogen produced.
6. Macroscopic samples are huge collections of atoms (on the order of Avogadro's number), so we can assume that we have an average sampling of isotopes. In certain samples (especially those that are human-made), the natural abundance of a particular isotope may not be present and so it would not be acceptable to use the average mass.
7. If both isotopes have the same charge upon ionization, then U-238 will be heavier and therefore will be deflected less (or move slower) by the magnet in the experiment.
8. No, 1 g does not equal 1 amu. That they are numerically equivalent means that the number of amu/atom is equal to the number of g/mol; this is a case of simply scaling up the numbers and using different units.
9. Carbon is arbitrarily defined to be exactly 12 amu, but protons and neutrons do not have a mass of exactly 1 amu and electrons do not have zero mass. These small deviations result in noninteger masses.
10. Macroscopic samples are huge collections of atoms (on the order of Avogadro's number), so we can assume that we have an average sampling of isotopes. In certain samples, the natural abundance of a particular isotope may not be present, and so it would not be acceptable to use the average mass.

Chapter 2:

Fill in the Blank:

1. diffraction
2. magnetic quantum number
3. Electromagnetic radiation
4. frequency
5. node
6. more
7. threshold frequency
8. gamma rays
9. Constructive interference
10. moves to a lower-energy level
11. wavelength
12. quantum-mechanical model
13. The Heisenberg uncertainty principle
14. emission spectrum
15. complementary
16. three; −1, 0, and +1.

17. vacuum
18. principal quantum number
19. indeterminate
20. wave properties
21. probability of finding an electron at a given location
22. quantum numbers
23. angular momentum quantum number
24. photoelectric effect
25. decreases
26. energy difference

Problems:

1. a.

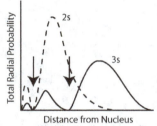

    b. Nodes are indicated with arrows on the figure.
    c. An electron in the $3s$ orbital can be closer to the nucleus, but it is less likely to be closer.
2. a. 284 nm
    b. 1.06×10^{15} Hz
    c. Since 500 nm light is a longer wavelength (lower energy) than the wavelength of light that will eject an electron, no electrons will be ejected from the metal.
    d. When the intensity of the light is increased tenfold, there will be ten times as many electrons ejected with zero kinetic energy.
    e. 7.00×10^{-19} J
    f. The difference in energy between the $n = 1$ and $n = 4$ levels is greater than the energy required to ionize an electron from the $n = 4$ level.
3. 32 electrons total
    $n = 4$; $l = 3, 2, 1, 0$; $m_l = -3, -2, -1, 0, +1, +2, +3$; and $m_s = +1/2, -1/2$
4. a. 3.63×10^{-35} m.
    b. The wavelength is so small that it could never be detected.
    c. Since the transition really comes down to what can be measured, measuring the smallest possible wavelength is the most quantum system that can be detected.
    d. No interference pattern would be observed since the wavelength is too small and the pattern will be too compressed to see anything.
    e. The zero-point energy of a baseball will actually be infinitesimally small (~zero) because the ball is too large to measure a smaller amount—we can think of the baseball as a nonquantum mechanical system because of its size.
5. a. 6.45×10^{14} Hz
    b. 257 kJ/mol
    c. The rate of photon emission has changed.
6. The wavelength of the photons will be 400 nm, and the energy will be 4.97×10^{-19} J.

Concept Questions:

1. a. Satellites are very large, and the uncertainty is very small, so it will not have an effect on the measurement of its position.
    b. If an electron were in an orbit, it would follow a trajectory that would allow for precise knowledge of the position and velocity of the electron.

2.

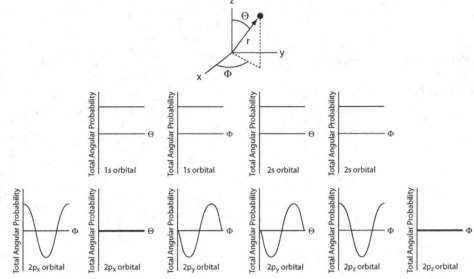

3. a. If I were to turn on a radio, I would be able to pick up a radio station.
    b. Radio waves have very long wavelengths and low energy, which makes them safe for general use.
4. a. When the slits move apart, the bright spots become more closely spaced.
    b. A longer wavelength will result in a pattern with the bright spots spaced further apart.
    c. Making the slits smaller increases the certainty in the position so there will be an increased uncertainty in the momentum in the y direction. The smaller slits will result in a pattern with the bright spots spaced further apart.
5. a. UV light has more energy, but since the intensity is unchanged, the number of incident photons is the same.
    b. Since UV light has greater energy per photon, each ejected electron will have absorbed more energy and will therefore have greater kinetic energy.
    c. Greater intensity will mean that more photons will be ejected.
    d. The kinetic energy of the particles is determined by the wavelength (energy) of the incident photons. Since the wavelength of light didn't change, the kinetic energy of the particles won't change.
6. Once the spacing becomes very small, it will not be detectable. Once the levels are no longer detectable, they become essentially meaningless.
7. The clicks correspond to a single photon since it is emitted from a single decay process. The single photons relate directly to single particles.
8. Electrons must have a smaller wavelength than visible light if using electrons allows one to image smaller things.
9. Green light is lower in energy than violet light, so the substance with the smaller work function, sodium, will be able to absorb green light.

Chapter 3:

Fill in the Blank:

1. ground state
2. Cations
3. paramagnetic
4. decreases; increases
5. $n - 1$
6. always larger
7. adding an electron
8. spin up; spin down
9. quantum numbers
10. shielding
11. spins
12. $[X]ns^1$
13. periodic

14. xenon; krypton
15. average bonding radii
16. ionization energy
17. Valence electrons
18. not bonded to another atom
19. decreases; increases
20. energy
21. anions
22. Metals

Problems:

1. a. $1s^22s^22p^2$
   b. $n = 1, 1 = 1; m_l = -1, 0,$ or $+1; m_s = +1/2$ or $-1/2$
   c. Yes, there are two unpaired electrons in the $2p$ orbital and unpaired electrons lead to magnetism.
   d. Carbon should have a larger radius as it has a smaller Z_{eff}.
   e. The ionization energy of carbon should be less than that of nitrogen since the Z_{eff} is less.
   f. The carbide ion will have a larger radius as it has more electrons but the same Z_{eff}.
   g. The electron affinity should be negative as the energy will be released when an electron is added to the neutral atom. This release of energy is a result of the attraction between the protons in the nucleus and the added electron.

2. a. B: $1s^22s^22p^1$
      O: $1s^22s^22p^4$
      F: $1s^22s^22p^5$
      He: $1s^2$
      Na: $1s^22s^22p^63s^1$
      K: $1s^22s^22p^63s^23p^64s^1$
      Rb: $1s^22s^22p^63s^23p^64s^23d^{10}4p^65s^1$
   b. B: 3
      O: 6
      F: 7
      He: 2
      Na: 1
      K: 1
      Rb: 1
   c. B, O, F, Na, K, Rb
   d. Rb, K, Na, He, B, O, F
   e. He, F, O, B, Na, K, Rb
   f. Rb, K, Na, B, O, F, He
   g. F, O, B, Na, K, Rb

3. a. P: [Ne]$3s^23p^3$

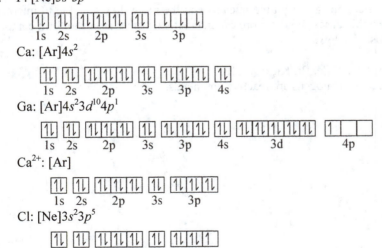

   Ca: [Ar]$4s^2$

   Ga: [Ar]$4s^23d^{10}4p^1$

   Ca^{2+}: [Ar]

   Cl: [Ne]$3s^23p^5$

303
Copyright © 2018 by Pearson Education, Inc.

   b. P: 5
      Ca: 2
      Ga: 3
      Ca^{2+}: 0 or 8
      Cl: 7

   c. P, Ga, Cl

   d. P: 5
      Ca: 2
      Ga: 3
      Ca^{2+}: N/A
      Cl: 7

   e. Ca^{2+}, Cl, P, Ga, Ca

   f. Ca^{2+}, Cl, P, Ga, Ca

   g. Ca^{2+}, Cl, P, Ga, Ca

   h. The smaller the atom/ion is, the closer the electrons are to the positively charged nucleus. According to Coulomb's law, the closer the electrons are to the nucleus, the more energy will be required to remove the electron.

   i. Ca^{2+} will have the highest ionization energy since there is an excess of protons in the nucleus, which pulls the electron density closer to the nucleus.

4. a. S^{2-}: $1s^2 2s^2 2p^6 3s^2 3p^6$
      Cl^-: $1s^2 2s^2 2p^6 3s^2 3p^6$
      Ar: $1s^2 2s^2 2p^6 3s^2 3p^6$
      K^+: $1s^2 2s^2 2p^6 3s^2 3p^6$
      Ca^{2+}: $1s^2 2s^2 2p^6 3s^2 3p^6$

   b. Ca^{2+}, K^+, Ar, Cl^-, S^{2-}

   c. Ca^{2+}, K^+, Ar, Cl^-, S^{2-}

   d. S^{2-}, Cl^-, Ar, K^+, Ca^{2+}

   e. These species are expected to have very similar chemical reactivity since they have the same number of valence electrons (the same electron configuration).

5. a. $1s^2 2s^2 2p^6 3s^2 3p^6 4s^2 3d^{10}$

   b. Zn will be diamagnetic since all of its electrons are paired in the ground state.

   c. Zinc will form a +2 ion since it will lose electrons from the $4s$ orbital.

   d. $n = 3$, $l = 2$, $m_l = +2$, $m_s = +\frac{1}{2}$ (m_l and m_s can have multiple values)

   e. $1s^2 2s^2 2p^6 3s^2 3p^6 4s^2 3d^9 4p^1$

   f. The ionization energy of the ground state will be higher than the ionization energy from the excited state. In the excited state configuration, the highest energy electron is further away from the nucleus (lower force is felt by the electron) and the energy of the electron is closer to zero (the ionized electron).

6. The electron affinity of Li^+ is −520 kJ/mol.

7. The larger the atom is, the further the electrons are from the nucleus and the more they are able to interact with other species. Kr, Xe, and Rn are large noble gases and therefore can interact with other atoms to form compounds.

8. Metals: All elements other than those listed below.
   Metalloids: B, Si, Ge, As, Sb, Te, At
   Nonmetals: H, He, C, N, O, F, Ne, P, S, Cl, Ar, Se, Br, Kr, I, Xe, Rn

9. Halogens are group 7A on the periodic table. Halogens are reactive nonmetals.

10. a. Na^+

    b. O^{2-}

    c. Mg^{2+}

    d. Al^{3+}

    e. S^{2-}

    f. Cl^-

Concept Questions:

1. See radial distribution plots in textbook.
    a. The higher value of n, the further away from the nucleus the electron density is, meaning that the atom is larger when higher-energy orbitals are occupied.
    b. The further away from the nucleus that an electron is, the smaller/higher the energy of the electron will be (Coulomb's law). The higher the energy of the electron, the closer it is to the ionized state (zero energy). So, as electron density is pushed out from the nucleus, the electron energy increases and ionization requires less energy.
    c. It is possible for an electron in the $3s$ orbital to be closer, but it is not very probable.
2. Transition metals have electrons in the ns and $(n-1)d$ orbitals, which are very close in energy. When electrons are removed, multiple stable configurations are available to the system.
3. a. Ionization energy for sodium is positive, but the electron affinity of chlorine is negative. In addition, the positive and negative ions form a crystal structure in which there are strong attractions between the ions, which contribute favorably (negative enthalpy) to the formation of sodium chloride.
    b. The magnesium ion will have to lose two electrons, which will require more energy, but two chlorine ions will form and the interaction energy between the +2 ions and the −1 ions will be stronger (Coulomb's law). The formation of $MgCl_2$ is, therefore, more exothermic.
4. a. The distance between the valence electrons and the nucleus changes more dramatically from $n=1$ to $n=2$.
    b. The d orbital is half full for Mn, so there is a jump.
    c. As you get further from the nucleus, the stability associated with a half-full shell becomes less important as the energy differences are smaller.
5. a. They will be the same in that discrete energies will be associated with transitions from one level to another, but they will have many more lines and those lines will be much closer together.
    b. The difference in energy becomes smaller because the additional electron/proton causes a smaller relative energy change. For example, going from +1 to +2 is a doubling of the Coulombic force, while going from +3 to +4 is a smaller change.
6. a. If the spins were randomly oriented, then there would be an increase in energy associated with the opposing magnetic fields (associated with the opposing spins).
    b. Two like charges, which repel each other, are in close proximity, and electrons in the same orbital more effectively shield one another.
    c. Since the s and d orbitals are close in energy, it is possible that the pairing energy is larger than the difference between the orbital energies (chromium has half-filled s and d orbitals). Also, as the electrons fill the orbitals, their energies shift and can actually switch order (copper has a filled d orbital and a half-filled s orbital).

Chapter 4:

Fill in the Blank:

1. formula mass
2. formula unit
3. ionic
4. molecular formula
5. electrostatic
6. Binary compounds
7. functional
8. mass percent composition
9. covalent
10. diatomic
11. hydrocarbons
12. octet (or duet in the case of He)
13. ionic
14. bond
15. Lewis structures
16. bonding pair
17. lattice energy
18. bond length
19. covalent

Problems:

1. H_2, O_2, N_2, F_2, Cl_2, Br_2, I_2
2. a. NaBr
   b. MgS
   c. S_2O_4
   d. H_2CO_3
   e. HI
   f. $HClO_4$
   g. SeN_2
   h. Fe_2O_3
   i. AgCl
   j. $SiCl_4$
   k. $Ca(HCO_3)_2$
   l. $CoPO_4 \cdot H_2O$
3. a. diphosphorus pentoxide
   b. rubidium oxide
   c. titanium(IV) oxide
   d. beryllium iodide
   e. dinitrogen tetraoxide
   f. iron(III) sulfate
   g. xenon tetrafluoride
   h. phosphorous acid
   i. lithium hydroxide
   j. copper(II) fluoride
   k. copper(II) sulfate pentahydrate
   l. chlorous acid
4. a. 141.94 amu
   b. 186.94 amu
   c. 79.87 amu
   d. 262.81 amu
   e. 92.02 amu
   f. 399.91 amu
   g. 207.29 amu
   h. 81.99 amu
   i. 23.95 amu
   j. 101.55 amu
   k. 249.70 amu
   l. 68.46 amu
5. 68.13%
6. 2.6×10^{22} sodium atoms
7. 1.0×10^{23} C atoms
8. $CuCl_2 \cdot 6H_2O$
9. 22.1% $CaCO_3$
10. a. 48.6% C, 8.16% H, and 43.2% O
    b. 5.50×10^3 g O
    c. $C_3H_6O_2$
    d. $C_6H_{12}O_4$
11. 27.37% Na, 1.200% H, 14.30% C, and 57.14% O. 2.9 g of oxygen in a 5.0 g sample of $NaHCO_3$.
12. CH_3 is the empirical formula of the hydrocarbon, and 107.95 g of oxygen were used.
13. You will need to consume 3.5 mg of NaF.
14. a. 215.69 g
    b. 2.79×10^{22} molecules
    c. 3.91×10^{23} H atoms
    d. 44.55% C

15. a.   BaCO₃

Ba^{2+} $\left[\begin{array}{c} \ddot{O}: \\ \| \\ C-\ddot{O}: \\ | \\ :\ddot{O}: \end{array}\right]^{2-}$

The bond between the barium ion and the carbonate ion is ionic. The bonds between carbon and oxygen atoms are covalent.

b.   KNO₃

K^+ $\left[\begin{array}{c} \ddot{O}: \\ \| \\ N-\ddot{O}: \\ | \\ :\ddot{O}: \end{array}\right]^{-}$

The bond between the potassium ion and the nitrate ion is ionic. The bonds between nitrogen and oxygen atoms are covalent.

c.   Na₂SO₄

$2Na^+$ $\left[\begin{array}{c} \ddot{O}: \\ \| \\ :\ddot{O}-S-\ddot{O}: \\ \| \\ :\ddot{O} \end{array}\right]^{2-}$

The bonds between the sodium ions and the sulfate ion are ionic. The bonds between sulfur and oxygen atoms are covalent.

d.   LiIO

Li^+ $\left[:\ddot{I}-\ddot{O}:\right]^{-}$

The bond between the lithium ion and the hypoiodite ion is ionic. The bond between iodine and oxygen is covalent.

e.   Mg(ClO₃)₂

Mg^{2+} 2 $\left[\begin{array}{c} \ddot{O}: \\ \| \\ :Cl-\ddot{O}: \\ \| \\ :\ddot{O} \end{array}\right]^{-}$

The bonds between the magnesium ion and the chlorate ion are ionic. The bonds between chlorine and oxygen atoms are covalent.

Concept Questions:

1. The chemical formula indicates the number of each type of atom, not the grams of each type of atom.
2. In ionic compounds, electrons are transferred, resulting in positively charged cations and negatively charged anions. The charged species interact with each other in all directions. Conversely, in molecular compounds, the electrons are shared between atoms and their localization results in discrete units.
3. In an ionic compound, the formula represents the smallest whole-number ratio of ions that will have a neutral unit. The formula doesn't represent the number of atoms in an independent group as molecular formulas do.
4. Molecular compounds combine in various ways, whereas ionic compounds form specific formula units based on the charges of the ions.
5. Functional groups: alcohol, ether, and amine.
    One of the OH groups in structure (a) is replaced with an ether group in structure (b); since this is the only difference, it must be this difference that results in the different reactivity.

6. a.

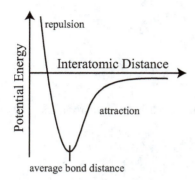

    d. As the distance between the atoms gets very large, they have a smaller effect on one another and the potential energy between them goes to zero.

Chapter 5:

Fill in the Blank:

1. ionic
2. formal charge
3. bond energy
4. bond length
5. polar covalent
6. Fluorine
7. Radicals
8. third
9. triple bond
10. trigonal planar
11. electrostatic
12. uneven
13. <109.5°

Problems:

1. a. $BaCO_3$

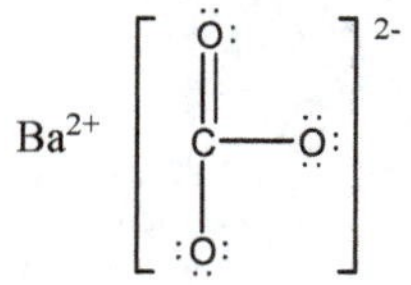

    The bond between the barium ion and the carbonate ion is ionic. The bonds between carbon and oxygen atoms are covalent.

    Formal Charges: Ba: 2+, C: 0, double bonded O: 0, and single bonded O: −1

    b. KNO_3

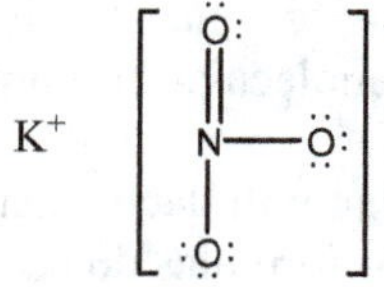

    The bond between the potassium ion and the nitrate ion is ionic. The bonds between nitrogen and oxygen atoms are covalent.

    Formal Charges: K: +1, N: +1, double bonded O: 0, and single bonded O: −1

c. Na_2SO_4

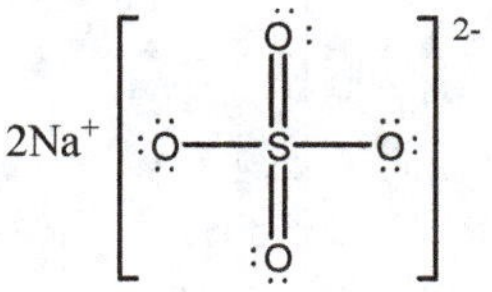

The bonds between the sodium ions and the sulfate ion are ionic. The bonds between sulfur and oxygen atoms are covalent.

Formal Charges: Na: +1, S: 0, double bonded O: 0, and single bonded O: −1

d. LiIO

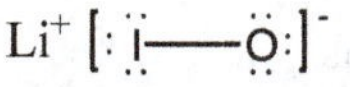

The bond between the lithium ion and the hypoiodite ion is ionic. The bond between iodine and oxygen is covalent.

Formal Charges: Li: +1, I: 0, O: −1

e. $Mg(ClO_3)_2$

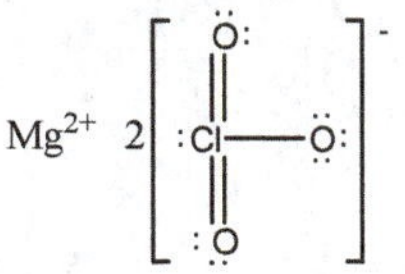

The bonds between the magnesium ion and the chlorate ion are ionic. The bonds between chlorine and oxygen atoms are covalent.

Formal Charges: Mg: +2, Cl: 0, double bonded O: 0, single bonded O: −1

2. a. ΔEN C–O: 1.0, dipole moment ~0.88, ~25%

   b. ΔEN N–O: 0.5, dipole moment ~0.44, ~12%

   c. ΔEN S–O: 1.0, dipole moment ~0.88, ~25%

   d. ΔEN I–O: 1.0, dipole moment ~0.88, ~25%

   e. ΔEN Cl–O: 0.5, dipole moment ~0.44, ~12%

3. a.

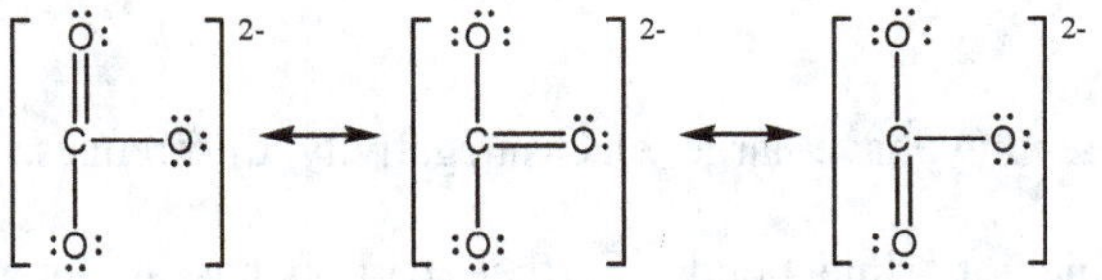

   b.

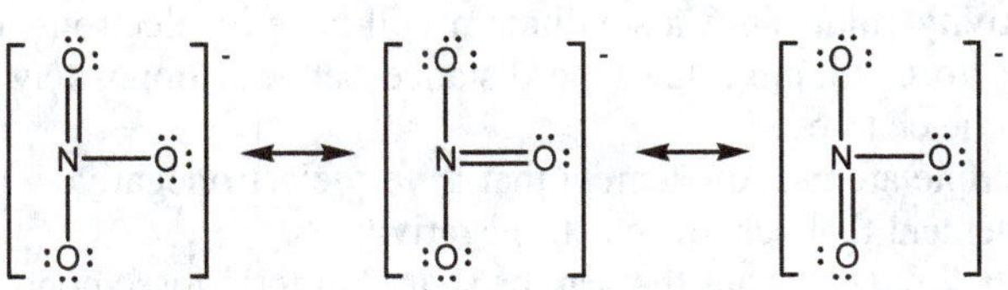

   c.

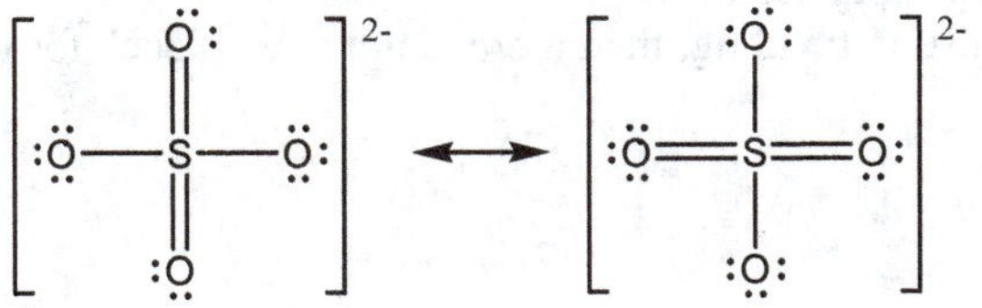

d. No resonance structures

e.
$$\left[\begin{array}{c} \ddot{O}: \\ \| \\ :Cl{=}\ddot{O}: \\ \| \\ :\ddot{O}: \end{array}\right]^{-} \longleftrightarrow \left[\begin{array}{c} \ddot{O}: \\ \| \\ :Cl{-}\ddot{O}: \\ \| \\ :\ddot{O} \end{array}\right]^{-} \longleftrightarrow \left[\begin{array}{c} :\ddot{O}: \\ \\ :Cl{=}\ddot{O}: \\ \| \\ :\ddot{O} \end{array}\right]^{-}$$

4. a. $\Delta H = -142$ kJ/mol; this reaction is exothermic.
    b. $\Delta H = -170$ kJ/mol; this reaction is exothermic.
    c. $\Delta H = 0$ kJ/mol; this reaction is neither endothermic nor exothermic.
5. a. C = 0 and H = 0
    c. The bond length should be between 154 pm and 134 pm: ~144 pm since the bonds are not double bonds or single bonds, but an average of the two.
6. a. Carbon is the least electronegative atom and is therefore most willing to share electrons.
    b. [S=C=N]⁻ makes the greatest contribution since it has the smallest formal charges with nitrogen (the most EN element) assigned a −1 formal charge.
7. a. Tetrahedral, trigonal pyramidal
    b. Linear, linear
    c. Tetrahedral, trigonal pyramidal
    d. Tetrahedral, tetrahedral
    e. Tetrahedral, tetrahedral
    f. Linear, linear
    g. Octahedral, square planar
    h. Linear, linear
    i. Tetrahedral, tetrahedral
    j. Linear, linear
    k. C: tetrahedral and O: tetrahedral, C: tetrahedral and O: bent
    l. C: tetrahedral and O: tetrahedral, C: tetrahedral and O: bent
    m. Trigonal bipyramidal, t-shaped

8.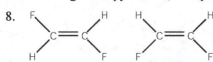

    a. The geometry of both carbon atoms is trigonal planar.
    b. The hybridization of both carbon atoms in both structures is sp^2.
    c. The first molecule shown will not have a net dipole, while the second molecule shown will have a dipole moment. The difference is due to the difference in symmetry.

Concept Questions:

1. When we look at metals and nonmetals, we see that they have large electronegativity differences, so both definitions apply.
2. Carbon forms four bonds and can form single, double, or triple bonds. Further, carbon has an intermediate electronegativity value, meaning that it can form strong, covalent bonds with many other elements.
3. The smaller the atom is, the larger the electronegativity value since a small atom will have its electrons relatively close to the positively charged nucleus. According to Coulomb's law, the distance between oppositely charged species is a determining factor in the force between the charges.
4. Central atoms tend to share their electrons with multiple atoms. An element that is very electronegative will not be as likely to share electrons with other atoms as an element that has low electronegativity.
5. Boron has only three valence electrons, so it has only three electrons that can be shared to form three bonds.
6. a. Any element that has accessible d orbitals can have an expanded octet.
    b. If an element does not have accessible d orbitals to use in bonding, then there is not any "place" for the extra electrons to be.

7. $2H_2 + O_2 \rightarrow 2H_2O$
    Since the reaction is exothermic, we can conclude that the energy released upon bond formation is greater than the energy needed to break the reactant bonds. We see that we must break two H–H bonds and one O–O double bond while making four H–O bonds. Since the oxygen molecule has a strong oxygen–oxygen double bond, we can conclude that the H–O bond is much stronger than the H–H bond.
8. Fluorine does not have energetically accessible *d* orbitals with which it can form an expanded octet. Further, fluorine is very electronegative and so is not as willing to share electrons with other atoms.
9. A polar bond will result in an overall uneven distribution of charge (molecular polarity) only if a similar polar bond that extends equally in the opposite direction does not cancel it out. If a bond is nonpolar, it will not have an uneven distribution of charge associated with it no matter what geometry the bonds adopt.

Chapter 6:

Fill in the Blank:

1. sp^2; three; one
2. a radical
3. five
4. head-to-head; side-to-side
5. destructive
6. equals
7. spread (or delocalized)

Problems:

1. a. tetrahedral, trigonal pyramidal, sp^3, polar NH bonds, polar molecule, <109.5°; all bonds are sigma bonds.
    b. Linear, linear, sp, polar bonds, polar molecule, 180°; the C–H bond is a sigma bond, and the CN bond is one sigma and two pi bonds.
    c. Tetrahedral, trigonal pyramidal, sp^3, polar bonds, polar molecule, <109.5°; all bonds are sigma bonds.
    d. Tetrahedral, tetrahedral, sp^3, polar bonds, nonpolar molecule, 109.5°; all bonds are sigma bonds.
    e. Tetrahedral, tetrahedral, sp^3, polar bonds, nonpolar molecule, 109.5°; all bonds are sigma bonds (the actual structure has one PO double bond, which corresponds to one sigma and one pi bond).
    f. Linear, linear, C: sp and S: sp^2, polar bonds, nonpolar molecule, 180°; the CS bonds have one sigma and one pi bond each.
    g. Octahedral, square planar, sp^3d^2, polar bonds, nonpolar molecule, 90°; all bonds are sigma bonds.
    h. Linear, linear, sp, polar bonds, polar molecule, 180°; the CO bond is one sigma bond, while the CN bond is one sigma bond and two pi bonds.
    i. Tetrahedral, tetrahedral, sp^3, polar bonds, polar molecule, 109.5°; the SCl bonds are one sigma bond, and the SO bonds are one sigma and one pi bond.
    j. Linear, linear, sp, polar bonds, polar molecule, 180°; the SC bond and the CN bond each have one sigma and one pi bond.
    k. C: tetrahedral and O: tetrahedral, C: tetrahedral and O: bent, polar bonds, polar molecule, HCH: 109.5°, COH: <109.5°; all bonds are sigma bonds.
    l. C: tetrahedral and O: tetrahedral, C: tetrahedral and O: bent, polar bonds, polar molecule, HCH: 109.5°, COC: <109.5°; all bonds are sigma bonds.
    m. Trigonal bipyramidal, t-shaped, polar bonds, polar molecule, 90°; all bonds are sigma bonds.
2. a.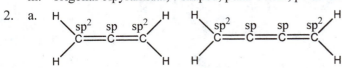
    b. CH_2CCH_2 has six sigma bonds and two pi bonds. CH_2CCCH_2 has seven sigma bonds and three pi bonds.

    c.

d. As can be seen in the drawings, the CH_2CCH_2 molecule will have CH_2 groups that are perpendicular to one another, while the CH_2CCCH_2 molecule will have CH_2 groups that are parallel.

3.

σ^*_{2p}	☐	σ^*_{2p}	☐	σ^*_{2p}	☐
π^*_{2p}	☐☐	π^*_{2p}	☐☐	π^*_{2p}	↑☐
σ_{2p}	↑↓	σ_{2p}	↑	σ_{2p}	↑↓
π_{2p}	↑↓ ↑↓	π_{2p}	↑↓ ↑↓	π_{2p}	↑↓ ↑↓
σ^*_{2s}	↑↓	σ^*_{2s}	↑↓	σ^*_{2s}	↑↓
σ_{2s}	↑↓	σ_{2s}	↑↓	σ_{2s}	↑↓
N_2		$N_2{}^+$		$N_2{}^-$	

   a. N_2 is the most stable.

   b. $N_2{}^+$ and $N_2{}^-$ are both equally stable and are less stable than N_2.

   c. $N_2 - 3$, $N_2{}^+ - 2\frac{1}{2}$, $N_2{}^- - 2\frac{1}{2}$

   d. $N_2{}^+$ and $N_2{}^-$ are both paramagnetic.

4. a. S_2 has a bond order of 2, while Cl_2 has a bond order of 1.

   b. S_2 will have a stronger bond since it has fewer antibonding interactions.

   c. S_2 will have a shorter bond length since it has fewer antibonding interactions.

   d. S_2 will be paramagnetic since it has unpaired electrons.

5. a. CO has the same electron configuration (same arrangement of electrons in the molecular orbitals) as N_2 does.

   b. I would expect N_2 to have a stronger bond since the orbitals will overlap more fully.

   c. The CO bond energy is 1072 kJ/mol, and the N_2 bond energy is 942 kJ/mol. The CO bond is polar, which results in a stronger bond.

Concept Questions:

1. When all bonds are single bonds, they form sigma bonds, which do not change in energy as they rotate. In a double bond, however, a pi bond forms and this side-to-side overlap is broken upon rotation. The breaking of the pi bond requires energy and is therefore not favorable.

2. Delocalized electrons have more space over which they can be, which leads to a lower-energy configuration. In pi bonding, the electrons will be delocalized and will therefore be lower in energy.

3.

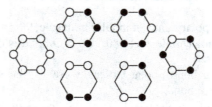

Chapter 7:

Fill in the Blank:

1. physical
2. chemical
3. limiting reagent
4. percent yield
5. stoichiometry

Problems:

1. a. physical change

   b. physical change

   c. chemical change

   d. chemical change

2. a. $N_2O_4 \rightarrow 2NO_2$

    b. $2Fe + 3Cl_2 \rightarrow 2FeCl_3$

    c. $2NaOH + H_2SO_4 \rightarrow 2H_2O + Na_2SO_4$

    d. $2NaCl + BeF_2 \rightarrow 2NaF + BeCl_2$

    e. $4Mg + Mn_2O_4 \rightarrow 4MgO + 2Mn$

    f. $2HCl + Na_2CO_3 \rightarrow 2NaCl + H_2O + CO_2$

    g. $CH_3CH_2OH + 3O_2 \rightarrow 2CO_2 + 3H_2O$

    h. $2C_2H_6 + 7O_2 \rightarrow 4CO_2 + 6H_2O$

    i. $Fe_2(SO_4)_3 + 12KSCN \rightarrow 2K_3Fe(SCN)_6 + 3K_2SO_4$

    j. $CuO(s) + C(s) \rightarrow Cu(s) + CO(g)$

    k. $2Co(NO_3)_3(aq) + 3(NH_4)_2S(aq) \rightarrow Co_2S_3(s) + 6NH_4NO_3(aq)$

3. a. $2NH_3 + CO_2 \rightarrow CH_4N_2O + H_2O$

    b. 240.8 kg CH_4N_2O

    c. The ammonia is completely consumed, and 34.9 kg of CO_2 remain.

    d. 69.9%

4. a. $2H_2(g) + O_2(g) \rightarrow 2H_2O(l)$

    b. 342 g of air

    c. 11.3 g of H_2O, 0.0 g of O_2, and 9.2 g H_2

5. a. $2Na(s) + 2H_2O(l) \rightarrow 2NaOH(aq) + H_2(g)$

    b. 0.10 g H_2

    c. 0.061 g H_2

    d. 14%

6. a. $C_6H_{12}O_6(s) + 6O_2(g) \rightarrow 6CO_2(g) + 6H_2O(l)$

    b. Either answer can be justified. Glucose is the limiting reactant overall as there is an "unlimited" supply of oxygen in the atmosphere. Oxygen is the limiting reactant in a given breath as we have reserved glucose and only a single breath's worth of oxygen.

    c. 16 g

7. 0.78%

Concept Questions:

1. This is a physical change since the species will be unchanged chemically—they will just be partitioned into different areas of space.

2. The number of molecules is not conserved—only the number of each type of atom is conserved on each side of the equation.

3. a. A side reaction occurs, resulting in the formation of another product. Some of the product is lost when it is transferred to a weighing vessel.

    b. It is unlikely that you would make more product than you calculate theoretically since the theoretical yield assumes that all of one reactant is used up.

    c. If a precipitate is wet when it is weighed, then water is being included in the product mass.

Chapter 8:

Fill in the Blank:

1. Oxidation-reduction
2. nonelectrolyte
3. polyprotic
4. concentrated; dilute
5. precipitation reaction
6. decreases
7. acid–base indicator
8. the ion charge
9. Spectator ions
10. strong electrolyte
11. stock solution

Problems:

1. a. H_2CO_3
   b. HI
   c. $HClO_4$
2. a. phosphorous acid
   b. chlorous acid
3. a. $Na_2CO_3(aq) + 2HCl(aq) \rightarrow H_2O(l) + CO_2(g) + 2NaCl(aq)$
   b. $CO_3^{2-}(aq) + 2H^+(aq) \rightarrow H_2O(l) + CO_2(g)$
   c. 741 mL
   d. 0.31 g
4. a. $Pb(NO_3)_2(aq) + (NH_4)_2S(aq) \rightarrow PbS(s) + 2NH_4NO_3(aq)$
   b. $Pb^{2+}(aq) + S^{2-}(aq) \rightarrow PbS(s)$
   c. 0.369 g PbS
   d. 0.106 M NH_4^+
   e. The original solution has a greater concentration of ions (0.462 M) than the final solution does (0.363 M), and so it will have greater conductivity.
5. a. 1.67 M
   b. 0.167 M
   c. $2NaOH(aq) + H_2SO_4(aq) \rightarrow 2H_2O(l) + Na_2SO_4(aq)$
      $2Na^+(aq) + 2OH^-(aq) + 2H^+(aq) + SO_4^{2-}(aq) \rightarrow 2H_2O(l) + 2Na^+(aq) + SO_4^{2-}(aq)$
      $OH^-(aq) + H^+(aq) \rightarrow H_2O(l)$
   d. 0.0510 M
6. a. The concentration of Na_2CO_3 will be 0.0 M since sodium salts are completely soluble.
   b. $Mg(C_2H_3O_2)_2(aq) + Na_2CO_3(aq) \rightarrow 2NaC_2H_3O_2(aq) + MgCO_3(s)$
   c. 3.2 M Na^+
   d. 3.2 M Na^+
   e. 0.84 g $MgCO_3$
   f. The light bulb would be brighter before the reaction took place since there will be more aqueous ions in the solution before the formation of the solid (which removes ions from solution).

Concept Questions:

1. A homogeneous mixture is independent of the size of the sample, meaning that an expression that gives the amount of material in a liter will be a scaled-up quantity from the milliliter sample.
2. If one is looking at qualitative results, such as whether or not a precipitate forms in a reaction, then simply knowing an approximate solution concentration would be sufficient.
3. In the solution, the two species (Ag^+ and Cl^-) are able to move freely and can physically come into contact and react to form a solid precipitate (AgCl).
4. The actual nature of the reactants is the formula of the substance when it is prepared; for example, sodium chloride is always found as a solid, neutral salt, even though it is completely soluble in water and sodium ions are often spectator ions. The net ionic equation, however, gives information about species that change in the reaction.
5. Sulfates might not interact with water as strongly as they will interact with the cation of the salt.

Chapter 9:

Fill in the Blank:

1. Kinetic energy; potential energy
2. joule (J)
3. calorie (cal)
4. constant
5. state
6. positive
7. on
8. thermal
9. heat capacity
10. negative

11. heat
12. endothermic
13. Hess's law
14. a pressure of 1 atm
15. pure elements in their standard state
16. lattice energy
17. Born–Haber cycle
18. decreases

Problems:

1. a. Heat is negative. Work is negative. The energy change is negative.
    b. Heat is zero (unless you count friction). Work is negative. The energy change is negative.
    c. Heat is zero. Work is negative. The energy change is negative.
    d. Heat is positive. Work is zero. The energy change is positive.
2. a. The reaction is exothermic since the enthalpy change is negative.
    b. The temperature in the calorimeter will increase since heat is released in the reaction.
    c. The internal energy is higher for the reactants. Energy is released in the reaction, so the products will be at lower energy than the reactants are.
3. a. The enthalpy change is –2658 kJ.
    b. The internal energy change is –2661 kJ.
    c. The work that is done is the products pushing against the atmosphere.
    d. Yes, it is reasonable since the energy and the enthalpy are almost the same, and to three significant figures, the two are equal.
4. a. $C_3H_8(g) + 5O_2(g) \rightarrow 3CO_2(g) + 4H_2O(g)$
    b. –2043.9 kJ
    c. Yes, combustion reactions release heat, which is reflected by the fact that the enthalpy change is negative.
    d. The reaction releases heat (has a negative enthalpy change), which means that it's an exothermic reaction.
    e. The internal energy of the reactants was higher. Heat was released, and work was done as the reaction occurred, and this lowered the internal energy of the system.
    f. $\Delta H = \Delta E + PV$. The system did work ($\Delta n > 0$) so $w < 0$ and $P\Delta V > 0$. Since the enthalpy change is negative, the value of ΔE must be a larger (magnitude) negative number.
5. a. For C(s): –393.5 kJ/mol
        For $CH_4(g)$: –802.5 kJ/mol
        For $C_8H_{18}(g)$: –5074 kJ/mol
    b. For C(s): –393.5 kJ/mol CO_2
        For $CH_4(g)$: –802.5 kJ/mol CO_2
        For $C_8H_{18}(g)$: –634.3 kJ/mol CO_2
    c. Methane is the most environmentally friendly because it releases the most heat with the smallest amount of carbon dioxide.
6. a. 24.2 J/mol °C
    b. 24.5 J/mol °C
    c. 25.4 J/mol °C
    d. 24.8 J/mol °C
    e. 25.3 J/mol °C
    f. 25.1 J/mol °C
7. 59 g
8. a. 0.897 J/g °C
    b. Aluminum
    c. If the metal was placed in the boiling water when it was transferred, hot water would also be transferred. The mass of the hot substance will be higher than measured, and the temperature change will be different due to the high heat capacity of the water.
9. a. The enthalpy change for this reaction should be negative.
    b. $\Delta H_{rxn} = -285.8$ kJ/mol
    c. 100 mol H_2 and 50 mol O_2

10. a. $H_2(g) + I_2(g) \rightarrow 2HI(g)$ $\Delta H° = -9.4$ kJ/mol
    b. -9.4 kJ/mol since there is no PV work done in the bomb calorimeter and the pressure is constant.
    c. 0.57 °C
11. $CH_4(g) + 2O_2(g) \rightarrow CO_2(g) + 2H_2O(g)$ $\Delta H = -803$ kJ/mol
12. $CH_4(g) + 2O_2(g) \rightarrow CO_2(g) + 2H_2O(l)$ $\Delta H = -847$ kJ/mol
13.

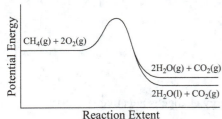

14. a. $\Delta H = -142$ kJ/mol; this reaction is exothermic.
    b. $\Delta H = -170$ kJ/mol; this reaction is exothermic.
    c. $\Delta H = +222$ kJ/mol; this reaction is endothermic.
15. a. -3836 kJ/mol
    b. -692 kJ/mol
    c. -606 kJ/mol

Concept Questions:

1. Gas stove tops are the least efficient: first, the burning of gas heats the air around the pot significantly. Second, some of the energy is lost as work. Electric stoves are more efficient because there is direct contact between the heating element and the pan; there is also not as much energy lost as work. Finally, the induction cooktops are the most efficient because the energy is transferred directly from the stove top to the pan, with no heat being generated except through the pan.
2. A gold pot would be best because gold has the smallest specific heat capacity, meaning that it requires the least amount of heat in order to change the temperature.
3. a. The molecular complexity of all of the metals is approximately the same since they all have approximately the same molar heat capacity value.
    b. This is a result of differences in the molar mass that are specific to a specific metal.
    c. 18.0 cal/mol °C is equal to 75 J/mol °C, which, compared to the metals, shows that liquid water is much more complex than the metal samples.
4. a. Liquid water has a higher heat capacity since the temperature of the ice changes more.
    b. The water molecules in the solid will have a larger change in molecular motion, assuming that the temperature of the solid water changes more than the liquid water temperature changes.
    c. Heat will be transferred from the ice to the liquid, and heat will also be absorbed from the surroundings. As heat is transferred, the molecular motion of the ice increases as expected for all processes.
5. Most processes in the laboratory are carried out open to the environment, which is a constant pressure condition.
6. This is true by definition; it is an arbitrary zero point that is used as a comparison point for all other enthalpy values.
7. Enthalpy changes are usually very close to the changes in internal energy for all reactions that do little or no PV work. Enthalpy changes are much easier to measure as they are equal to the heat exchanged at constant pressure (easily maintained in an open container).
8. Looking at the values in the book, we see that the distance is not much different across a row, so the large change in charge has a much larger impact.

Chapter 10:

Fill in the Blank:

1. Dalton's law of partial pressures
2. mean-free-path
3. temperature; Kelvin
4. Pressure
5. inversely

6. Charles's
7. repulsions
8. faster
9. mmHg
10. pascal (Pa)
11. 22.4 L
12. vapor pressure
13. 8.3145 L·atm/J·K

Problems:

1. 1 mmHg = 13.22 mm saltwater
2. a. 0.30 atm, 4.4 psi, 230 torr, and 230 mmHg
   b. 6.3 kPa
   c. 9.5 kPa
   d. 34 kPa
3. 2.1 atm
4. 2.6 atm
5. a. 63 g He
   b. 0.17 g/L
   c. 1.2 g/L
   d. 1.0×10^4 hrs
6. a. 0.85 atm
   b. 33 L
7. a. 3.90 atm
   b. 3140 mmHg
8. 140 atm
9. a. $2HCl(aq) + Na_2CO_3(s) \rightarrow H_2O(l) + CO_2(g) + 2NaCl(aq)$
   b. 8.98 mL CO_2
   c. 31.4 mmHg
10. 2.8 L
11. a. $Zn + 2HCl \rightarrow ZnCl_2 + H_2$
    b. 767 mmHg
    c. 97.7% Zn
12. Particle size is negligible. Average kinetic energy is proportional to temperature in kelvins. Collisions are perfectly elastic.
13. 576.2 m/s
14. Hydrogen gas will effuse 3.7 times as fast, and therefore nitrogen will take 3.7 times as long.
15. a. 2.9 g
    b. 2.9 g
    c. Methane is appropriately described as an ideal gas under these conditions, as demonstrated by the fact that both calculations gave the same numerical result.

Concept Questions:

1. a. The pressure inside your tires is greater than atmospheric pressure. You can tell because when a tire pressure gauge is put on the valve stem, air releases spontaneously.
   b. The tires don't collapse or explode because the rubber of the wheels exerts a force against the air pressure.
   c. Underinflated tires are dangerous because there will be pressure against the side of the tires, which weakens them. Underinflated tires also get hotter, which weakens them.
2. a. The two gases have the same kinetic energy since they are both at the same temperature.
   b. The two gases exert the same partial pressure since there are the same numbers of moles of each.
   c. Gas A is lighter than gas B, which means that if the two gases have the same kinetic energy, gas A moves faster.
   d. Gas B contributes more to the average density since it is heavier.
3. a. The two gases have the same kinetic energy since they are at the same temperature.
   b. The average speed of the H_2 gas is greater. Since the two gases have the same kinetic energy and H_2 is much lighter, it must be moving more quickly.

c. 10 g of H_2 has more particles than 10 g of F_2 does, so the pressure in the flask containing H_2 will be greater.

d. The volume of the two gases will be the same since gases occupy the entire volume of their container and the volumes of the containers are the same.

4. As the temperature of the gas increases, it moves faster. The faster the gas moves, the more often it hits the walls of the container. Since pressure is directly related to the number of times the gas hits the walls of the container, the pressure increases as the temperature increases.

5. When the container size increases, the gases will have further to travel before they hit the sides of the container. Since the particles are moving at the same speed, the longer distance results in fewer collisions and, consequently, a decrease in pressure.

6. Carbon tetrachloride has a much larger value of *a* than carbon dioxide does. This indicates that carbon tetrachloride has stronger intermolecular forces than carbon dioxide does. Carbon tetrachloride has a *b* value that is larger than the value for carbon dioxide, meaning that carbon tetrachloride has a larger volume correction term than carbon dioxide does.

Chapter 11:

Fill in the Blank:

1. critical point
2. increased
3. heat of fusion
4. decreases
5. N, F, and O
6. surface tension
7. capillary action
8. network covalent atomic solids
9. permanent dipoles
10. positive; negative
11. boiling temperature
12. sublimation
13. greater
14. increase

Problems:

1. a. Dispersion
    b. Dispersion
    c. Ion–Dipole
    d. Dipole–Dipole
    e. Dipole–Dipole and H-bonding
    f. Dispersion
    g. Dipole
    h. Ion–Dipole
2. a. $NaCl < NH_3 < NF_3 < He$
    b. $NH_3 < NF_3 < NCl_3 < NBr_3$
    c. $HF < ClF < F_2$
    d. $CH_3CH_2CH_2CH_2CH_2CH_2CH_2CH_3 < CH_3CH_2CH_2CH_2CH_3 < CH_3CH_2CH_3$
3. a. The gas at 123 °C has the largest kinetic energy, which is determined by the temperature.
    b. The gas at 123 °C has the largest internal energy since the most heat has been transferred at this point without any input or output of work.
4. a. NH_3 is polar and therefore has dipole–dipole interactions that cause the pressure to be less than the ideal gas law predicts.
    b. The gas with the highest boiling point is the gas with the strongest intermolecular forces, which, in this case, is ammonia.
    c. H_2 will have the lowest critical temperature since it is an ideal gas at the temperature used in the tabulated data.

5. a. 24.2 kJ/mol
    b.

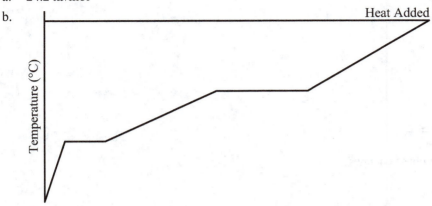

    c. 21.1 kJ
6. a. 33 kJ
    b. 780 kJ
    c. The values calculated are for 15 g of water, so we must calculate the heat per gram: 2.2 kJ/g for the hydrogen bonds and 52 kJ/g for the oxygen–hydrogen bonds. We see that the heat of vaporization is very close to the heat required to disrupt all of the hydrogen bonds but much smaller than the heat required to break the covalent bonds within the water molecules.
7. a. The gas at 123 °C has the largest kinetic energy, which is determined by the temperature.
    b. The gas at 123 °C has the largest internal energy since the most heat has been transferred at this point without any input or output of work.
    c.

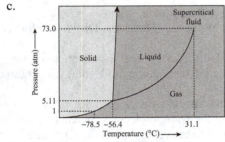

    d. At –60 °C, the liquid phase is never the most stable phase (no matter the pressure) as can be seen in the phase diagram.
    e. The vapor pressure of solid CO_2 at –78.5 °C is 1 atm. We know this because this temperature is the normal sublimation point, which is the point at which the vapor pressure of the solid is equal to the normal pressure (1 atm).
    f. No; the sample has changed to a gas.
    g. In this case, the sample will go from solid to liquid instead of going from a solid to a gas.
8. a.

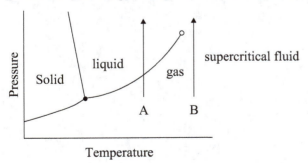

b.

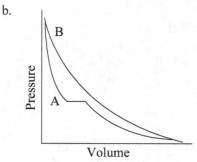

c. Ideal gases do not undergo phase changes.

Concept Questions:

1. a. The two compounds should have roughly the same size of dispersion forces since they have the same molar mass and the same general shape.
    b. The butanol should have a higher boiling point since it has an OH bond, which can form hydrogen bonds with the OH bonds on other butanol molecules. The stronger intermolecular forces will mean that more energy (higher temperature) will be required to vaporization.
    c. The butanol should have a higher heat of vaporization since it has stronger intermolecular forces, as explained in part b.
2. a. The pentane beaker will empty out first since the intermolecular forces between pentane molecules are weaker than the intermolecular forces between pentanol molecules.
    b. Since the pentane has a higher vapor pressure due to the weaker intermolecular forces, the concentration of vapor over pentane will be higher.
    c. Due to the weaker intermolecular forces, pentane will boil at a lower temperature.
3. Covalent bonds are too strong, and DNA must be able to separate in replication processes. Dispersion forces are too weak, and DNA could come apart and be susceptible to damage.
4. A higher heat of vaporization corresponds to stronger intermolecular forces. When vaporization occurs, the intermolecular interactions are effectively destroyed, so the stronger the interaction, the more energy is required to break that interaction.
5. The molecules that leave the liquid and go into the gas phase are the fastest moving particles since they have enough kinetic energy to overcome the intermolecular forces that hold the water molecules together in the liquid phase. As the fastest particles leave the liquid, the distribution shifts to a lower average speed (lower temperature).
6. A sodium chloride solution should have a higher boiling point since the interactions between the ions of sodium chloride and the polar water molecules are very strong.
7. The first region of the heating curve corresponds to the solid region of the phase diagram. The line between the solid and liquid phases is represented by the second region of the heating curve. The third region of the heating curve corresponds to the liquid region of the phase diagram. The line between the solid and liquid phases is represented by the fourth region of the heating curve. The fifth region of the heating curve corresponds to the gas region of the phase diagram.

Chapter 12:

Fill in the Blank:

1. addition polymer
2. p-type conductor
3. decreases
4. network covalent atomic solids
5. packing efficiency
6. crystalline
7. ionic solids; high
8. band gap
9. $CaCO_3$ and SiO_2
10. Copolymers

Problems:

1. a. AlCl₃ < NaBr < NaCl < MgS
   b. CO < CO₂ < CCl₄
   c. O₂ < CO₂ < SO₂
   d. SO₂ < P₂O₃ < SiO₂ < Al₂O₃

2. Simple cubic: 48% empty   Example: Polonium   Density: 9.32 g/mL
   Body-centered cubic: 32% empty   Example: Sodium   Density: 0.968 g/mL
   Face-centered cubic: 26% empty   Example: Silver   Density: 10.5 g/mL
   The more empty space there is, the lower the density will be if the mass of the atom is kept constant. Since polonium is the most massive (209 g/mol), it is more dense than sodium (with the lowest molar mass).

3. a.

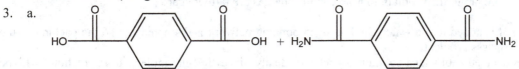

   b. Kevlar is formed in a condensation reaction with the elimination of a water molecule as each monomer adds to the polymer.
   c. Hydrogen bonds between the oxygen atoms and the hydrogen of the NH group will hold the strands together.
   d. At high temperatures, the strands move more freely and the hydrogen bonding network is disrupted.
   e. Polymers do not have specific molar masses associated with them since the strand length can vary widely.

Concept Questions:

1. a. This substance is expected to be a conductor since the higher-energy band is occupied by electrons.
   b. If all of the electrons are removed from the conduction band, the material will become a semiconductor since the band gap is small.
   c. The conductivity is expected to increase with increasing temperature since more electrons would have sufficient energy to overcome the band gap.

2. a. HDPE does not have branches while LDPE does. This means that more strands can pack together in the HDPE form; as the strands are closer together and can interact over a longer range, the intermolecular forces (dispersion forces) are stronger and the substance is more rigid and hard.
   b. The HDPE melts at a higher temperature since the intermolecular interactions are stronger.
   c. Polyethylene is probably formed in an addition reaction because there is a double bond in the molecule and there is no water molecule to expel as needed in a condensation reaction.

Chapter 13:

Fill in the Blank:

1. miscible
2. Henry's law constant
3. aqueous
4. Osmotic pressure
5. higher
6. Entropy
7. less
8. saturated
9. homogeneous
10. ideal solution
11. negative manner
12. heat of solution
13. van't Hoff factor
14. decreases; increases
15. molality

Problems:

1. a. Since this is ionic, it will dissolve in water.
   b. Since CO_2 is nonpolar, it does not dissolve significantly in water. A nonpolar solvent would be more suitable.
   c. Since hexane is nonpolar, it does not dissolve significantly in water. A nonpolar solvent would be more suitable.
   d. Lithium hydroxide is ionic, so it will dissolve in water.
   e. The nitrate ion is charged and should therefore dissolve in water.
2. a. The original solution was unsaturated since more ammonium chloride dissolved. After the addition, the solution is saturated since there is solid present.
   b. The heat of hydration must have a smaller absolute value than the lattice energy.
   c. Ammonium ions are larger than sodium ions are, so the charges are farther apart.
3. 0.33 M
4. a. When the ethanol is added to the water, the hydrogen bonding network is disrupted and so the packing of the water molecules is less efficient.
   b. I do not expect an ethanol/water solution to behave ideally since the intermolecular interactions will be different. It will deviate in a positive fashion since the interactions between the ethanol and water will be weaker than the interactions between water molecules.
   c. 3.4 M
   d. 4.3 m
   e. 16%
   f. 0.072
   g. 0.93
5. a. 0.22 atm
   b. 0.0306 atm
   c. I expect that the actual value will be less than the value calculated since ammonia and water should have strong interactions.
6. a. P = 58.8 torr
      Freezing point = −114.5 °C
      Boiling point = 78.5 °C
      Osmotic pressure = 3.60 atm
   b. P = 55.6 torr
      Freezing point = −118.7 °C
      Boiling point = 81.1 °C
      Osmotic pressure = 43.00 atm
   c. P = 59.0 torr
      Freezing point = −114.1 °C
      Boiling point = 78.3 °C
      Osmotic pressure = 0.103 atm
   d. P = 56.7 torr
      Freezing point = −115.8 °C
      Boiling point = 79.4 °C
      Osmotic pressure = 17.0 atm
   e. P = 25.8 torr
      Freezing point = −169.8 °C
      Boiling point = 112.4 °C
      Osmotic pressure = 540.5 atm
      Note: The values calculated for this problem are unrealistic since the methanol solution has a mole fraction that is greater than 50%.
7. 52 g of NaCl per kg of water, which is a large number. Based on the size of this number, we can assume that salt is added to water in order to flavor food, not to change the temperature of the water.

Concept Questions:

1. Solids do not have particles that move freely, and therefore they will not mix completely on a molecular level with gas particles.
2. Once there are more than six carbon atoms, the dispersion forces become more important than the potential hydrogen bonding that can occur with the OH group.
3. As the temperature increases, the gas particles move faster. These fast-moving particles access the surface of the liquid more often, where they can escape into the surroundings. The fast-moving particles also have more kinetic energy, which allows them to overcome the intermolecular forces holding them in the solution phase.
4. a. Carbon tetrachloride will have a larger freezing point change.
   b. The intermolecular forces will affect the values. The stronger the intermolecular forces of the solvent, the more the freezing point will be altered upon the addition of a solute.
   c. The heat of vaporization is much larger than the heat of fusion, so there is a smaller impact on the boiling point than there is on the freezing point.
5. a. If the solvent–solvent interactions are about the same as the solvent–solute interactions, then when the mixing occurs, there will be no real contraction (stronger interactions) or expansion (repulsions dominate).
   b. No, this is not always a valid assumption. A solution that has a negative deviation of vapor pressure is one that has strong solute–solvent interactions and we could expect that the solution volume will deviate from the additive volume.
   c. The solvent–solute interactions will be strong, so the vapor pressure will be lowered.
6. The lattice energy increases with increasing charge and decreasing ion size. The greater the lattice energy, the less soluble the substance, so from least soluble to most soluble, $MgO < MgCl_2 < RbBr_2$.
7. a. I would expect more NaCl to dissolve in the pentanol because the OH group of pentanol can interact with the sodium and chloride ions.
   b. The boiling point changes because the vapor pressure changes. A lower vapor pressure results in a higher boiling point.
   c. The boiling point changes by a different amount for two reasons: first, the two solvents have different boiling point elevation constants, and second, the NaCl will dissociate to a different extent. Based on the dissolution of the NaCl, we can expect that the pentanol solution will have a higher boiling point because there will be more dissolved particles.
8. This process is not favorable since the sodium ions are going from low to high concentration or from a more disordered state (with mixing of solvent and sodium) to a more ordered state (with less mixing).
9. a. A water–ethanol solution will deviate in a positive fashion because the hydrogen bonding network of the water will be disrupted with the introduction of the ethanol.
   b. It will be less than the ideal solution because the vapor pressure will be higher.
   c. The nitric acid will deviate in a negative fashion because it will dissociate into ions.
   d. It will be higher than the ideal solution because the vapor pressure will be lower.
   e. There will be a different change in the boiling point elevation because of the dissociation of the acid and because of the differences between the interactions between solvent and solute.
10. Soap molecules align themselves in a sphere, with the nonpolar side oriented inward where they can trap the dirt. The nonpolar sides of the molecules are on the outside of the sphere, where they allow for the soap sphere to be dissolved in water.

Chapter 14:

Fill in the Blank:

1. rate law
2. first
3. slow step
4. doubles
5. tangent
6. proven true
7. catalyst
8. time
9. intermediate
10. Chemical kinetics
11. decreases

12. increases
13. overall reaction order
14. elementary step
15. activation energy
16. Enzymes
17. p; z
18. inverse concentration
19. concentration (or molarity)
20. molecularity
21. proportional to
22. mechanism
23. kinetic energy of the sample
24. activation energy divided by R
25. decrease

Problems:

1. Rate = (0.0082 M-2s-1)[H$_2$][NO]; the rate for experiment 4 is 1.31×10^{-5} M/s.
2. Mechanism 1 is the best choice as it has a rate law that is consistent with the experimentally derived one.
3. rate = k[H$_2$][I$_2$]

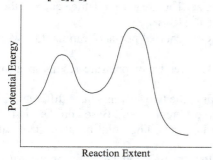

4.

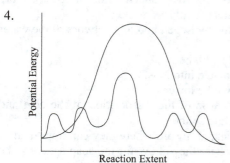

5. a. [phenol acetate] vs. time

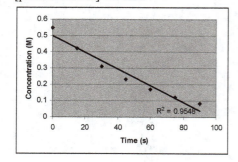

b.   1/[phenol acetate] vs. time

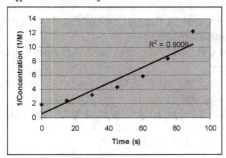

c.   ln[phenol acetate] vs. time

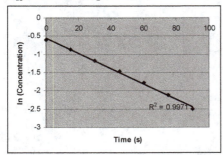

The plot of the natural log of concentration versus time is most linear. This corresponds to a first-order rate law.

6. a. 4.9×10^{-9}
    b.   2.86
    c.   497 K
7. a.   130.8 kJ
    b.   The activation energy is very large, and so the rate will be significantly affected by even small changes in temperature.
    c.   129.2 kJ
8. 1.1×10^{11} atoms
9. a.   1.9 minutes
    b.   No, it is not expected that this reaction would occur in a single elementary step since that would require four reactive species to come together at the same time (not observed) and the rate law is first-order, not fourth-order.
10. a.   Rate = $k[C_3H_6]^2$
    b. $C_3H_6^*$
    c.   0.18 will remain
    d.   The concentration does not affect the rate *constant*, which depends on the reaction itself and the temperature.
11. 2.6 atm

Concept Questions:

1. An increase in temperature will result in an increase in the number of collisions because the particles will move faster; it will have no impact on the geometry criteria; it will increase the number of particles with sufficient energy to overcome the reaction barrier.
2. Experiments 2 and 3 have a different initial concentration. Experiment 1 was carried out at a different temperature from experiments 2 and 3.
3. The glow stick is saved because the reaction is slowed down considerably by putting it in the freezer.
4. The reaction order is the sum of the exponents for the overall rate law, which is a mathematical construct. The molecularity reflects the number of reactive species that come together in a single reaction step, which has a physical basis.
5. This reaction must occur in multiple steps because there are nine particles reacting together. Plus, there are no known reactions that are more complicated than termolecular.

6. The slope of the reverse reaction has a larger absolute value than the slope of the forward reaction because the activation energy of the reverse reaction is greater than the activation energy of the forward reaction. The rate constant of the reverse reaction will therefore be more sensitive to temperature than the rate of the forward reaction.
7. The particle that removes excess energy does not have any orientational restrictions.
8. If NO and/or NO_3 are observed, then mechanism 1 can be ruled out. The rate laws of mechanisms 2 and 3 are different and therefore can be tested using an initial rates experiment.
9. We can write $[A]^2$ as $[A][A]$ and think of an increase in the concentration of A as an increase in two reactant species, so the rate will quadruple when the concentration of A is doubled.
10. Energy is not consumed in the generation of a large supply of enzymes, and there is no need to maintain a large supply of them in the body.
11. Substance X will have the highest concentration if it decays via second-order kinetics.

Chapter 15:

Fill in the Blank:

1. do not change
2. number of gas particles
3. reactants
4. reaction quotient
5. equal
6. products
7. inverted
8. equilibrium constant
9. minimizes
10. reversible

Problems:

1. $K_c = 1.2 \times 10^{-25}$
2. The reaction is not at equilibrium; it will proceed in the forward direction in order to reach equilibrium.
3. a. 0.109
    b. 0.95
    c. 0.91
    d. An increase in the pressure will result in a shift toward the side of the reaction with fewer gas particles, toward the reactants in this case. This shift is reflected in the smaller fraction of reactant particles that have decomposed in part c as compared with part b.
4. a. $P_{NH_3} = P_{H_2S} = 0.332$ atm
    b. 1.73%
    c. $P_{NH_3} = 2.490$ atm
       $P_{H_2S} = 0.044$ atm
5. a. Reactants are favored at equilibrium since $K < 1$.
    b. 0.048 mol
    c. The reaction will proceed in the forward direction since $Q < K$.
    d. 7.1×10^{-4}
    e. 2.2×10^{-6}
6. a. 0.0097 atm
    b. 0.347 atm
    c. In part b, the amount of starting material is so large that the amount that reacts will be a very small fraction. Note that the starting quantity is more than 100 times the value of K_P.
7. a. The reaction will shift toward reactants.
    b. No shift occurs.
    c. The reaction will shift toward products.
    d. No shift occurs.
    e. No shift occurs.
    f. No shift occurs.
    g. The reaction will shift toward reactants.

8. a.   The reaction will shift toward reactants.
    b.   The reaction will shift toward reactants.
    c.   The reaction will shift toward products.
    d.   No shift occurs.
9. a. $AgHCO_3$, H_2O, Ag^+, HCO_3^-, H_3O^+, and CO_3^{2-}
    b.   The addition of hydrochloric acid would increase the quantity of the silver bicarbonate present.
    c.   Adding NaOH would result in a decrease in the concentration of H_3O^+.

Concept Questions:

1. We could use isotopes and prepare an equilibrium mixture. If the isotopic distribution equalized itself on both sides of the reaction, equilibrium must be dynamic.
2. The concentrations are not always equal when $K = 1$. The concentrations will only be equal when they are equal to 1; otherwise, the concentration of C is equal to the product of the concentrations of A and B.
3. Definition of equilibrium: $rate_1 = rate_{-1}$

    $rate_1 = k_1 [A]^a [B]^b$

    $rate_{-1} = k_{-1} [C]^c$

    Combining these: $k_1 [A]^a [B]^b = k_{-1} [C]^c$

    Rearranging give: $\dfrac{k_1}{k_{-1}} = \dfrac{[C]^c}{[A]^a [B]^b}$

    $K_{eq} = \dfrac{[C]^c}{[A]^a [B]^b}$ therefore $K_{eq} = \dfrac{k_1}{k_{-1}}$

4. When the concentration of a reactant or product is changed, the rate of the forward or reverse reaction is changed. The rate constants, however, do not change. So, the rates will change until equilibrium is reestablished and that equilibrium will have a new set of concentrations associated with it, but the ratio of the product concentrations raised to their stoichiometric coefficients to reactant concentrations raised to their stoichiometric coefficients will remain unchanged.
5. Using the Arrhenius plot, we can see that a given temperature change will result in a change of the rate constant that is larger with a larger activation energy (slope of the graph). For an endothermic reaction, the activation energy of the forward reaction is larger than the activation energy of the reverse reaction. Since the equilibrium constant is a ratio of the rate constants, we see that the numerator changes more than the denominator and the equilibrium constant value increases with an increase in the temperature.
6. Let's call the number of particles with sufficient energy to overcome the barrier in the forward direction N_1 and the number of particles with sufficient energy to overcome the barrier in the reverse direction N_2. For an endothermic reaction, the activation energy of the forward reaction is greater than the activation energy of the reverse reaction. As can be seen in the distribution, the value of N_2 will be less affected than the value of N_1 by a given temperature change.

Chapter 16:

Fill in the Blank:

1. polyprotic
2. percent ionization
3. Lewis
4. proton acceptor
5. 1.0×10^{-7} M; 7.00
6. ionizes
7. weaker
8. acid ionization constant
9. stronger
10. basic
11. substance that increases $[H^+]$ in solution
12. acid
13. acidic

14. increasing
15. amphoteric

Problems:

1. a. $HCHO_2(aq) + H_2O(l) \rightarrow H_3O^+(aq) + CHO_2^-(aq)$
   b. $HCHO_2/CHO_2^-$ and H_2O/H_3O^+
   c. 2.56
   d. 11.44
   e. 6.1%
   f. i. 2.99
      ii. 14.9%
   g. The preparation of a dilute stock solution as that outlined in the original problem is not useful; it is more useful to use a concentrated stock solution to prepare all dilute solutions.

2. 4.3×10^{-10}

3. 3.4e-6

4. 4.4×10^{-5} g

5. $[HClO] = 0.0015$ M; $[H_3O^+] = [ClO^-] = 7.2 \times 10^{-6}$ M; pH = 5.14

6. 3.44

7. a. 1.1×10^{-3}
   b. 23%

8. a. $NaHCO_3(aq) + HC_2H_3O_2(aq) \rightarrow NaC_2H_3O_2(aq) + CO_2(g) + H_2O(l)$
   b. $NaHCO_3/H_2CO_3$ and $HC_2H_3O_2/C_2H_3O_2^-$
   c. Yes, a reaction will occur because there is no product present when these solutions are combined. The value of Q is therefore zero and less than the value of K for the reaction.

9. a. The solution will be basic since the base ionization constant is greater than the acid ionization constant.
   b. Assuming a 1.0 M solution, pH = 5.13, and percent ionization = 0.00075%
   c. pH = 9.62, and percent ionization = 0.0041%
   d. We can see that the percent ionization of the base is higher than the percent ionization of the acid; this is consistent with the answer given in part a.
   e. $NaHCO_3(aq) + HC_2H_3O_2(aq) \rightarrow NaC_2H_3O_2(aq) + CO_2(g) + H_2O(l)$
      When carbonic acid forms, it decomposes into carbon dioxide and water. The formation of carbon dioxide is observed as fizzing.

10. a. $[OH^-] = 1.2 \times 10^{-4}$ M and $[H_3O^+] = 8.1 \times 10^{-11}$ M
    b. 6.05×10^{-7}
    c. The base is moderately weak since the base ionization constant is a number to the power of −7.
    d. 0.49%
    e. Yes, the answer in part d is consistent with the answer in part c since the percent ionization is so small.

11. a. $H_2C_6H_6O_6$, $HC_6H_6O_6^-$, $C_6H_6O_6^{2-}$, H_3O^+, H_2O, OH^-
    b. 2.86

12. 21 mL

13. a. $Cr^{3+}(aq) + 7H_2O(l) \rightarrow Cr(H_2O)_5OH^-(aq) + H_3O^+(aq)$
    b. 2.10

14. The oxygen atom in HOCN makes the hydrogen atom more positive and, therefore, more acidic.

15. Ethene can serve as a Lewis acid because the pi orbitals involved in the double bond can be used to accept a lone pair upon rearrangement.

Concept Questions:

1. Water dissociates to form hydronium ions and hydroxide ions, which means that it's an acid and a base according to the Arrhenius definition. Water accepts a proton and donates a proton in accordance with the Brønsted–Lowry definition. The transferred proton accepts a lone pair from the water molecule in accordance with the Lewis definition.

2. The percent ionization is larger because the reaction goes further forward, but since there is a smaller amount of the acid, fewer hydronium ions will be produced.

3. The percent ionization will be greatest in the 1000 mL water. The less concentrated the weak acid is, the greater the percent ionization, so 1000 mL water will be greater than 500 mL of water. The NH_3 is a weak base, and there will be some NH_4^+ present in the solution, which will suppress ionization of the ammonium ion so the water will have a greater ionization than the NH_3 solution.

4.

The resonance structures shown above indicate that the proton is relatively easy to remove (as in an acid).

5. H_3PO_4 is the strongest acid, and PO_4^{3-} is the weakest acid. Since the strongest acid has the weakest conjugate base and the weakest acid has the strongest conjugate base, PO_4^{3-} will have the largest K_b value.

6. CO_2 will form carbonic acid when dissolved in water, resulting in an acidic solution.

7. H_3PO_4 is the weakest acid listed because the phosphorous atom is the least electronegative (as compared with S and Cl). The weakest acid will ionize the least and therefore give the highest pH.

8. Any other acid, when dissolved in water, will produce hydronium ions. In water, therefore, the other acids don't exist.

9. a. All of these acids dissociate completely to produce hydronium ions, and therefore they will all have the same strength.

   b. Using a different solvent will allow the acids to be compared with one another.

   c. $HI > HBr > HCl$

   $HClO_4 > HCl$

   $HClO_4 > H_2SO_4$

10. Water contributes 0.0000001 MH^+; the precision of this number is in the ten millionth place. In the concentration 0.0010 M, the precision is in the ten thousandth place. Since the rules of significant figures say that, in addition, the answer has the same precision as the least precise term in the problem, the addition of 0.0000001 M and 0.0010 M will give 0.0010 M.

11. The numbers associated with the hydronium ion concentration for weak acids are very small and must be reported as a factor of ten. The pH scale allows the numbers to be reported in a much simpler form.

12. Amines have a lone pair that can accept a proton.

13. Borax is a base: $Na_2[B_4O_5(OH)_4]\cdot 8H_2O$.

Chapter 17:

Fill in the Blank:

1. pH; added acid or base
2. Lewis acid; Lewis bases
3. its conjugate base; their conjugate acid
4. solubility product constant
5. equivalence point
6. increases
7. pK_a
8. molar solubility
9. Henderson–Hasselbach
10. a pH of the $pK_a \pm 1$
11. amphoteric
12. polyprotic acid
13. smaller
14. buffering capacity
15. greater than
16. endpoint
17. a factor of ten
18. formation constant
19. decreases

Problems:

1. a. 33.1 g $NaC_2H_3O_2$
    b. 4.61
    c. 4.74
    d. Because the molar mass of sodium acetate is greater than the molar mass of acetic acid, adding equal masses of the two will result in a solution that has a greater quantity of acid than base.
    e. 4.75
    f. 14 L
2. a. 4.58
    b. 0.032
    c. In part a, the change in pH is huge as compared with that in part b. This is because in part b, there is an equilibrium system that contains a weak base that is able to react with the strong base and thereby remove hydronium ions from solution.
3. a. 25.0 mL of HCl
    b. The titration will not have a buffer region because only strong acids and bases will ever be present.
    c. pH = 13.16
    d. pH = 12.55
    e. pH = 7.00
    f. pH = 1.39
4. a. As the equivalence point, the pH will be dominated by the conjugate acid of hydrazine and the solution will therefore be acidic with a pH < 7.
    b. 4.05 mL
    c. 10.75
    d. 10.61
    e. 4.46
    f. 1.094
    g. 0.864
    h. 0.749
5. a. 9.246
    b. The pH at half equivalence is approximately equal to the pK_a of the acid (or pK_b of the base).
6. a. pH = 2.88
    b. pH = 2.48, so $\Delta pH = -0.40$
    c. pH = 2.15, so $\Delta pH = -0.73$
    d. The pH in part c is much lower because more acid has been added to the solution. In part c, the effect is essentially to dilute the initial acid solution.
    e. 7.98
    f. Phenol red is the best choice because its pKa is very close to the pH at the equivalence point.
    g.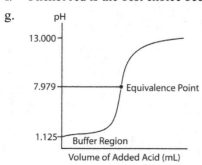
7. a. 6.6×10^{-5} M
    b. 9.8 mg
    c. No, $MgCl_2$ is soluble, and the hydronium ion can combine with the carbonate ion to increase the solubility of the magnesium ion.
    d. Yes, $Mg(OH)_2$ is a sparingly soluble salt with a solubility that is two orders of magnitude less than the solubility of magnesium carbonate.
8. Fe^{2+}, then Ca^{2+}, then Mg^{2+}. The iron carbonate salt has the smallest solubility product constant (it is least soluble), and the magnesium carbonate salt has the largest solubility product constant (it is the most soluble).

9. a. 7.31×10^{-7} M
    b. 1.2×10^{-7} M
    c. 3.57×10^{-11} M
    d. 0.125 M
10. a. 1.8×10^{-7} M
    b. HCl will react with the OH^-, which will drive the solubility reaction forward.
    c. NaOH will dissociate into Na^+ and OH^-. The presence of OH^- in solution will decrease the solubility of copper(II) hydroxide.
    d. 7.2×10^{-4} M

Concept Questions:

1. They are added in such small amounts that their contribution to the acidity or basicity of the solution is negligible.
2. HInd will be dominant before the equivalence point, and Ind^- will be dominant after the equivalence point.
3. The pH at half the equivalence point will be equal to the pK_a of the acid, so the pH will be 2.39. The $[OH^-]$ at the equivalence will be equal to the square root of the K_b of the conjugate base multiplied by the concentration of the conjugate base initially present. For a 1 M solution, the pOH at the equivalence point is then equal to the negative log of the square root of the K_b value, 5.81, and the pH is 8.19.
4. a. The result is the same whether we allow it to happen in one step or in multiple steps. Splitting it up just allows the mathematical manipulations to be simpler.
    b. No, in actuality, the reaction occurs in both directions all at the same time.
    c. The math is more complicated, and the calculations will need to be repeated multiple times if we try to mimic reality.
5. The region above the equivalence is not a buffer since it is not representative of a weak acid and conjugate base pair, which are needed for a buffer. It looks flat because the pH is dominated by the strong base—as more base is added, the concentration of the acid doesn't really change.

Chapter 18:

Fill in the Blank:

1. free energy
2. decreases
3. nonspontaneous process
4. zero
5. Entropy
6. zero
7. smaller than 1
8. state function
9. maximum work available
10. negative; negative
11. increases
12. smaller; the size of the molecule (molar mass)
13. multiplied by the same number
14. Thermodynamics; kinetics
15. negative; positive
16. adding tabulated values
17. negative

Problems:

1. d. $COF_2(l)$ will have the highest entropy. Liquids have larger entropy values than solids and COF_2 is less symmetric than BF_3 while their molar masses are similar.
2. $\Delta S > 0$ (two aqueous species from one solid species), $\Delta H > 0$ (adding heat will increase product formation), and $\Delta G > 0$ (this is a sparingly soluble salt—$K < 1$).
3. a. ΔS_{vap}(methane) $= -21.3$ J/°C mol
        ΔS_{vap}(hexane) $= -47.0$ J/°C mol

b. Intermolecular forces between hexane molecules are greater than intermolecular forces between methane molecules; as a result, the liquid phase of hexane is more ordered than the liquid phase of methane. When the phase change occurs, a larger change in order occurs with the hexane.

4.
a. $2H_2(g) + O_2(g) \rightarrow 2H_2O(l)$ $\Delta H = -571.6$ kJ/mol $\Delta S = -326.6$ J/mol K
$C(graphite) + O_2(g) \rightarrow CO_2(g)$ $\Delta H = -393.5$ kJ/mol $\Delta S = 2.9$ J/mol K
$CH_4(g) + 2O_2(g) \rightarrow CO_2(g) + 2H_2O(l)$ $\Delta H = -890.5$ kJ/mol $\Delta S = -242.9$ J/mol K

b. $2H_2(g) + O_2(g) \rightarrow 2H_2O(l)$ $\Delta G = -474.2$ kJ/mol
$C(graphite) + O_2(g) \rightarrow CO_2(g)$ $\Delta G = -394.4$ kJ/mol
$CH_4(g) + 2O_2(g) \rightarrow CO_2(g) + 2H_2O(l)$ $\Delta G = -818.1$ kJ/mol

c. $2H_2(g) + O_2(g) \rightarrow 2H_2O(l)$ $\Delta G = -474.2$ kJ/mol
$C(graphite) + O_2(g) \rightarrow CO_2(g)$ $\Delta G = -394.4$ kJ/mol
$CH_4(g) + 2O_2(g) \rightarrow CO_2(g) + 2H_2O(l)$ $\Delta G = -818.1$ kJ/mol

d. $\Delta G = -50.5$ kJ/mol

e. 7×10^8

f. The reaction will be spontaneous at all temperatures.

5.
a. $2O(g) \rightarrow O_2(g)$ $\Delta S < 0$
$H_2O(l) \rightarrow H_2O(g)$ $\Delta S > 0$
$2H(g) + O(g) \rightarrow H_2O(g)$ $\Delta S < 0$
$C(graphite) + 2O(g) \rightarrow CO_2(g)$ $\Delta S < 0$
$C(graphite) + O_2(g) \rightarrow CO_2(g)$ $\Delta S < 0$
$C(graphite) + 2H_2(g) \rightarrow CH_4(g)$ $\Delta S < 0$
$2H(g) \rightarrow H_2(g)$ $\Delta S < 0$

b. $2O(g) \rightarrow O_2(g)$ $\Delta S° = -117.0$ J/mol K $\Delta H° = -498.4$ kJ/mol
$H_2O(l) \rightarrow H_2O(g)$ $\Delta S° = 118.8$ J/mol K $\Delta H° = 44.0$ kJ/mol
$2H(g) + O(g) \rightarrow H_2O(g)$ $\Delta S° = -201.7$ J/mol K $\Delta H° = -927.0$ kJ/mol
$C(graphite) + 2O(g) \rightarrow CO_2(g)$ $\Delta S° = -114.1$ J/mol K $\Delta H° = -891.9$ kJ/mol
$C(graphite) + O_2(g) \rightarrow CO_2(g)$ $\Delta S° = 2.9$ J/mol K $\Delta H° = -393.5$ kJ/mol
$C(graphite) + 2H_2(g) \rightarrow CH_4(g)$ $\Delta S° = -80.8$ J/mol K $\Delta H° = -74.6$ kJ/mol
$2H(g) \rightarrow H_2(g)$ $\Delta S° = -98.7$ J/mol K $\Delta H° = -436.0$ kJ/mol

c. $2O(g) \rightarrow O_2(g)$ $\Delta G° = -463.5$ kJ/mol
$H_2O(l) \rightarrow H_2O(g)$ $\Delta G° = 8.6$ kJ/mol
$2H(g) + O(g) \rightarrow H_2O(g)$ $\Delta G° = -866.9$ kJ/mol
$C(graphite) + 2O(g) \rightarrow CO_2(g)$ $\Delta G° = -857.9$ kJ/mol
$C(graphite) + O_2(g) \rightarrow CO_2(g)$ $\Delta G° = -394.4$ kJ/mol
$C(graphite) + 2H_2(g) \rightarrow CH_4(g)$ $\Delta G° = -50.5$ kJ/mol
$2H(g) \rightarrow H_2(g)$ $\Delta G° = -406.6$ kJ/mol

d. $2O(g) \rightarrow O_2(g)$ $\Delta G° = -463.4$ kJ/mol
$H_2O(l) \rightarrow H_2O(g)$ $\Delta G° = 8.5$ kJ/mol
$2H(g) + O(g) \rightarrow H_2O(g)$ $\Delta G° = -866.9$ kJ/mol
$C(graphite) + 2O(g) \rightarrow CO_2(g)$ $\Delta G° = -857.8$ kJ/mol
$C(graphite) + O_2(g) \rightarrow CO_2(g)$ $\Delta G° = -394.4$ kJ/mol
$C(graphite) + 2H_2(g) \rightarrow CH_4(g)$ $\Delta G° = -50.5$ kJ/mol
$2H(g) \rightarrow H_2(g)$ $\Delta G° = -406.6$ kJ/mol

e. -801.1 kJ/mol

f. K is huge and will cause an overflow error on most calculators.

g. The reaction will be spontaneous at all temperatures.

6. A reaction will be spontaneous at all temperatures if the entropy change is positive and the enthalpy change is negative.

7.
a. $\Delta G° = -174.1$ kJ/mol

b. 1.02 atm

c. 5.02%

d. At temperatures above 314 K.

e. At low temperatures, the reaction becomes more spontaneous and the equilibrium constant value increases. At high pressures, the reaction will also favor product formation with no change in the equilibrium constant value.

8. a. −22.7 kJ
    b. 9520
    c. −22.7
    d. −34.1 kJ
    e. The situation in part c is closer to equilibrium as the calculated ΔG_{rxn} value (which represents how far a system is from equilibrium) is closer to zero.
9. a. $\Delta G° = -199.5$ kJ/mol
    b. The reaction is product-favored.
    c. 9.297×10^{34}
    d. $\Delta S° = -4.4$ kJ/mol
    e. The entropy change is almost zero, reflecting the fact that there is the same number of gas particles on either side of the reaction and that the number of particles per molecule remains constant.
    f. The reaction is enthalpically driven since the free energy change is such a large negative number and the entropy change is such a small negative number.
10. a. 30%
    b. Isobutane is more stable at room temperature.

Concept Questions:

1. The entropy of the universe increases even though the entropy of the system decreases.
2. At absolute zero a system has no motion or no entropy, which is not possible.
3. The entropy change is negative for the condensation, and the reaction is exothermic, so it will be favorable only at low temperatures.
4. A gas can fill a container in more ways (configurations) than it can occupy only a portion of the container. The increased number of configurations represents a greater entropy, so the process is spontaneous.
5. a. $Ag(s)$ will have a larger standard entropy value since it has more protons, neutrons, and electrons; thus, it has more energetically equivalent ways to arrange its components.
    b. Br_2 at 398 K will have a larger standard entropy value since at higher temperature the particles have more random (thermal) energy.
    c. $2F(g)$ will have a larger standard entropy value since it has more independent degrees of freedom.
6. The reaction is carried out more slowly and more reversibly, which allows more energy to be extracted; this is in accordance with the concept that the maximum available work is equal to the free energy change of the reaction.
7. This means that neither the product nor the reactant is favored at equilibrium.
8. The change is relatively small because of the hydrogen bonding network, which gives a complex system. Methanol also forms H-bonds and therefore also has a small entropy change of vaporization.
9. The higher the entropy, the more the molecular complexity and the more places for energy to be dispersed when heat is applied. When heat is stored as internal energy, the temperature change is small—this is a large heat capacity.

Chapter 19:

Fill in the Blank:

1. Nernst equation
2. electrolytic
3. force associated with electron motion
4. increase; loss
5. mass; charge
6. coulombs
7. voltaic (or galvanic)
8. anode
9. electrical potential; volts
10. dry-cell
11. salt bridge
12. Overvoltage
13. positive
14. oxidizing; reduced
15. below
16. positive

17. fuel cell
18. zero
19. anode; cathode
20. concentration
21. platinum
22. water

Problems:

1. a. $4Al(s) + 3O_2(g) \rightarrow 2Al_2O_3(s)$
   b. $23H_2O(l) + 3NO_3^-(aq) + 8Al(s) \rightarrow 3NH_3(aq) + 8Al(OH)_4^-(s) + 5H^+(aq)$
   c. $3H_2O(l) + 3Cl_2(g) \rightarrow 5Cl^-(aq) + ClO_3^-(aq) + 6H^+(aq) + 4e^-$
   d. $2HNO_3(aq) + 3H_3AsO_3(aq) \rightarrow 2NO(g) + 3H_3AsO_4(aq) + H_2O(l)$
   e. $2NO_2(g) + 7H_2(g) \rightarrow 2NH_3(g) + 4H_2O(l)$
   f. $3Cu(s) + 8HNO_3(aq) \rightarrow 3Cu(NO_3)_2(aq) + 2NO(g) + 4H_2O(l)$
   g. $2Cr(OH)_3(s) + ClO_3^-(aq) + 4OH^-(aq) \rightarrow 2CrO_4^{2-}(aq) + Cl^-(aq) + 5H_2O(l)$
   h. $2Au^{3+}(aq) + 6I^-(aq) \rightarrow 2Au(s) + 3I_2(s)$
   i. $FeO(s) + CO(g) \rightarrow Fe(s) + CO_2(g)$
   j. $TiO_2(s) + 2Cu(s) + 4H^+(aq) \rightarrow 2Cu^{2+}(aq) + Ti(s) + 2H_2O(l)$

2. a. The cell does not have a salt bridge to maintain electrical neutrality. Without a salt bridge, current will not flow.
   b.
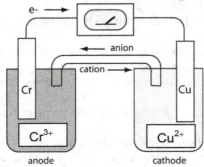
   c. Oxidation: $Cr(s) \rightarrow Cr^{3+}(aq) + 3e^-$
      Reduction: $Cu^{2+}(aq) + 2e^- \rightarrow Cu(s)$
   d. $2Cr(s) + 3Cu^{2+}(aq) \rightarrow 3Cr^{3+}(aq) + 2Cu(s)$ $E° = 1.07$ V
   e. $Cr(s) \mid Cr^{2+}(0.010$ M$) \parallel Cu^{3+}(0.20$ M$) \mid Cu(s)$
   f. $\Delta G° = -619$ kJ
   g. $K = 10^{108}$
   h. No, the potential that is read will not be the same as the standard cell potential because the concentrations of the reagents are not standard.
   i. $E = 1.08$ V

3. a. $[Pb^{2+}] = 50.4$ M
   b. $[Pb^{2+}] = 0.02$ M
   c. The concentration of lead(II) nitrate could be increased, or the concentration of nickel(II) nitrate could be decreased.

4. a. 92,800 min
   b. 2.4 A
   c. 0.64 A
   d. It will take the same amount of time in a more dilute solution. The concentration of the solution is not a factor in calculating the current.

5. a. Yes, the reaction occurs spontaneously since the voltage is a positive number.
   b. Oxidation: $Ti(s) + 2H_2O(l) \rightarrow TiO_2(s) + 4H^+(aq) + 4e^-$
      Reduction: $Cu^{2+}(aq) + 2e^- \rightarrow Cu(s)$
   c. $Ti(s), TiO_2(s) \mid H^+(aq) \parallel Cu^{2+}(aq) \mid Cu(s)$
   d. $E° = 1.2110$ V
   e. $E = 1.291$ V
   f. A cell potential of zero will be measured when the reaction reaches equilibrium.

6. a. $2Ag^+(aq) + Cd(s) \rightarrow 2Ag(s) + Cd^{2+}(aq)$ $E° = +1.20$ V
   b. The potential will increase since there will be less Cd^{2+} in solution.
   c. The potential will stay the same since the size of the electrode does not matter.

7. a. Anode: $2Br^-(aq) \rightarrow Br_2(g) + 2e^-$
   Cathode: $Na^+(aq) + e^- \rightarrow Na(s)$
   The reactions that occur at the anode and cathode are the reactions that have a smaller absolute value of the potential. The closer a negative potential is to zero, the easier it is to carry out that reaction.
   b. The reactions will be different because the electrolysis of water is easier to carry out than the oxidation of bromine or reduction of sodium.

8. anode: $4OH^-(aq) \rightarrow O_2(g) + 2H_2O(l) + 4e^-$ cathode: $2H_2O(l) + 2e^- \rightarrow H_2(g) + 2OH^-(aq)$

9. a. $C_3H_5O_2^- + 2H_2O \rightarrow C_2H_3O_2^- + CO_2 + 6H^+ + 6e^-$
   b. $E° = 0.190$ V
      $\Delta G° = -110,000$ J
   c. $\Delta G° = 61800$ J
   d. The conversion of lactate to pyruvate is spontaneous, while the conversion of NAD^+ to $NADH$ is nonspontaneous.
   e. When these reactions are carried out together, NAD^+ can be produced.
   f. 5.56 mol of NAD^+ will be consumed when 1 mole of lactate is converted to acetate and CO_2.

Concept Questions:

1. a. $B < D < A < C$
   b. B will be the cathode (reduced), and C will be the anode (oxidized).

2. $K_c > 1$, $\Delta G° < 0$, and $E_{cell}^o > 0$.

3. a. The standard cell potential equals the measured cell potential when the cell is prepared at standard conditions.
   b. The standard hydrogen electrode has a cell potential of zero by definition; other half-cells will have a reduction potential that is in reference to the zero and cannot, therefore, be exactly zero.

4. Half-cell potentials are relative values as they are measured against the standard hydrogen electrode, which is arbitrarily assigned a value of 0 V.

5. $K_c = 1$, $\Delta G° = E_{cell}^o = 0$. A concentration cell has the same reaction at the anode and cathode but in different amounts. At equilibrium, as well as under standard conditions, the two sides will be identical.

6. The electrons in a battery are traveling in the opposite direction as they would spontaneously travel in an electrolytic cell.

7. A negative cell potential represents the minimum amount of work that must be put into a galvanic cell in order for it to run, while the actual voltage represents the actual amount of work required. The difference between the minimum and actual amounts gives the overvoltage.

Chapter 20:

Fill in the Blank:

1. nitrogen-14
2. Alpha emission
3. Fusion
4. cyclotron
5. fission
6. parent; daughter
7. strong force
8. neutrons; beta decay
9. beta emission
10. nuclide
11. Geiger–Müller counter
12. mass defect; energy; mass
13. positron

Answer Key

Problems:

1. a. $^{210}_{84}\text{Po} \rightarrow \, ^{0}_{-1}\text{e} + \, ^{210}_{85}\text{At}$

    b. $^{222}_{88}\text{Ra} \rightarrow \, ^{4}_{2}\text{He} + \, ^{218}_{86}\text{Rn}$

    c. $^{234}_{90}\text{Th} \rightarrow \, ^{0}_{-1}\text{e} + \, ^{234}_{91}\text{Pa}$

    d. $^{138}_{57}\text{La} \rightarrow \, ^{0}_{1}\text{e} + \, ^{138}_{56}\text{Ba}$

    e. $^{185}_{79}\text{Au} \rightarrow \, ^{4}_{2}\text{He} + \, ^{181}_{77}\text{Ir}$

    f. $^{40}_{19}\text{K} + \, ^{0}_{-1}\text{e} \rightarrow \, ^{40}_{18}\text{Ar}$

2. a. Beta decay
    b. Beta decay
    c. Sequential decay
    d. Stable
    e. Positron emission
    f. Positron emission
    g. Beta decay

3. a. $^{19}_{10}\text{Ne} \rightarrow \, ^{0}_{1}\text{e} + \, ^{19}_{9}\text{F}$

    b. 10%, 4.18×10^{23} F atoms
    c. 1 half life – 17.22 s.
    d. 27.3 s

4. a. Beta decay

    b. $^{131}_{53}\text{I} \rightarrow \, ^{0}_{-1}\text{b} + \, ^{131}_{54}\text{Xe}$

    c. A small dose can result in mutations that can lead to cancer. A large dose can result in cell death, which is less dangerous.

5. a. 15,500 years
    b. The age of the wood will be the same. The rate of decay determines the age, not the quantity, of the wood.

6. a. 0.340407 amu
        8.570 MeV/nucleon
    b. 0.664429 amu
        8.717 MeV/nucleon
    c. 0.958434 amu
        8.584 MeV/nucleon
    d. 1.44296 amu
        8.146 MeV/nucleon

7. a. 0.78 MeV
    b. 3.7095×10^{8} m/s

8. 120 J

Concept Questions:

1. Alpha decay is the least dangerous because it has the smallest penetrating ability, even though it is the most energetic type of decay. Gamma radiation emission is the most harmful because it has the largest penetrating ability.

2. Magic numbers must be related to a physically relevant phenomenon such as the energetics of the subatomic particles in the nucleus.

3. There is no collision that needs to occur between the radioactive element and another element.

4. C-14 has a half-life of about 5,000 years. After about 10 half-lives (50,000 years), there is such a small amount of C-14 that it is practically undetectable.

5. Since zircon rejects lead, it can be assumed that any lead found in the sample is the result of uranium decay and so the quantities of N_0 and N_t will be simple to find.

Chapter 21:

Fill in the Blank:

1. Ethers
2. the most hydrogen atoms
3. structural formula
4. carbonyl
5. isomers
6. enantiomers; a clockwise or counterclockwise direction
7. *para*-dichlorobenzene
8. racemic mixture
9. double bond
10. cis–trans isomers
11. Stereoisomers
12. hydroxyl
13. oxidation–reduction reactions
14. saturated
15. phenyl

Problems:

1.

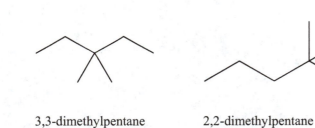

3,3-dimethylpentane   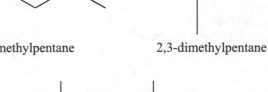 2,2-dimethylpentane 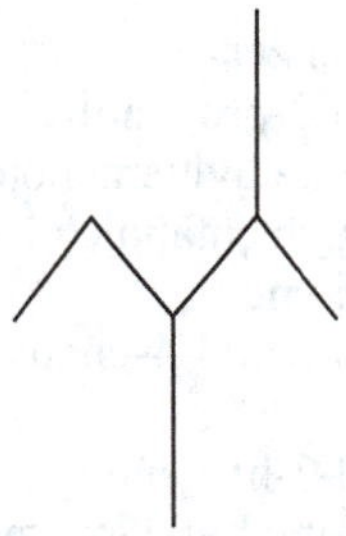   2,3-dimethylpentane

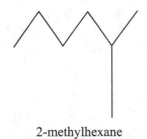

2-methylhexane   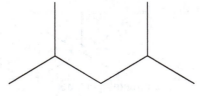 2,5-dimethylpentane   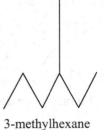 3-methylhexane

heptane

2.

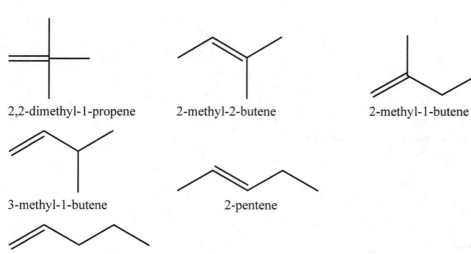

2,2-dimethyl-1-propene   2-methyl-2-butene   2-methyl-1-butene

3-methyl-1-butene   2-pentene

1-pentene

3. a. Ether: ethylpropyl ether
    b. Amine: ethyldimethylamine
    c. Ester: ethyl ethanoate
    d. Benzene: 1,2-difluorobenzene
    e. Alcohol: isopropanol or 2-propanol
    f. Carboxylic acid: 2-3-dimethylpentanoic acid
    g. Carboxylic acid: 2-propylpentanoic acid
    h. Alkyne: 4-methyl-2-hexyne
    i. Carboxylic acid: 2,3-dimethyl-4-phenylbutanoic acid
    j. Aldehyde: 3-ethylpentanal
    k. Alkene: 2,5,6-trimethyl-2-heptene
    l. Alcohol: 3-isopropyl-4-methyl-1-pentanol

4.

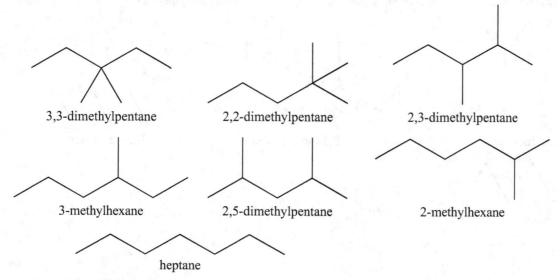

3,3-dimethylpentane   2,2-dimethylpentane   2,3-dimethylpentane

3-methylhexane   2,5-dimethylpentane   2-methylhexane

heptane

5. a. F_2 reacts with benzene using iron(III) chloride as a catalyst

    +HF

    b. $2C_6H_{14}(g) + 19O_2(g) \rightarrow 12CO_2(g) + 14H_2O(l)$

338

c.

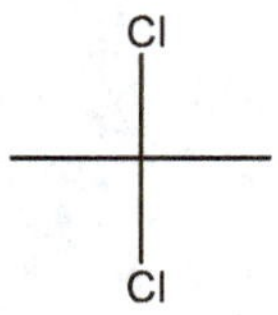

d.

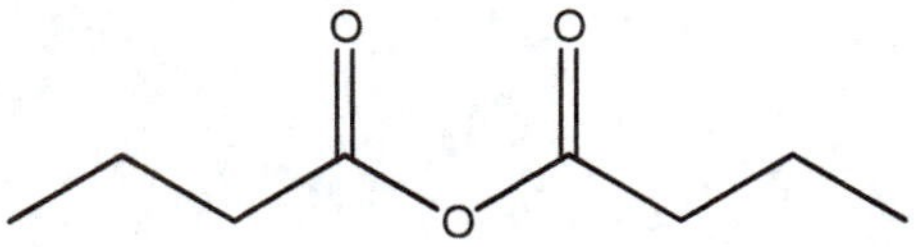

+ H$_2$O

e.

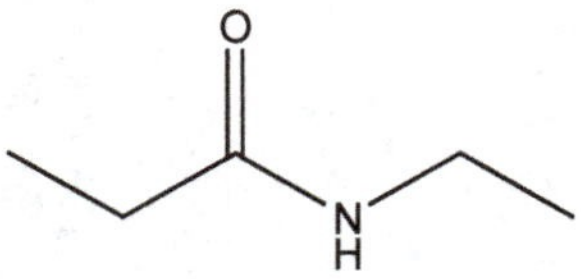

+ H$_2$O

f.

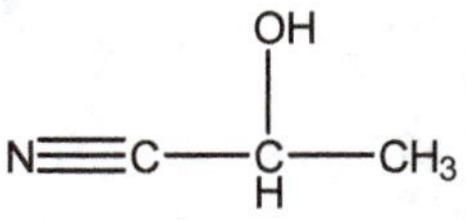

g.

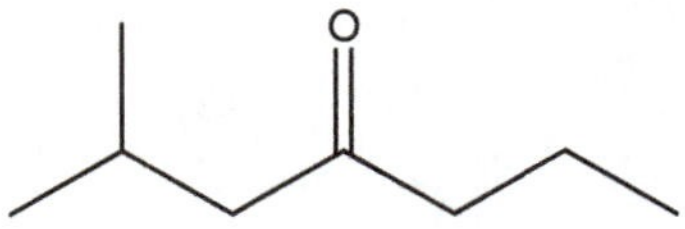

h.

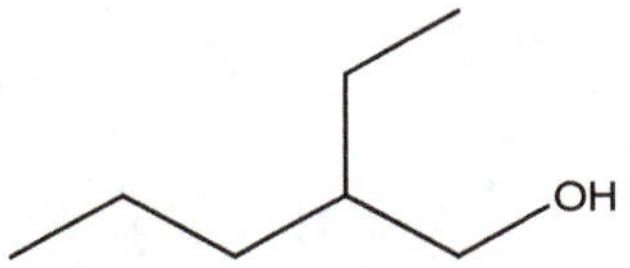

i.

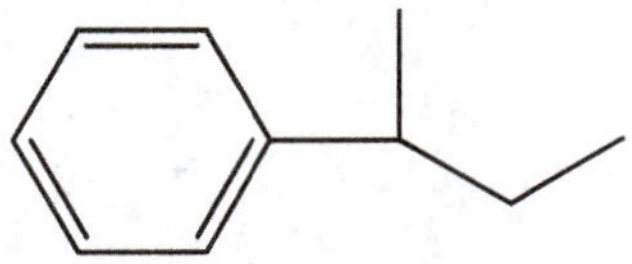

j.   (CH$_3$CH$_2$)$_2$NH$_2^+$NO$_3^-$

Concept Questions:

1. a.   C$_3$H$_8$ will boil first since it has the weakest intermolecular forces. The dominant IMFs are dispersion forces that are weakest when the molecule size is smallest.
    b.   C$_{40}$H$_{82}$ will be a solid at room temperature. As the largest molecule, it will have the strongest IMFs and will therefore undergo a phase change at a higher temperature.
2. The cis isomer has a dipole moment so that the dominant intermolecular force is a dipole–dipole interaction. The trans isomer has no dipole and therefore has only dispersion forces between the molecules. Since dipole–dipole interactions are stronger than dispersion forces are, the cis isomer will have a higher boiling point.
3. The more substituted carbon atom will be better able to accommodate the negative charge by spreading it out over a larger area.

Chapter 22:

Fill in the Blank:

1.   strong-field; weak-field
2.   increases
3.   ligands
4.   Coordination
5.   spectrochemical series
6.   oxidation state; coordination number
7.   lanthanide contraction
8.   Geometric; cis–trans and fac–mer.

9. Low-spin
10. polydentate ligands; chelates
11. z^2 and $x^2 - y^2$; xy, xz, and yz
12. Lewis acid; a Lewis base

Problems:

1. a. Potassium hexacyanoferrate(II)
   b. Diamminetetrachloroplatinum(IV)
   c. Hexacarbonylchromium(0)
   d. Triamminetriaquachromium(III) chloride
   e. Tetraamminedichlorocobalt(III) chloride
   f. Amminetrichloroplatinate(II) ion
   g. Trisethylenediamminenickel(II) chloride
   h. Tetraamminebromochlorocobalt(II)
2. a. $[Co(NH_3)_5Br]SO_4$
   b. $[Co(NH_3)_5SO_4]^+$
   c. $[CoF_6]^{3-}$
   d. $[Cu(H_2O)_2(CN)_4]Cl_2$
   e. $[Co(en)_3]Cl_3$
   f. $[Cr(NH_3)_5Cl]SO_4$
   g. $Pt(NH_3)_2Cl_2$
3. a. Coordination isomer
   b. Linkage isomer
   c. Coordination isomer
4. a.

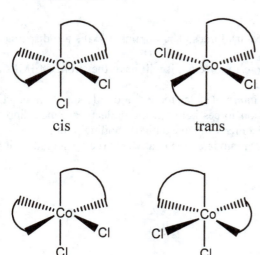

   b. No isomers exist.
   c. No isomers exist.
   d.

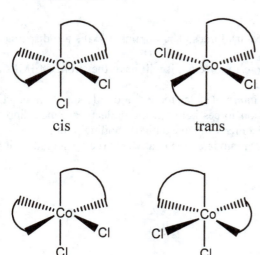

e.

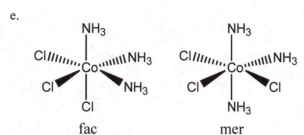

fac   mer

  f. No isomers exist.

5. a. 0
  b. 0
  c. 0
  d. 3
  e. 0
  f. 4

6. $a = b = c = e < d < f$

7. A green compound absorbs red light, which is low in energy. The compounds with the smallest crystal field splitting will absorb red light, so d and f will appear green.

Concept Questions:

1. a. Br and SO_4 have different field splitting properties, so the transition from the lower- to higher-energy state will have different energies.
  b. Isomers with ligands that have similar crystal field splitting abilities will look the same.
  c. The violet compound has a smaller field splitting, so it is more likely to be magnetic.
  d. SO_4^{2-} must exert a stronger ligand field, since the energy difference between the levels is greater.
  e. Green light: ~520 nm or 4×10^{-19} J. Yellow light: ~575 nm or 3×10^{-19} J. And we see that the green light is higher in energy, and so the red complex has a larger crystal field splitting.

2. a. pentaammineaquacobalt(III) ion
  b. The complex containing only NH_3 absorbs a higher-energy light (violet) than the complex containing and H_2O ligand. The higher-energy light corresponds to a larger field splitting.
  c. The weaker splitting of $[Co(NH_3)_5(H_2O)]^{3+}$ means that the higher-energy d orbitals are closer in energy than those in the $[Co(NH_3)_6]^{3+}$ complex. As electrons fill the higher-energy d orbitals rather than pairing up, $[Co(NH_3)_5(H_2O)]^{3+}$ will be more magnetic.

3. See Figures 22.13, 22.17, and 22.18. The tetrahedral configuration appears to promote magnetism more because the energy between the levels is always smaller.

4. a. In the original problem, electrical energy is used to excite electrons; when these electrons relax back down, they emit light with an energy that corresponds to the crystal field splitting. The compound that appears green has a larger crystal field splitting, so Y exerts a stronger field.
  b. The magnetic compound is MX_6. Since the field splitting is smaller, electrons can occupy the higher-energy level.
  c. When illuminated with white light, the compounds will appear as their complementary colors: MY_6 will appear red and MX_6 will appear green.
  d. X donates more electron density to the metal since the CO bond energy changes more in the metal compound with X than the one with Y.